THE COLLECTED PAPERS OF

# Albert Einstein

VOLUME 10

*THE BERLIN YEARS:*

*CORRESPONDENCE, MAY–DECEMBER 1920*

*and*

*SUPPLEMENTARY CORRESPONDENCE, 1909–1920*

DIANA KORMOS BUCHWALD

GENERAL EDITOR

# THE COLLECTED PAPERS OF ALBERT EINSTEIN

*English Translations Published to Date*

VOLUME 1
The Early Years, 1879–1902
*Anna Beck, translator; Peter Havas, consultant* (1987)

VOLUME 2
The Swiss Years: Writings, 1900–1909
*Anna Beck, translator; Peter Havas, consultant* (1989)

VOLUME 3
The Swiss Years: Writings, 1909–1911
*Anna Beck, translator; Don Howard, consultant* (1993)

VOLUME 4
The Swiss Years: Writings, 1912–1914
*Anna Beck, translator; Don Howard, consultant* (1996)

VOLUME 5
The Swiss Years: Correspondence, 1902–1914
*Anna Beck, translator; Don Howard, consultant* (1995)

VOLUME 6
The Berlin Years: Writings, 1914–1917
*Alfred Engel, translator; Engelbert Schucking, consultant* (1997)

VOLUME 7
The Berlin Years: Writings, 1918–1921
*Alfred Engel, translator; Engelbert Schucking, consultant* (2002)

VOLUME 8

The Berlin Years: Correspondence, 1914–1918

*Ann M. Hentschel, translator; Klaus Hentschel, consultant* (1998)

VOLUME 9

The Berlin Years: Correspondence, January 1919–April 1920

*Ann M. Hentschel, translator; Klaus Hentschel, consultant* (2004)

VOLUME 10

The Berlin Years: Correspondence, May–December 1920

and

Supplementary Correspondence, 1909–1920

*Ann M. Hentschel, translator; Klaus Hentschel, consultant* (2006)

THE COLLECTED PAPERS OF

# Albert Einstein

VOLUME 10

*THE BERLIN YEARS:*

*CORRESPONDENCE, MAY–DECEMBER 1920*

*and*

*SUPPLEMENTARY CORRESPONDENCE, 1909–1920*

ENGLISH TRANSLATION
OF SELECTED TEXTS

Diana Kormos Buchwald, Tilman Sauer, Ze'ev Rosenkranz,
József Illy, and Virginia Iris Holmes, Editors

Ann M. Hentschel, TRANSLATOR
Klaus Hentschel, CONSULTANT

Princeton University Press
Princeton and Oxford

Published by Princeton University Press, 41 William Street, Princeton, New Jersey 08540
In the United Kingdom: Princeton University Press, 3 Market Place,
Woodstock, Oxfordshire OX20 1SY
pup.princeton.edu



ISBN-13: 978-0-691-12826-9

ISBN-10: 0-691-12826-X

Library of Congress catalog card number: 87-160800

This book has been composed in Times.

The publisher would like to acknowledge the editors of this volume for
providing the camera-ready copy from which this book was printed.

Princeton University Press books are printed on acid-free paper
and meet the guidelines for permanence and durability of the
Committee on Production Guidelines for Book Longevity
of the Council on Library Resources.

Printed in the United States of America

1 3 5 7 9 10 8 6 4 2

# CONTENTS

# PUBLISHER'S FOREWORD

We are pleased to be publishing this translation of selected documents of Volume 10 of *The Collected Papers of Albert Einstein,* the companion volume to the annotated, original-language documentary edition. As we have stated in all earlier volumes, these translations are not intended for use without the documentary edition, which provides the extensive editorial commentary necessary for a full historical and scientific understanding of the source documents. The translations strive first for accuracy, then literariness, though we hope that both have been achieved. The documents were selected for translation by the editors of *The Collected Papers of Albert Einstein.*

We thank Ann M. Hentschel and Klaus Hentschel for their continuing good work and dedication to making Einstein's correspondence available to the English-speaking world; Diana Kormos Buchwald for her attention to the translation project; Rosy Meiron for stylistic suggestions; and Osik Moses for indexing, typesetting the manuscript, and producing the camera-ready copy, assisted by Jennifer Nollar, Rudy Hirschmann, and Linny Schenck. Many thanks to Alice Calaprice for her thorough copyediting of the manuscript.

Finally, we are most grateful to the National Science Foundation, Grant No. SBR-9710507, for its support of our project.

Princeton University Press
July 2006

# LIST OF TEXTS

SUPPLEMENTARY CORRESPONDENCE, 1909–1920

CORRESPONDENCE, MAY–DECEMBER 1920

# SELECTED TEXTS

# Vol. 5, 161a. To Vladimir Varićak[1]

Bern, 19 May 1909

[Not selected for translation.]

# Vol. 5, 197a. To Vladimir Varićak

Zurich, 15 February 1910

[Not selected for translation.]

# Vol. 5, 197b. To Vladimir Varićak

Bern [Zurich], 28 February 1909 [1910][1]

[Not selected for translation.]

# Vol. 5, 202a. To Vladimir Varićak

Zurich, 5 April 1910

[Not selected for translation.]

# Vol. 5, 202b. To Vladimir Varićak

Zurich, 11 April 1910

Highly esteemed Colleague!

Your two letters have given me great joy, as has your interesting treatise on the transformation.[1] As regards the rotating rigid body, my view of the matter is about as follows.

First of all, it cannot be excluded that the abstraction of the freely moving rigid body does not fit at all into the theory of relativity.[2] Take, e.g., the case that a rigid rod which at first hovers freely at rest in space suddenly receives a momentum during an infinitely short time. The end in *B* can experience a change in position, or acquire a velocity, as a consequence

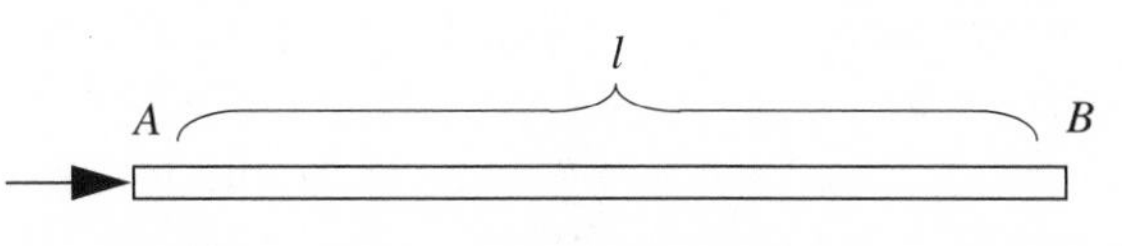

of this momentum at the earliest after a time period $\frac{l}{c}$, because otherwise superluminal signals would exist, with severe absurdities. Hence, the rod will either be deformed, or it will move, as a consequence of this momentum only after a certain time. Both consequences are so absurd (even the first assumption, if you think about it more closely). It rather seems to make more sense to do without the notion of a rigid body of finite size at all, especially if one only needs the infinitely small rigid body for the definition of space and time.

It seems that I have not yet explained to you clearly enough the core of the difficulty that in my opinion stands against the treatment of the rotating rigid body. First, we need to note that one is not obliged to deal with the problem of the *coming into being* of the rotation; here even worse difficulties lurk than in the case of uniform rotation. As regards the latter, it obviously does not suffice that radii and peripheral lines deform in the Lorentzian way. *This must moreover be valid for each material element of the rotating disc*. But satisfying this requirement seems to be impossible—this seems to have been proven in particular by Herglotz.[3]

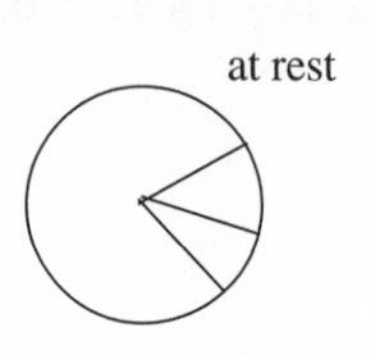


I have laid out the issue to myself by means of the following consideration.[4] Let radii be drawn in a material disc at rest. These must, for the Lorentzian contraction to take place in a rotating disc, be deformed in the manner that you indicated, if the disc rotates, judged from the coordinate system $K$ at rest. Hence about thus:

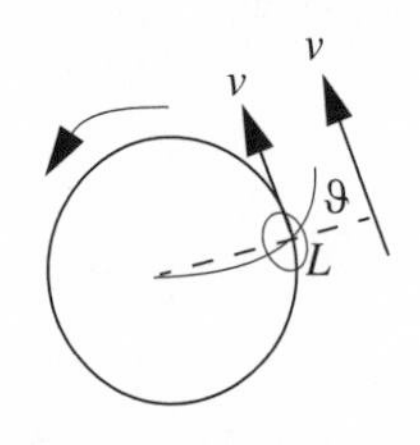


Consider now a small part of the rotating disc, that is enclosed in the line $L$ in the figure. Here the radial line, as seen from $K$, forms the angle $\vartheta$ with the peripheral line. Now one introduces a second coordinate system $K'$ which also does not rotate, but whose origin moves with the velocity $v$ (mean velocity of the particle under consideration). Judged from this system, the element under consideration does not have a translational velocity but only an angular velocity and an acceleration. It seems difficult to assume that, with respect to the coordinate system $K'$, the angle $\vartheta$ ($\vartheta'$) should differ from 90°. At least one can see immediately that neither acceleration alone nor rotation can have such an effect. Perhaps you might succeed in finding an expression for $\vartheta'$, in which somehow the product of acceleration & angular velocity enters.

But for the time being it seems absurd to me to assume that the angle $\vartheta'$ differs from 90°. But if $\vartheta' = 90°$, then we also have $\vartheta = 90°$. Then what Ehrenfest says is true, what is known to me already for years.[5]

That paper in the ⟨Journal⟩ Archive des Sciences ...[6] is of no interest for you. It is not a reprint of that paper in the Jahrbuch ...[7] but a kind of explicit exposition of the epistemological foundations of the theory of relativity, to which I obligated myself by giving an uncautious promise. I did not do a good job with the thing by the way.[8]

With best greetings your devoted

A. Einstein

My wife sends best greetings, too.

# Vol. 5, 203a. To Vladimir Varićak

[Zurich, 23 April 1910]

[Not selected for translation.]

# Vol. 5, 235a. To [Otto Lehmann][1]

Zurich, 1 December 1910

Highly esteemed Colleague,

First of all, many thanks for your kind letter, your paper, and above all for the papers you sent me earlier.[2] Now to your example!

1) Your consideration must still take into account that the lines of force emanating from a rod become denser as a result of the Lorentz contraction. The electric field strength is thus increased by the ratio $\frac{1}{\sqrt{1-\frac{v^2}{c^2}}}$ $\left(= 1 + \frac{1}{2}\frac{v^2}{c^2}\right)$, whereby the electrostatic energy is increased by this ratio.

2) The relative lessening of the repulsion through electrodynamic forces that you observed has the value $\frac{v^2}{c^2}$; it is therefore only halfway compensated by the increased repulsion mentioned under (1). The energy is therefore in fact ⟨larger⟩ smaller than if the rods were at rest.

3) This does not indicate a violation of the principle of relativity, however, because with moving rods one must distinguish between the force $K$, acting between them, from the standpoint a frame of reference not moving along with them, and the force $K'$, from the standpoint of a frame of reference moving along with the rods. If $K'$ differed from the force between the rods at the same distance while they were at rest, there would be a contradiction with the principle of relativity. But the force $K$ between the rods from the point of view of the non-moving system can

certainly differ from it; a priori it is not clear at all how $K$ should be defined. If one defines the force on moving bodies by the relation between force and the magnitude of motion, it can be shown that

$$K = K'\sqrt{1-\frac{v^2}{c^2}} \text{ must be true,}$$

just as it also results from your *special* example.

Thank you very much for your kind invitation to give a talk in Karlsruhe.[3] However, whenever possible I avoid giving talks before larger audiences.

With great respect, yours very sincerely,

A. Einstein.

# Vol. 5, 242a. To Heinrich Zangger[1]

[Zurich, 1 January 1911]

Dear Mr. Zangger,

Cordial New Year's greetings to you and your wife![2] One doesn't write to a prolific writer the way you do to me. I carefully rechecked my calculations concerning the viscosity of suspensions[3] but deemed everything in order. I also requested Mr. Hopf to check it over.[4] Who knows whether Perrin didn't have some strong swelling of the particles (1.4-fold in diameter).[5] Bredig thinks it very well possible. He considers it very difficult to make well-defined suspensions.[6]

I believe I found a new kind of influence by a magnetic field on electrons, but the magnitude of the effect still needs to be calculated.[7] In radiation theory the evil spirit is constantly leading me around by the nose.[8] Did you eventually get the Tammann?[9] I looked for you everywhere before your departure but was unable to track you down.

I hope your stay will do you & your wife a lot of good. With cordial greetings, yours,

A. Einstein

Best regards & and greetings from my wife.[10]

# Vol. 5, 255a. To Vladimir Varićak

Zurich, 24 February 1911

[Not selected for translation.]

# Vol. 5, 257a. To Vladimir Varićak

Zurich, 3 March 1911

Highly esteemed Colleague!

I thank you very much for your letter and the paper.[1] I have now read the beautiful study by Lewis and Tolman,[2] but I cannot understand how you can draw from this an ⟨support⟩ endorsement of your opinion. I want to justify my opposite opinion explicitly.[3]

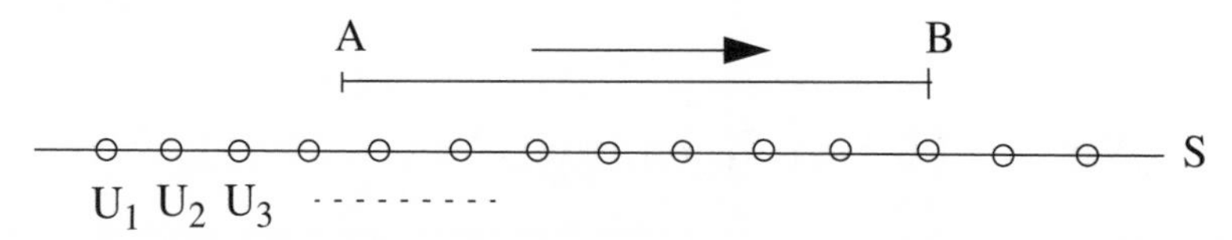


Let *S* be a nonaccelerated frame of reference, in which there are clocks of the same kind at rest with it. Let these be synchronized, e.g. by means of light rays, so that they show the time of *S*. Let the rod *AB* be in uniform motion relative to *S*. Its "real" length, i.e., the length measured by itself, be *l*. Then it follows from the rel. theory in the well-known way that its length with respect to *S* is $l\sqrt{1-\frac{v^2}{c^2}}$. This means: if one determines those clocks in *S*, which show the same state of hands, when the points *A* and *B* are passing by them, then the distance of these points measured in *S* is $l\sqrt{1-\frac{v^2}{c^2}}$. The contraction is observable by measurement, hence "real." In order that you see that the contraction is not simply affected by the definition of simultaneity in *S*, i.e. of a purely conventional nature, I add: it is impossible, to reset the clocks in such a way, that even after this resetting the rod always has the length $l'$ with respect to S, if it has the velocity $\pm v$ measured by means of the clocks. From this one can conclude with Ehrenfest that a rotation without *elastic* deformation is excluded in the theory of relativity, if you assume in addition that a transversal contraction does not take place.[4] One cannot ask whether one has to conceive of the contraction as a consequence of the modification of the molecular forces or as a kinematic consequence from the foundations of the theory of relativity.[5] Both points of view are justified side by side. The latter point of view corresponds roughly to the one of Boltzmann, who treats the dissociation of gases *in a molecular-theoretic manner*; this is completely justified, although one can derive the laws of dissociation from the second law without kinetics.[6] A ⟨principal⟩

difference exists not with respect to the *result* but only with respect to the foundation, on which one grounds the consideration.

If you are going to publish your note, it is my duty to also express my own point of view publicly, because my silence might be interpreted as agreement, and because I believe that your note could create confusion. I therefore ask you to let me know whether you still want to publish it, and in which journal I should publish my response.[7]

With all high respect & and with best greetings your

A. Einstein

# Vol. 5, 267a. To Heinrich Zangger

Prague, Thursday [before 1 June 1911][1]

Dear Mr. Zangger,

You are a splendid fellow to take up the telegraph administration matter so promptly. I am prepared to do whatever might make me useful in this affair. So use me as you see fit in this regard. I am mainly informed about the goings on in Chavan's office & am persuaded that things were done with a complete lack of circumspection and conscientiousness, so any state controls of the material delivered was quite illusory & the work by the office useless.[2] Good suggestions advanced on the part of Chavan always sailed straight into the wastepaper basket, and he personally was harassed and neutralized. All evidence points to lack of expertise on the part of Chavan's superiors, since Chavan saved everything in writing for fear that the facts might be distorted. It will be argued against me as an expert that I could not pass as being objective, owing to personal connections. But they will surely be allowed to interview me; I wouldn't hesitate to travel to Switzerland (from mid-July onward, when my vacation starts).[3] Your faint hints, which you give me in place of more precise information, only make me cross. That's not how one dangles bait. But I earnestly ask you please to visit me in Prague at Pentecost and be our guest.[4] Then I hope to worm the secrets out of you and, on the side, show you the wonderfully beautiful city, the city of these barbarians. The culture of these people really is backward. I haven't discovered any true scientific interest among my colleagues yet, only a kind of arrogance. I close in the hope of seeing you here very soon!

Cordial regards, yours,

Einstein.

# Vol. 5, 344. To Heinrich Zangger

Prague, 27 January [1912][1]

[Not selected for translation.]

# Vol. 5, 349a. To Heinrich Zangger

[before February 1912][1]

Dear Mr. Zangger,

You recently showed me a relation on heat of vaporization. Wasn't it $\frac{D-pv}{v^{2/3}}=$ independent of temperature? If this isn't approximately right, then my thing on capillarity is rubbish.[2] Please jot down for me, if possible, where that article (I think it was in English) can be found. Do you remember what it was that I grumbled about? Youth ends quickly...

With cordial regards, yours,

Einstein.

# Vol. 5, 349b. To Robert Heller[1]

Prague, 1 February 1912

Dear Mr. Heller,

Your little card gave me great pleasure. In the summer I shall soon be breathing the free air of Zurich again! I am enormously happy & will *never* forget that I owe all this solely to my dear friend Zangger. What is bothering him?[2] I note from his letters that he is depressed.[3] How might one do him a little favor? He is surely very isolated and is being aggravated a lot. He will not be given due credit until he turns his back on Zurich sometime.[4]

I am glad that we shall soon be able to resume our relaxed relations again.[5] In the meantime, best regards from your

Einstein

Best regards also from my wife.

Hearty greetings to Mr. Zangger.

# Vol. 5, 374a. To Heinrich Zangger

Prague, 17 March [1912]

Dear Friend,

Don't be angry because I am writing so little. I am up to my ears in the problem of gravitation, so much so that I cannot gather the energy to write a letter. Aren't you familiar with this condition? I can hardly believe it! You are such an energetic fellow and did not indulge yourself that way. But the thing is going well; there is some light at the end of the tunnel.

It will be nice when we are both living on Zurich Hill. I can't come to Zurich before my move to look for an apartment.[1] When we come to Zurich we shall put our things in storage and stay in a guest house until we have found a suitable apartment. (*Nummen nüt gesprengt* [Look before you leap], as the Bernese say.)[2]

I am so sorry that you are still being irritated by the pack of physicians.[3] If only some of my indifference could rub off on you. Right now I am in the middle of an ugly dispute with Nernst, who is simultaneously very offended and shameless, mainly because I dare to raise doubts about his sacred heat theorem.[4] He wrote me, e.g., that he and his "highly talented" pupils were wondering about how superficial my last papers were and that as a senior colleague he would give me his fatherly advice etc. etc. I replied with the advice that he and his highly talented pupils need not waste their time on this stuff, but that I would write it anyway. At the beginning of April I am going to Berlin in order to discuss this matter scientifically with several people there, if possible also with Nernst (but only in the presence of third parties).[5] At the same time I shall be visiting an old uncle there.[6]

Cordial regards from your

Einstein

Best regards to your wife and to Heller, also best regards from my wife to you, your wife & Heller. I congratulate the latter very much on passing successfully his examination. I read it here in "Bohemia." Our colleague Ehrenfest was visiting me.[7] He is a highly intelligent theoretical physicist. He is submitting his Habilitation thesis at the Polytechnic. If Debye should leave, he would be an excellent staff member for the university.[8] I would most like to see him as my successor here. But his fanatical atheism makes this impossible.[9]

# Vol. 5, 439a.To Vladimir Varićak

[Zurich, 14 May 1913]

[Not selected for translation.]

# Vol. 8, 5a. From Heinrich Zangger

[between ca. 14 April and 1 July 1914][1]

Dear friend Einstein,

[Text unintelligible]... Only when you were standing in Berlin as if in front of a cage... and Switzerland in summer is nice when Dahlem is broiling[2] [I thought [then] that you still have examinations to hold at the Poly[3] in July [and since it was July] I didn't want to do anything foolish and miss you by chance. I have finally obtained the proceedings of the [Prussian] Academy, it must be nice... for people who [...] [don't] like to read books like you and me.

You have not yet unlearned to apologize, if you don't write, but neither you nor I believe in an improvement.

Varicac[4] wanted to help the renters in the Hofstrasse, its seems that one battle follows another [there] But there was nothing to do.

They say Abraham will come to Zurich[5] [who will go] to Göttingen, Frankfurt, Debye?[6]

On the 1st of August you come to Zurich!

first because the Berliners will do something out of envy, if there are rays around the sun for your sake[7] [and] don't do it in the photographic [camera] because then it is better one can enjoy the short sleeves [...]

If you know when you are coming, please send a postcard a few days beforehand.

# Vol. 8, 16a. To Heinrich Zangger

[Berlin,] 27 June [1914][1]

Dear friend Zangger,

You're furious about my unbelievable silence and scold me unjustifiably calling me a Berliner; but a Berliner always has his mouth open and he can manage a lot with that![2] Whereas I have a semipathological inhibition toward writing, as generally with any action that requires any definite decision. That's how tiredness and age manifest themselves with me.

Life appeals to me very much here, I have to admit. The sheer amount of competence and glowing interest in science one finds here! I'm repeatedly fascinated by the colloquium and the phys. society.[3] And the people, you ask? They are basically the same as everywhere. In Zurich they feign republican probity, here military rigidity and discipline, but here, as there, they're governed by the same drives, and here, as there, only a few rise above raw instinct.

Despite being in Berlin, I am living in tolerable solitude. But here I have something that makes for a warmer life, namely, a woman whom I feel closely attached to, ⟨namely⟩ a cousin of roughly the same age. She was the main reason for my going to Berlin, you know.[4] Surely I told you so already at the time.

At the academy things vary, sometimes boring, sometimes highly interesting. People there are highly heterogeneous; everyone is one of its kind. Everything I used to think about Berlin people has vanished into thin air. One could think, no doubt influenced by political sentiments, that they are conceited imperialistic Berliners, who want to gobble up the world, know it all, etc. Such do exist, along with others, as everywhere. Only in matters open to change do local traits emerge: dutifulness, an almost mind-numbing need to follow the herd, authoritarianism, lack of taste, respect for acknowledged achievements.

My work is moving within a rather modest scope. Just now I'm writing about quantum theory and Nernst's theorem. There is still very much left of value in the latter, although it cannot be generally valid. The man struck lucky; he's mad at me because I'm not a silent worshipper.[5] The theory of gravitation encounters as much respect as it does suspicion.[6] Lorentz offered a detailed lecture on it.[7] The solar eclipse is being received by the astronomers with good weapons. I am absolutely convinced of the correctness of my theory. I haven't worked on it here yet, though I've recovered physically to do so, mainly through the solicitude of my relatives.[8] I was quite seriously worn out!

I'm happy that you have been able to resume your work again. If I can, I'll come to visit you in the summer. A bit of Berlin, as I have it, would do you good as well: no obligations and worries! I didn't understand your insinuation about Varicak.[9] He is—as far as I can judge, a good sort of fellow. He had a kind of relationship with my wife, which can't be held against either of them. It only made me feel my sense of isolation doubly painfully: I am all the more grateful now that fate has granted me the affection of a good woman after all.

Don't be angry at me anymore, but write me in comfort during a genuinely leisurely spare time, yours,

Einstein.

Best regards to Heller!

# Vol. 8, 34a. To Heinrich Zangger

[Berlin,] 24 August 1914

Dear friend Zangger,

My deepest condolences on the heavy loss you suffered. I knew your dear mother as an excellent mother.[1] I saw many a tear in her eye when you were so sick and an indescribable joy on her face when you were slowly regaining your strength.[2] It never became clear to me where your refinement and sensitivity and your quick reflexes came from, besides the Germanic blood coursing through your veins.

What a horrific picture the world is now offering! Nowhere is there an island of culture where people have retained human feeling. Nothing but hate and a lust for power![3] The question, where can justice be found? is becoming sheer mockery. One lives the life of a stranger on this planet, happy when one isn't done in for outmoded sentiments. I feel so strangely drawn to early Christianity and feel as acutely as never before how much nicer it is to be anvil than hammer. What galls me most is that now even the best talent is being forced into this senseless butchery and henchman's service. I have blind luck to thank that I was spared this.

I sit all day long peacefully at work in my lodgings. Yesterday I discovered a pretty method for deriving the fundamental formulas for the absolute differential calculus indispensable to gravitational theory.[4] My wife and my boys are in Zurich. In the future they will live apart from me, hard though it is for me to be without my boys. I could not stand the wife any longer. It is hardly conceivable to me now how I wasn't able to summon the moral energy to come to this decision.[5] In part it depended on the fact that my means could not have permitted living separately.

The observations of the solar eclipse have surely been suppressed by Russia's floggers, so I won't live to see the decisive results about the most important finding of my scientific wrestling.[6]

I hear that Debye will be Kleiner's successor. An invaluable acquisition for your university![7] Abraham seems not to have made it to the Poly, mainly because of Weiss.[8] This will surely have grim repercussions. I fear that my success⟨ion⟩or in Zurich is not going to be chosen with sufficient expertise and objectivity.

Cordial regards from your

Einstein.

# Vol. 8, 41a. To Heinrich Zangger

[Berlin, after 27 December 1914][1]

Dear friend Zangger,

Many thanks for your friendly lines. I almost came to see you to make arrangements about caring for a son of Planck's who was wounded and captured in France.[2] But he is now out of danger & on his way toward a recovery.

The world is like a madhouse now. What drives people to kill and maim each other so savagely?[3] I think, in the end, it is the sexual character of the male that leads to such wild explosions from time to time, *if it's not prevented by careful organization*.[4] The special calamity of our times, however, is that bestial instincts together with existing tools are leading to veritable destruction. The real betterment of the masses moves so slowly ahead in relation to the rapid development of technology that there is now a disparity of the worst kind. We must therefore, in my opinion, strive toward a large-scale organization that acts toward individual states the way the latter act toward individual thieves. But the *bestia masculina* still resists this again one by one. That is how a completely dispassionate person like me can appear *lacking* to others. It's all the same to me. As long as I am left in peace, I calmly continue my work with the usual pleasure, without letting myself be infected by mass psychosis. What hurts me most bitterly, though, is that people like me are prevented by the force of circumstances from acting the way I do.

I have had the particularly good fortune of separating from my wife.[5] I know very well that, from the point of view of others, this looks like unparalleled brutality. But for me it was a question of survival; my nerves could not have withstood any longer the pressure that had been bearing down on me for years, which this barbaric character had been exerting on me. It is an act of defense in the fullest sense. You can imagine what a difficult decision it was to forgo watching my beloved boys grow up! For them it is better not to grow up in a home in which the father and mother confront each other like enemies. It is also beneficial for the development of their social awareness and habits if they grow up in socially relatively well-balanced Switzerland, where the teaching in schools is executed without ulterior motives.

Of late I have been working incessantly on gravitation and have now arrived at wonderful clarity about the interrelations. The matter is now so compelling that anyone really gaining profound insight must find it difficult to shake himself free from the spell of this construct. If all physical events are completely determined by other perceptible physical events, then the chosen path becomes a necessity.[6]

The philosophical quotes you relayed to me do not rouse an appetite for more. These fellows haven't the least feel for the narrow constraints of the terms they are fumbling around with. When I come and visit you sometime, let's take a closer look at one of them and tear it apart; it would be very interesting for me to experience such a thing with you, but not flaccid Bergson.[7]

The size of the city here gives occasion that one gets to know highly interesting people.[8] The advantage is that any given nature is empowered to the point that one can choose one's activities and the people that affect one.

Be embraced by your old friend,

Einstein.

Do write me again soon but slowly and leisurely, simplistically, and with *legible* writing. You know well I am a *nonmedical man*.

I live here very cozily all by myself in a comfortable small apartment;[9] the housekeeper tidies up. As soon as better times arrive, you must turn up in my den.

# Vol. 8, 45a. To Heinrich Zangger

[Berlin,] 11 January 1915

Dear friend Zangger,

Your letter and your postcard arrived. As a man with faith in society you are seeking to assuage the terrible suffering that people are now falling victim to, and are finding solace in action.[1] The likes of us, however, are so disgusted by the abundance of incomprehensible, puzzling, and unspeakable ugliness that we crawl even deeper inside our snail shell of contemplation. You probably received the paper on general relativity;[2] it is a successful end to my struggles in this area. Accept this booklet as a sign of amicable feelings, not as a presumption that you should immerse yourself in such ponderings! At the present time I am collaborating with de Haas (a young Dutch man, Lorentz's son-in-law) on a very interesting experimental matter about the nature of magnetism.[3] The issue in question is whether paramagnetism can really be attributed to revolving electrons; the goal is surely attainable. As soon as the analysis is completed, I'll send you an offprint. Here scientific life is almost at a standstill. Everyone is working and suffering for the state, in part voluntarily, in part involuntarily.[4] If only one could do something about it so that the former relatively harmonious conditions were reinstated! But in this state of agitation such lack of passion is offensive to those who are more deeply stuck in this affair.

Recently Edgar Meyer wrote and asked me to do something for him with regard to obtaining Kleiner's position in Zurich.[5] But I believe that I would rather harm than help his prospects if I put in a word for him without being asked. I tell you, though, that Meyer is an excellent physicist, who would be more useful to the university than most others. Couldn't you perhaps suggest that I be asked?

It's good that among the fine minds strikingly many are considered cripples by the state, especially among theoreticians. Debye, Born, and Laue were all deemed absolutely unfit.[6] The latter told me recently that he had insurmountable difficulties learning the rifle drills—a pretty illustration of the manifesto published by my colleagues, according to which "we owe our scientific prowess not least to our military education."[7] In recent days I made the acquaintance of our colleague

Natanson from Cracow, a fine theoretical mind.[8] He is a Polish Jew and grew up in Russia, now 50 years of age. I quickly took a liking to him as I rarely do with people; blood runs thicker than water! Dear Zangger, make sure that no wimps are called to Zurich as physicists, either to the Polytechnic or to the university; once you have them, you never get rid of them. Take care that Keesom does not come to the Polytechnic as a theoretician;[9] he knows a lot, has original ideas to a certain degree, but cannot teach people how to think critically.

With cordial greetings and best wishes for the so sadly started 1915, your old

Einstein.

# Vol. 8, 69a. From Hans Albert Einstein

[Zurich, before 4 April 1915][1]

Dear Papa,

We got the postcard; I don't know what you meant to say by it, though. But I hope you were thinking: "Well, it would, admittedly, be nice, we'll see."[2] Imagine, Tete[3] can already multiply and divide, and I am doing gometetry (geometry),[4] as Tete says. Mama[5] assigns me problems; we have a little booklet; I could do the same with you then as well. But why haven't you written us anything lately? I just think: "At Easter[6] you're going to be here and we'll have a Papa again." Yours,

Adu!

# Vol. 8, 69b. From Hans Albert Einstein

[Zurich, before 4 April 1915][1]

Dear Papa,

Today we told each other our dreams. Tete suddenly said:[2] "I dreamed that Papa was here!" Then I thought: "It really would be much nicer if you were with us." I can tinkle away on the piano much better now already; not long ago I played a Haydn and a Mozart sonata and some sonatinas.[3] In short, I could also play with you. The examination is approaching now; but at the same time, so is Easter.[4] Last Easter we were alone;[5] do we have to spend this Easter alone as well? If you were to write us that you are coming,[6] that would be the finest Easter bunny for us. We can live here quite well, you know, but if Mama[7] gets ill one day, I don't know what to do. Then we would have no one but the maid. Also for this reason it would be better if you were with us. Yours,

Adu.

# Vol. 8, 91a. From Hans Albert Einstein

[Zurich, 28 June 1915]

Dear Papa,

You should contact Mama[1] about such things, because I'm not the only one to decide here.[2] But if you're so unfriendly to her, I don't want to go with you either.[3] We have plans for a nice stay that I'd only give up very reluctantly. We are going at the beginning of July and are staying the whole vacation. It's very high up.[4] Yours,

A. Einstein.

# Vol. 8, 96a. To Heinrich Zangger

[Sellin,] 16 July 1915

My dear friend Zangger,

Your friendly lines greatly impressed me, not by their content but because I see what an active interest you are taking in my fate.[1] In the matter itself you are mistaken. My fine boy had been alienated from me for a few years already by my wife, who has a vengeful, ordinary disposition, but also is so sly that outsiders and particularly men are always deceived by her. If you only knew what I had to live through with her, you would hold it against me that I did not find the energy for so long to separate myself from her. The postcard I received from little Albert had been inspired, if not downright dictated, by her. It said: "As long as you aren't friendlier with Mama, I don't want to go with you. Anyway, we're going into the countryside in July and I don't want to give up that stay."[2] Where they wanted to go was not conveyed to me, not even the new address, which I only learned about from you. When I write to Albert, I get no response at all.

Under these circumstances it appeared as if I couldn't see the children at all if I came now to Zurich in July, as I was firmly resolved to do. So at the last minute I decided, while I was at Göttingen giving talks about the general theory of relativity,[3] to relax here in Sellin, where my cousin had rented lodgings with her children.[4] I'm going to stay here until August 1st, because I need that much rest. From the 1st of August I am ready to come to Zurich, even if my children are so incited against me that they don't want to have anything to do with me; then I'll come to see you again. Give me a time period between 1 August and 1 October; I will certainly come. I would surely have been there on the 15th of July if I hadn't been deterred by the ugly postcard.[5] I left the children to my wife; she shouldn't fill them with animosity toward me, less for my sake than for the children's, whose moods are dampened by it.

In your reply please also write me what's wrong with my little boy.[6] I'm particularly fondly attached to him; he was still so sweet to me and innocent.

Do answer soon, your truly grateful

A. Einstein.

# Vol. 8, 122a. To Heinrich Zangger

[Eisenach, 24 September 1915]

Dear friend Zangger,

See over there the man of strong conviction, who always knew what he wanted and what he should do, the happy one![1] I just saw his small work room, that picturesquely overlooks the countryside. It was the last stop on the romantic trip home.[2]

Best regards! Your

Einstein

# Vol. 8, 124a. To Heinrich Zangger

[Berlin, 4 October 1915]

Dear friend Zangger,

I wrote to my Albert in detail. I want to *arrange* somehow that he stay with me here in my apartment for a whole month each year. Only then can he be relieved of all that pressure.[1] During that time I will devote myself to him to a large degree.

I do remember the fact that *an* American wrote me. Unfortunately I forgot to reply to him. I don't even remember what he wanted from me. What can be done? Did you receive the very interesting printed piece I sent to you, I wonder? Probably not. It was a very conciliatory manifesto that I would gladly not begrudge Romain R.; the best people in the country sig[ned].[2] Recently I visited Planck.[3] Such a sincere affection between coworkers in science as between us is rare indeed. How is your book doing?[4] When are you coming? 8 days ago I was with Stodola. He made a wonderful invention.[5]

Cordial regards from your

Einstein.

P. S. If my wife wants something in writing, I would file for a divorce at court.

# Vol. 8, 144a. To Heinrich Zangger

[Berlin, 15 November 1915]

Dear friend Zangger,

I thank you heartily for the detailed news. I was silent because I have been laboring inhumanly, but also with magnificent success.[1] I modified the theory of gravitation, having realized that my earlier proofs had a gap. Now the thing has become very simple and easy to understand.[2] *I have now derived from the theory the hitherto unexplained anomalies of planetary motion.*[3] Imagine my joy! I shall be glad to come to Switzerland at the turn of the year in order to see my dear boy. I am writing him at the same time.[4] I am gratefully obliged to you more than I can say for your conciliatory mediation.

Warmest greetings from your

Einstein.

My wife wrote me recently about the children in a manner that appears genuine. I am going to try to arrange my meeting with my Albert directly with her.[5]

# Vol. 8, 154a. From Hans Albert Einstein

[before 30 November 1915][1]

Dear Papa,

I'd like to answer you now:[2] I will come over New Year's, i.e., from the 31st to 2nd; I'm thinking of going on the Zugerberg.[3] I don't want to stay longer because Christmas is nicest at home. Besides, I got skis and would like to learn how to use them with my colleagues.

The ski equipment costs about 70 francs, and Mama[4] bought them for me on condition that you also contribute. I consider them a Chrismas present.

I'm also thinking of taking the skis along to the Zugerberg, in case it has snow.

Yours,

Adu.

# Vol. 8, 159a. To Heinrich Zangger

[Berlin, before 4 December 1915]

Dear friend Zangger,

Just now I received the enclosed letter from my Albert, which upset me very much.[1] After this, it's better if I don't take the long trip at all rather than experience new bitter disappointments. The boy's soul is being systematically poisoned to make sure that he doesn't trust me. Under these conditions, by attempting any approaches I harm the boy indirectly. Come, dear old friend, Lady Resignation, and sing me your familiar old song so that I can continue to spin quietly in my corner!

So we'll see each other again at Easter then. I have to attend a meeting of the Anti-War Council in Berne, the international council into which I let myself be elected.[2] In these times everyone must do whatever he can for the community as a whole, even if it is only slight and ineffectual.

Young Rohrer, who wrote his doctorate under you, sent me his dissertation. If he got the initial idea for the thesis from you, it is ugly of him not to breathe a word about it.[3] Currently I am also having quite a curious experiences with my dear colleagues. All but one of them is trying to poke holes in my discovery or to refute the matter, if only so very superficially; just one of them acknowledges it, insofar as he is seeking to "partake" in it, with great fanfare, after I had initiated him, with much effort, into the gist of the theory.[4] Astronomers, however, are behaving like an ants' nest that has been disturbed from its mindless humdrum by a walker's thoughtless misstep; they're biting away at the walker, without making the least impression on his shoes.[5] All of this is very droll without being unpleasant for me. If one is pressed into playing one's role as an actor in this farce, one is richly compensated for the pain and effort by being able to watch as a spectator the others' playacting.

As I read, Meyer has now become Kleiner's successor. I can't disapprove of the choice, despite the special argument we put forward against it. P[iccard]'s character appears to me in a somewhat tarnished light because of his unkind attitude toward his teacher and honest benefactor W[eiss], to whom he owes his knowledge and position.[6]

Heartfelt greetings, yours,

Einstein.

## Vol. 8, 161a. To Heinrich Zangger

[Berlin,] 9 December 1915

Dear friend Zangger,

Late yesterday your letter arrived, today the enclosed one from my wife.[1] This last letter, ⟨which I am enclosing,⟩ makes such a very convincing impression of honest good will that I also consider it right if I yield to my feelings and travel there—despite all the earlier bad experiences. So I'm really going to try and go somewhere with Albert (Zugerberg), in order to spend a few days with him undisturbed.[2] Then, to my great joy, I'll see you and Besso again as well and if possible also look up Dr. Zürcher to thank him for his kindness.[3]

I am quite overworked from the extraordinary exertions of the last few months. But the success is glorious.[4] The interesting thing is that now the initial hypotheses I made with Grossmann are confirmed, and the most radical of theoretical requirements materialized.[5] At the time we lacked only a few relations of a formal nature, without which the link between the formulas and already known laws cannot be attained. The matter is beginning to sink into my colleagues' minds as well. In 10 or 20 years it will be a matter of course. . . .

It seems to me almost better if all of you don't speak with my Albert. He could become accustomed to falseness far too soon if he sees that certain feelings are expected of him. It was a big mistake of mine to be annoyed about his postcard.[6] Such a thing shouldn't happen to a reasonable man of my age. Being overworked is, however, connected with heightened irritability; that excuses it somewhat. The one-sidedness of the relationship between parents and children is, after all, a natural law; the harshness of this arrangement is in general only softened by careful training—see the (fourth?) commandment of the Old Testam[ent].

I'm eagerly awaiting our next cozy chat. Best regards, also to your wife and your little ones, yours,

Einstein.

# Vol. 8, 185a. To Wilhelm Wirtinger[1]

[Berlin, 26 January 1916]

To the Dean of the Philosophical Faculty of the Imperial and Royal University of Vienna

Highly Esteemed Sir,

Regarding your inquiry of 19 January 1916, I am informing you that I would take under very serious consideration an offer of a position at your university.[2]

With great respect,

A. Einstein
Wittelsbacher St. 13
Berlin-Wilmersdorf

# Vol. 8, 196a. To Heinrich Zangger

[Berlin, 1 March 1916]

Dear friend Zangger,

My silence is a disgrace. But I'm coming at the beginning of April; then we can chat again. I've been working enormously on a final formulation of the general theory of relativity, which has now penetrated completely.[1] A simple experiment to demonstrate Ampère's molecular currents is now finished and tried out, too.[2] I'll demonstrate it to you all in Zurich as well. Besso will get a kick out of that.[3] I'm very pleased about the completion of his dissertation as well as about the fact that you seem to have become so much closer. Give him my warm regards. I'm feeling very well. I live in complete seclusion, work away and—hold my tongue. Soon we'll be sitting together, Areopagus-like, on your wonderful veranda. But this time you must allow me to live independently when I come to Zurich, if only for my boys' sakes, who are less hurt by it. I still have to work intensely right up to my trip.

Cordial greetings to you, your wife,[4] and the children from your

Einstein.

Fond greetings to Besso! Pray for me, a poor sinner, because I still haven't written [him.] Tell him that I decided on the formality of marriage with my cousin after all, because her grown-up daughters are otherwise seriously harmed by me.[5] It doesn't signify any injury either to me or my boys, but is my duty, which I mustn't dodge. Nothing in my life changes by it. Why should the original sin be even harder on these poor daughters of Eve?

# Vol. 8, 209a. To Elsa Einstein

[Zurich, 6 April 1916]

Dear Else,

I just arrived here[1] 2h 45 and have taken lodging at the "Gotthard."[2] I am writing to my boys as well as to you.[3] Few people crossed over the border. The inspection was very thorough, though, but entirely decent and polite (in Lindau).[4] Jacket & vest off, shirt opened; even trousers down, collar off. Every single piece was searched through. But the young official performed the procedures with much grace.[5] One can see that an experience depends more on *how* it's done than on *what* it is. The train passengers were lackluster until the Romanshorn–Zurich[6] stretch, when I conversed with a very worldly-wise young Swiss. The trip along Lake Constance was magnificent. I am enjoying myself and, in contrast to the last time, even my body is holding together well.[7]

A kiss from your

Albert.

Affectionate greetings to Ilse and Margot, likewise to Uncle & Aunt.[8]

# Vol. 8, 210a. To Elsa Einstein

[Zurich,] Saturday. [8 April 1916]

My Dear Else,

I hope my communications to you are reaching you better than yours reach me; for I haven't received anything yet from Berlin. Tomorrow I'm going with my Albert[1] on the excursion. We are walking Albis–Zuger Lake–Vierwaldstätter Lake–Seelisberg.[2] Today I received a letter from M[ileva][3] in which she denies having declared herself ready to file a divorce suit.[4] At the same time she requested discussing this in person. I rejected this and encouraged her by letter to file the claim.[5] Today I went on a sailing trip with Besso.[6] Life here is very unsettling for me, because I know so many who make a personal claim on my time. Today I had an interesting scientific idea. That means work again in Berlin. I wrote Ludwig Kraft a postcard.[7] This evening I was at Neter's, who is very reasonable.[8] I believe I've arranged everything well. He's not such a bad sort. Besso is very nice to me. He doesn't make the slightest effort to give me advice, nor do I encourage him to do so at all either.

Kisses from your

Albert

Greet Ilse and Margot affectionately for me. Also give my regards to Uncle & Aunt. Maja is at Lake Lugano but should be coming home soon.[9]

# Vol. 8, 211a. To Elsa Einstein

Zug. Monday. [10 April 1916]

My dear Else,

Yesterday I went on the hike[1] with the boy and am enjoying very much being with him. He is kindhearted, trusting, and surprisingly eager to learn and intelligent. My relationship with him is becoming very warm. Everything is fine except for the weather. That's why we're not going to Vierwaldstätter Lake for the time being.[2]

Kisses from your

Albert.

# Vol. 8, 232a. To Heinrich Zangger

[Berlin,] 11 July [1916][1]

Dear Friend,

Your long letter, in which you informed me about how my boys are faring, pleased me very much, but it also filled me with a certain concern in one respect.[2] Whenever my wife[3] confided in any one of my friends, I almost always had to give him up for lost. Besso alone counts as a famous exception.[4] So don't allow the slightest drop of venom into your subconsciousness. It would be such a pity on our fine relations. Surely not that I believed the woman would complain about me outright; it's a matter of indirect influence on the emotions, by which women so often get the better of us.—[5] My relations with the boys have frozen up completely again. Following an exceedingly nice Easter excursion, the subsequent days in Zurich brought on a complete chilling in a way that is not quite explicable to me.[6] It's better if I keep my distance from them; I have to content myself with the knowledge that they are developing well. How much better off I am than countless others, who have lost their children in the war! Planck also lost a son like that, the other one has been languishing in French captivity for almost 2 years.[7]

I'm extremely happy that Besso adjusted so well to his new profession.[8] He has your stabilizing influence largely to thank for that. The few days I spent with him during the Easter holidays were exceedingly fine.[9] He is one of the best minds I ever encountered and is at the same time an excellent fellow.

This vacation I'm not going to come to Switzerland, for one because I have confidence that peace is coming soon, so next spring I could travel there without those chicaneries; on the other hand,[10] because revisiting my boys would be rather more painful than enjoyable.

Concerning science, I'm only working on smaller things now, living a more contemplative life and appreciating the work of others. The general theory of relativity has now penetrated to the point where I can regard my task in this connection as completed.

I shut my eyes as best I can to the insane goings-on in the world at large, having completely lost my social consciousness. How can anyone merge in such a social monstrosity if one is a decent person? A fleeting glance at the newspaper is enough to make one disgusted with our contemporaries.[11] One can find solace only in certain individuals.

Cordial greetings, yours,

Einstein.

# Vol. 8, 237a. To Heinrich Zangger

[Berlin,] 19 July 1916

Dear friend Zangger,

In this upsetting affair I feel highly fortunate to have such genuine and good friends as you and Besso, who settle and mediate everything so calmly and surely under the hardest of circumstances.[1] Pardon me as well for not being impressed as earnestly as circumstances would have warranted upon first notification. At any rate, the symptoms seem to have been, and partly still are, very serious indeed. Isn't it possible that the nerves are behind it all?[2] It's good that the boys are now in the care of Mrs. Savić, a very good and intelligent woman (Viennese), married to a Serbian civil servant.[3] She is a former college classmate of mine and of my wife, her best friend, a person of much kindness and intelligence. If you have a chance, don't fail to discuss things thoroughly with her, especially considering that her impressions and experiences must be very interesting.

I considered whether I shouldn't take the boys myself during this period, but have arrived at the conclusion that this wouldn't be right. It would only add to my wife's irritability. Moreover, I would be subjecting the children to an inner conflict (mother or father?), no matter how careful my conduct. The experiences of last spring made me resigned to this relationship. First trust and warmth, then icy coldness.[4] The woman simply suffered under the impression that the bigger boy might become too attached to me and worked ⟨perhaps⟩ probably unconsciously so energetically against it that a very unfortunate abrupt change took place in the boy. Before the boys are grown up I'm not going to be allowed to play any substantial personal role in their lives, but only watch out that no particularly serious errors are made in their upbringing. I fear and believe that the worst deficiencies lie in the bodily area (cleanliness, teeth). I try in vain to make sure that the necessary things are done in this regard.

Dr. and Mrs. Zürcher also seem to be most solicitous toward my wife and the children, in the most selflessly generous way. I'm going to write them.[5]

Scientifically, I'm having a little breather. I examined gravitational waves,[6] most recently the quantum theory of light emission and absorption,[7] and the causes of lift in flight.[8] On the latter topic I also wrote a short popular article that was originally intended for the publication in honor of Kleiner;[9] I'll send you the article once it's printed so you can read it when you have time to sit back.[10] With Kleiner we lost a person who was closely connected to my life, at least as concerns its external development. He was an honest fellow, rare as they are to come by of late, but untalented as a scientist.[11]

The war is educating people here most propitiously, or better put, the food shortage is.[12] If it continues systematically like this, as seems to be the way things are going, the fellows will become quite likable yet. No person and no nation can take much outward success without damage.

Daily reports on my wife's condition are not necessary. I'm entirely satisfied if you or Besso inform me briefly on a postcard if there is anything urgent or if anything new is revealed about the condition.

Cordial greetings, yours,

Einstein.

# Vol. 8, 242a. To Emil Zürcher Jr. and Johanna Zürcher-Siebel

13 Wittelsbacher St., Berlin, 25 July 1916

Highly esteemed Mr. and Mrs. Zürcher,[1]

My friends Zangger and Besso inform me in every letter about the devoted and magnanimous way in which you are supporting my family.[2] I know that you are alleviating my wife's hard lot and thank you wholeheartedly.[3] I did not know how seriously ill she is until now, or better said, I *did not believe* in the severity of the situation.[4] But now I am shocked to see that the basis of the troubling letters is unfortunately quite real.[5]

I urge you please to apply your influence that (at my cost) an experienced and reliable nurse be engaged, if this hasn't already happened and if my wife does not prefer to be taken to a hospital.

The most difficult and perplexing problem for me now is the care of the children, during my wife's presumably lengthy illness, when they return from vacation. Most obvious would be that I come to Zurich then and take care of the boys during the day. Perhaps I'll even do so and ask you please to inform me in any event when they are coming back again.[6]

Unfortunately, however, grave reservations speak against my coming. You know that after a long, painful marriage we have been living separately for 2 years now at my instigation and that I have been strictly avoiding meeting my wife since then to avoid new upsets.[7] My trips to Zurich were only intended for my boys.[8] If I come to Zurich now, I fear that the wife will demand that I visit her. You can imagine how difficult it would be to decline such a request under these conditions. If ⟨this happens⟩ I go to see her, it will not only cause considerable agitation but I could find myself compelled to promise things that would separate me from my dear boys in the future as well. I tell you this so that you understand my anxious hesitancy. If, however, you and my friend Zangger deem my coming desirable in the interest of the little ones, then I would come as soon as you, with your better insight into the circumstances, consider it appropriate.

In sending you my best wishes, I remain in deep appreciation, yours,

A. Einstein.

# Vol. 8, 247a. To Heinrich Zangger

[Berlin,] 3 August [1916][1]

Dear friend Zangger,

I just received your long letter and am happy beyond words that such a kind spirit is hovering over my family there. But you really shouldn't take the children in, because a decent housemaid is there and my Albert really is quite reliable already, particularly in taking care of his little brother. The maid could keep a household account, and some trustworthy person could take a look at the book and give the maid housekeeping money and general instructions. Not only will it work that way, but Albert and the maid will even learn something in the process. My wife should be sent to the Theodosianum as soon as possible, if only to let you leave in peace on vacation.[2] The maid would then be relieved. She has had much to do in any case. *You shouldn't miss any of your hard-earned holidays.* Just so you know, I can tell you that in the past two years I have saved about 10,000 marks, so you don't need to worry about costs. The possibility of putting aside substantial sums for my children is one of the best things about my position here. If I live for a few more years, the children's education will be adequately provided for.– Do you have a clearer picture of the nature of my wife's illness yet?[3] I don't understand the condition in the least ("understand" meant in the simplest sense of the word).

Unfortunately, I again could not completely read many things in your letter. The regulations of August 1st don't bother me much.[4] I won't have to walk around naked, since I have a sufficient supply of clothes and my aesthetic demands in this regard are minimal. My insight into human nature does not teach me very much that's of interest. Taking events as they come, like hail heaven-sent, a helpless yearning for peace, a lack of understanding of the causal relations and of the psychology of others all the way up to the elite intellectual "know-it-alls" (and loudmouths), wistful conviction in one's own integrity and others' depravity, in short, stupidity and still more stupidity. Even the financial robber barons you are complaining about are no worse than the others. They are acting—like the others—in compliance with the training that has fallen to their lot, and consider themselves good, useful, and irreplaceable. Everywhere [there is] an unhealthy overrating of power and money and an indescribable ravaging of the soul.

I am living very tranquilly and reclusively and am content. I am on the best of terms with my colleagues and other people, partly because I understand them, partly because they—thanks to my reputation—find everything about me fine and good. This reputation holds up to vigorous tolerance tests. But beware when it swings the other way!

Scientifically, I have found a few nice things again. Now I've fully mastered airfoil theory. You'll soon receive a brief elementary article about this, which contains only the very basics, of course.[5] Furthermore, I found a nice addition to the quantum theory of radiation and gravitation theory (gravitational waves).[6]

Switzerland's problems are often the topic of discussion here as well.[7] Switzerland's situation is simply, by nature, very difficult. I do have the impression, though, that the authorities earned the mistrust of *both* parties involved from their lax supervision of the negotiations.

Soon you will get the offprints. Warm greetings from your

A. Einstein

# Vol. 8, 250a. To Heinrich Zangger

[Berlin,] Friday. [18 August 1916]

Dear Zangger,

Everywhere misery. Huguenin ailing and being operated on,[1] Besso sick, my wife fatally ill, my children isolated.[2] How did Huguenin's operation go? What's wrong with him? Send him my best regards. I hope he recovers very soon, likewise Besso, whose whereabouts I don't happen to know right now. He was on the Planalp near Brienz.[3] The sad fate of my wife affects me deeply. Your last letter together with the *Konversationslexikon* clarified the case for me as far as is practically necessary. At the age of 20 or 21 my wife had a gland operation. It was without a doubt a tuberculous infection.[4] Now the business is spreading further in the brain. I know very well that there is no hope left, but that the only question is how long the torment will last. I was so ignorant as not to know the link between glandular degeneracy and tuberculosis. Otherwise I would not have been so foolish as to have children with this woman. Up to now it looks as if they haven't contracted it. What luck that we can't look into the future; otherwise it would be intolerable.

It's possible that my wife's sister is going to Zurich for the children.[5] This would be a true consolation for me because the children like her very much. But when my wife's illness is advanced to the point that she has lost consciousness, I'll immediately take the children to live with me. I'll not send them to any school but will teach them privately, partly myself, and then send them to Switzerland for the last four years of schooling.[6] For I won't let them be Prussified on me, they have to become proper Swiss, who feel at home there and will later take up there lives there.

In the past few weeks I discovered a very nice bit of science. I'm sending you a short publication about it along with this letter.[7] The continuation will appear in the memorial issue for Kleiner by the Zurich Physical Society.[8] You might not be able to read it easily on your own. But Besso will explicate it to you. It deals with the underlying meaning and derivation of Planck's radiation formula. I'm also sending you a copy of the exposition of the general theory of relativity without the presumption of your really reading it.[9]

I'm looking forward to the time when I can shake your hand again. Without you and the Zürcher family[10] I would not have known what to do. It is hard knowing that one's loved ones are in such distress and not being able to help personally. Write me now candidly without the usual precautions about the facts and also about my children, and the way they are bearing their ill fortune, and accept my heartfelt greetings, yours,

A. Einstein.

Cordial thanks also to your wife.[11]

# Vol. 8, 261a. To Heinrich Zangger

[Berlin,] 26 September 1916

Dear Friend,

Now, after much patient effort, I managed to arrange that I be able to leave tomorrow morning to see my Dutch colleagues for a fortnight, in order to debate about the new scientific issues with them. Nordström is there, too.[1] I'm very much looking forward to it. But I can only send the 1400 francs upkeep for my family to Zurich around the 20th rather than the 1st of October.[2] Should I send the money to your address? For I don't know exactly who is holding the household reins.[3] My address over there is Prof. Ehrenfest, Leyden.[4] Please have the hospital and doctors' bills sent directly to me.[5] I hope you had a good and relaxing holiday and that you and your family and the whole of little Switzerland are experiencing good times, better than the people here.[6] I'm gradually beginning to understand now what a treasure people had with the Christian ideal of humility. Who is going to replace the power icon with a merciful God again?[7]

The tide is at ebb inside my brain case at the moment, but I'm completely satisfied with that. Life is very quiet. I hope I receive another card from you soon with a placating report.

Best regards, yours,

Einstein

# Vol. 8, 261b. To Elsa Einstein

[Leyden,] Thursday. [28 September 1916]

Dear Else,

I arrived here safely yesterday at ½ past 10 and then talked with Ehrenfest and Nordström until 1 o'clock,[1] at which time I was fed according to the rumors about the reception of travelers from the East.[2] The local people, with their simple, unassuming manner and high intellectual culture, appeal to me exceedingly. Such a huge difference! I'm blissfully happy during the discussions. It was really good that I took all that trouble[3] in order to be able to breathe in this air for a time. But I'm not going to forget you all, and especially you, because of it, and am not ungrateful.

Affectionate greetings also to the little minxes and the elders,[4] from your

Albert.

# Vol. 8, 261c. To Elsa Einstein

[Leyden,] Saturday. [30 September 1916]

Dear Else,

I haven't been able to think about nature and the arts until now, because physics and the physicists have been keeping me in breathless suspense.[1] Today I am going to Amsterdam to the Academy.[2]

More next time. Affectionate greetings also to the little minxes, from your

Albert.

Greetings to Uncle & Aunt. I have not been able to attend to the lard yet.[3]

# Vol. 8, 262a. To Elsa Einstein

[The Hague,] Thursday. [5 October 1916]

Dear Else,

Yesterday I was with Lorentz,[1] who lives in the countryside in Bussum near Amsterdam, all day long. We took a long walk to Zuider See, conducting very interesting conversations all the while. In the evening I drove to Haarlem to the De Haases, who gave me a touchingly warm welcome.[2] We chatted until ½ past 1 o'clock. Then I slept at their house, was in the laboratory there this morning,[3] and now (1–2) am traveling here to The Hague, where I'm arriving in the pouring rain and am quite tired. But I had to be here exactly today. It's a true test of the nerves.[4] I still want to have a breather the last few days before my trip homeward. Incidentally, I stayed completely healthy throughout. My new theories really have come alive here, because the best theoreticians are working on them.[5] The day after tomorrow I'm also going to Groningen with Ehrenfest,[6] where I'll be spending the night.

Tender greetings also to the little minxes and the elders,[7] from your

Albert.

# Vol. 8, 262b. To Elsa Einstein

[Leyden,] Saturday. [7 October 1916]

Dear Else,

This greeting is in haste. This evening the Groningen colleagues[1] are coming over here again to discuss relativity. So I don't need to travel to Groningen.[2] I can't get any lard. I have made inquiries at an official authority. Thus you will all have to receive me lardless but with kindness all the same.[3] The days spent here were unforgettably interesting and also very pleasant. My theory has found itself a true home here.[4] The personal culture of the local people here . . . . . . .[5] But even so, I am looking forward to you and our quiet life no less than before. Greetings & a kiss from your

Albert.

# Vol. 8, 263a. To Paul Bernays[1]

[Berlin,] Friday. [13 October 1916 or after 1920][2]

Dear Colleague,

The orientational sense of time exhibited by living organisms everywhere is intimately connected with the second law. It primarily involves processes of diffusion, irreversible chemical processes, heat conduction, viscous currents, etc. It is entirely correct that this temporal bias of events finds no expression in the fundamental laws we use as a basis. But the theory of relativity shares this circumstance with classical mechanics, likewise with conventional electrodynamics and optics. The second law is understood such that a very improbable state is set for one temporal limit (lower $t$-limit) of a four-dimensional region under consideration; for the region's upper $t$-limit the probability considerations then yield a state of greater probability. The puzzle is thus transferred into the boundary conditions and therefore avoids "explanation" by means of the equations.

There certainly is an unsatisfactory element to this procedure. But all previous theories offering an analysis of irreversible processes (e.g., kinetic gas theory) are the same as the theories of relativity in this regard. The fundamental equations are designed such that the equations are also satisfied by the temporally inverse process.–

Please pardon the sloppy drafting! A mountain of letters and other obligations awaited me here. For I returned yesterday evening from Holland. Wishing you bon voyage and a profitable semester, yours,

A. Einstein.

# Vol. 8, 263b. To Heinrich Zangger

[Berlin, 13 October 1916]

Dear friend Zangger,

Having returned last night from a 2-week trip to Holland, I found your postcard, from which I gather that there is slow improvement with oscillations.[1] I do, of course, agree to do what you deem right in this matter (Veraguth).[2] I'm sending 1400 francs for my family to your address (for October–December).[3] I'll pay *separately* the exceptional disbursements arising from my wife's illness to make sure there's no shortage. Please simply inform me *what* I should send *where*. Such payments shouldn't be delayed, because my ⟨interest⟩ conversion losses are enlarged by waiting. In general, in these uncertain times one should settle everything that's possible, since future difficulties are unforeseeable. This morning I received from my Albert a very sweet letter, which delighted me beyond words. I'll write him again immediately.[4] I had a very nice time in Holland among the local circle of my fellow colleagues.[5] What an enormous difference! . . . . . You should go there sometime as well. The theory of gravitation has grown powerful roots there; but this is certainly not the main source of my sympathies.[6]

Cordial greetings and many thanks for the continuing solicitude! Yours,

A. Einstein.

# Vol. 8, 269a. To Heinrich Zangger

[Berlin,] 25 October [1916][1]

Dear friend Zangger,

Cordial thanks for your warm and detailed letter. Albert is a gem of a boy. I replied to his good and honest letter about a week ago[2] and cannot refrain from writing to him right away again after seeing that he is still attached to me.

I am very glad that the wife is feeling better. Do let her stay long enough in the Theodosianum.[3] She shouldn't worry herself about the expense. I will pay extra for all the costs arising from the illness. She doesn't have to bother about that at all. It is, after all, too boisterous at home with the lively boys. It is fortunate beyond words that you and your wife are looking after my family in Zurich like this; otherwise I wouldn't know what to do. On the other hand, it is fortunate for more than one reason that they are in Zurich and not here.

In Holland I spent fine days with people of rare intellectual profundity and cultivation who have not swallowed ramrods and aren't suffering from any mass psychosis.[4] I read Treitschke with Ehrenfest.[5] You should do that as well. This is the name one could apply as the label for the upper class here. Otherwise, I am also reading Tolstoy's booklet, *A Confession*, whose train of thought is somewhat erratic but I feel much affinity with it while reading.[6] He recognizes after a hard struggle that a vivacious and thoughtful person cannot live without clinging onto something extrapersonal. But I cannot share his view that this extrapersonal thing must be the childish belief in God.

Since my last letter I have produced little on my own, mainly only a formal simplification of the general theory of relativity.[7] I am lecturing again this semester,[8] and there is a man wanting to condense this course into a book. This is commendable, because I myself lack the enterprising spirit to do it.

I hope you are completely healthy again and happily at work. Cordial greetings from your

Einstein.

# Vol. 8, 270a. From Heinrich Zangger

[between 31 October and 13 December 1916][1]

Dear friend Einstein,

You will have to explain the perihelion motion to me sometime, [I get] stuck at the mathematics.[2] I am likewise unclear about the gravitation-electricity relations you are implying.[3] Your visit is necessary, as you can see. The situation is getting progressively worse, many factories are at a standstill because we aren't getting any raw materials—so that one can overextend the goodwill of the Swiss, I read in a letter. If it wasn't an economic war, it's being made into one because they are saving up all exploitable elements, because they seem to need them in this fine world.

Justice Minister Klein[4] did [not?] respond, the 2000 tuberculars aren't coming to Switzerland.[5] They are kept in the prison camps. What will be the consequence? The standard for the [prospects?] of the future are, in the end, different over there [—], for otherwise I don't understand, [. . .]

Besso is depressed because he can't find a job; what can be done about it?[6] I don't know. I can't assess his abilities with regard to his teaching talent.

# Vol. 8, 276a. To Heinrich Zangger

[Berlin,] 16 November 1916

Dear friend Zangger,

It troubles me that you have to be in bed again. What's ailing you? Besso writes nothing about it.[1] Hearty thanks for the compilation of the bill.[2] The recently transferred 1400 francs were exclusively for the regular upkeep (October through December). The 1061.15 francs will follow separately in the coming days.[3] The wife's condition unfortunately still seems to be quite unstable; so at least they should not be short on funds. What you write about the boys makes me happy. I wrote Albert many times; but he inherited the writing diligence of his two parents. This is a natural process which I shouldn't complain about. Just like I am not writing a book, neither is he writing me. I shall be glad to write to R. Rolland when you relay his address to me. The recognition of his artistic achievements by the Nobel Committee is at least a small plaster on his wounds, but no more than a *small* one.[4] But maybe great things will come to pass so that he becomes reconciled with fate; I lay my trust in that.

Warm greetings and the best of health! Yours,

Einstein

I hope Huguenin is recovered.[5]

# Vol. 8, 278a. From Hans Albert Einstein

[Zurich, before 26 November 1916][1]

Dear Papa,

I received your postcard and did not reply to it sooner because I always have very much homework.[2] Mama has come home again in the meantime and has a nurse. We are tremendously happy. When Mama came home, we had a celebration. I had practiced a sonata by Mozart, and Tete had learned a song. We are very glad that Mama is here again, for we are not so alone.[3] Mama can at least listen while I'm playing the piano and does Latin a bit with me. It's much nicer this way. We are slaving away in Latin: we've already had the 1st, 2nd, 3rd + 4th conjugation and the 1st, 2nd, 3rd, 4th + 5th declension. The life insurance bill arrived a while ago and comes to *139.7 francs*.

Recently, I was working steadily on a ship I just made out of a simple wood carving.

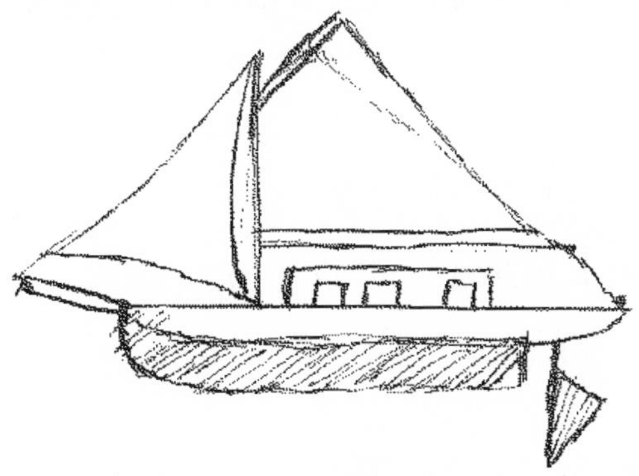

It can sail very well and I have lots of fun with it.

Many greetings, yours,

Albert.

# Vol. 8, 279a. From Hans Albert Einstein

[Zurich, after 26 November 1916][1]

Dear Papa,

I received your postcard today and thank you very much for the Christmas present you promised us. Tete is going to school in the spring.[2] I didn't write you earlier, because first Mama was not well, and later because there were a good 20 cm of snow and we went skiing.[3]

I made the ship like this:[4] At first I just had a block of wood that in cross section formed an almost equilateral, right-angled triangle. It was about 25 cm long. Then I hewed out the basic form with an ax. Then I carved it nicely into shape with a knife. All around the upper edge I nailed a thin little wooden ledge, and on top of that, I made a little cabin out of cigar-box wood.

- - - - - - - - - - - - - - - - - - - - - - - - - - -

I designed the sails so that I can raise and lower them.

With this method, the actual wood of the ship was not hollowed out but just an edge made with the little planks. The thing at bottom is a rudder.

The sonata goes like this: They are variations[5]

We are already looking forward to Christmas and are curious what the Christkindel is bringing us.

Christmas greetings from

Adu and Tete!

# Vol. 8, 282a. To Ejnar Hertzsprung

[Berlin,] 5 December 1916

Dear Colleague,[1]

I certainly am used to the sky not being the limit! You are right, of course, with $\sqrt[3]{\delta^2 M}$, likewise with the relatively strong influence of the density. This is even more unfortunate since density is only rarely obtainable and probably also quite uncertain as well. Besides, the magnitude that results from occultations is not the density but the *radius*, so the effect *vis-à-vis* the observation[s] is better defined by mass/radius than by the above formulation.[2]

From among the points you touched upon, I will just address the last two because they relate to actual physics (not stellar observation).[3]

An evaluation of the eccentricity of the Earth's orbit without an extraterrestrial light source is impossible, *in principle*, because according to the theory only the frequency depends on the gravitational potential, not the wavelength measured with a meter-stick.[4]

An evaluation of the differences of gravitational potential on Earth would be splendid, but these effects are *fabulously* tiny. If $\gamma$ is the gravitational acceleration, $h$ the difference in altitude, then the Doppler velocity equivalent to this is:[5]

$v = \gamma h/c (c =$ velocity of light).

For 3000 m that makes an altitudinal difference of

$v = 1/10$ millimeter/second.

I don't believe the thing about the Bunsen-burner flame, either. If the molecules generating the luminescence have a thermal velocity of something like at least 100 m/sec, then such tiny relative velocities of the flames cannot yield such a huge effect as to be visible to the naked eye. In any event, one ought to get to the bottom of this business. As far as I know, observations never yielded sharp lines that did not agree with kinetic gas theory; and this subject has been repeatedly and thoroughly treated.

Cordial greetings and wishing you full recovery soon, yours,

A. Einstein

# Vol. 8, 283a. To Michele Besso

[Berlin, after 6 December 1916][1]

Dear Michele,

Isn't it possible after all that Mileva's illness is of a nervous nature? From the description it seems to me to be like epilepsy. Its temporal nature also speaks for it.[2] I am very sorry for her and also for the boys, who are having a truly pitiful youth. But nothing can be done about it. I always designed my letters to Albert so that Mileva could read them without unpleasant feelings. So your suspicion in this regard cannot apply.[3] Besides, the environment of a patient always associates significant occurrences with external events.–

I have made myself more independent of my relatives insofar as I now almost always eat alone at home in the evenings. Then I always cook myself something small, because I have the good fortune of agreeing well with this most direct and honest form of service. I abandoned once and for all the idea of remarrying. You were absolutely right in this regard; in general one sees other people's situations more clearly than one's own! Pauli writes very enthusiastically about your lecture and about your article in the *E.T.Z.*, which I hope you won't hold back from me.[4] The shawm of peace is piping with increasing shrillness . . . it knows why.[5] I don't know yet exactly how I can see the children next Easter. In any case, I won't stay in Zurich, because that would cause agitation. I'd like most to have them come to Lucerne, but that would be hard to carry through. I am actually deprived of more rights than a jailbird. It appears that Weyl calculated with an incorrect Hamiltonian function. It is enough for me that the addition of gravitation unfortunately does not explain the electron.[6] Altogether, the connection between gravitation and electromagnetism has been quite superficial. ⟨On the other hand⟩ It is also hard for me to believe that God Almighty took the trouble to introduce two so very basically different states of space. An infernal joke is probably still behind this. I try in vain however to improve something.[7] I put Wiechert in a very bad light at the colloquium concerning Mercury.[8] Inertia of energy; fine. But adding a factor 10.7 is foolish. You also said so in your letter. What would you say if someone explained a mechanical phenomenon by converted heat, but set it equivalent to $10.7 \cdot 4.2 \cdot 10^7$?[9] The best thing about Gerber supposedly is that he derives the forces incorrectly from his already adventurous potential.[10] I owe this to Sommerfeld, who heard it from Seliger. Flamm's paper is quite nice. The light paths *are always* geodesic lines.[11]

There are pretty static and metrically isotropic solutions to the gravitational equations. I calculated them together with a fellow Russian tribal comrade.[12] At infinity the $g_{\mu\nu}$'s degenerate in such a way that the vel. of light becomes infinite and the inertia zero. Such a world necessarily has naturally measured infinitely large mass. It is apparently not possible for an *entire* universe of finite mass to exist, in this sense, even if you abandon the isotropic requirement. I cannot answer many of your questions, because I haven't analyzed them either.

The gravitational energy does not become negative—at least to first approximation. It is hard to say whether it is "merely a counting chip." The concept is just as valid as that of kinetic energy in classical mechanics.[13] I'll send you the wave paper; it's quite pretty.[14]

I hope, by the time you receive this letter, not just Anna but you too have the flu behind you.[15] I am sweating profusely over a popular book I am writing about the theories of rel.[16] The first part about the special theory is already finished.

My brain is in quite an ebb, which isn't at all unpleasant, though. Someone in Munich has apparently found a quite fabulous way to prove the bending of light rays from existing observations. About that another time!

Warm greetings to both of you, yours,

Albert.

# Vol. 8, 287a. To Heinrich Zangger

[Berlin,] Monday, 8 January [1917][1]

Dear friend Zangger!

That's quite a calamity about the money. I just found out from my bank that they still could not issue my check made out to you in the amount of 1500,[2] because they have not been able to obtain the consent to send out the money.[3] This can lead to quite considerable difficulties in future. No dams exist that could withstand such a difference of economic levels in the long run. This period must be splendid for a theoretical political economist, likewise for a social psychologist, who would have liked to have cast all things human aside. I am terribly eager to set out to see you again on the island of the relatively uninvolved who are weathering this storm under the cheerful motto "more luck than wits." I am living here somewhat like a drop of oil on water, as you can easily imagine, knowing me.[4] The discrepancy also in the unexpressed views of life separates people. Contact is always upheld, however, through the purely intellectual, especially physics, of course.– You'll be receiving the money within a few weeks, in the course of which the exchange rate is threatening to shoot up extraordinarily. But I would not like to incur any debts in Switzerland as long as ever possible and also not touch my wife's dowry,[5] because I'm wary that the money I'm sending into Switzerland under current conditions is still being used to best advantage. Meanwhile the local bourgeoisie are basking in the surplus of money here, which they interpret as wealth. Right now I am happily relishing the fact that my Albert is now kindly disposed toward me again. He already wrote me two genuinely childish and cheerful letters, which downright touched me.[6] Your influence has surely contributed strongly. From the last one I gather that my wife recuperated from the most recent relapse.[7] I am not going to bore you anymore with meddlesome medical questions but will take everything as it comes. When I come to Switzerland for Easter, or perhaps a little later, I probably won't want to stay right in Zurich, because it could offend my wife and children again if I don't visit them in the apartment. Best would be if I stayed in the countryside near Zurich so that we can see each other often.[8] Then, if at all possible, I would also like to look up Romain Rolland, if he doesn't feel visits by strangers are bothersome.[8] I cannot bring myself to write, because, strangely enough, I cannot find the right way out. Scientifically speaking I haven't accomplished anything special but am pleased with the enthusiastic reception that the general theory of relativity is encountering among my professional colleagues, also in England and America. No war or human delusions penetrate into this sanctum. I won't leave Switzerland before you have really absorbed this matter, as far as is possible without mathematics.

Hearty greetings to you from your

A. Einstein.

# Vol. 8, 287b. To Heinrich Zangger

[Berlin,] Tuesday, 16 January [1917][1]

Dear friend Zangger!

Today I approach you with a request that perhaps only you can fulfill for me. Recently my stomach has been bothering me more and more frequently and my appearance became so desperate that I gave in to my cousin's[2] insistence and, contrary to my most ingrained principles, went to the doctor.[3] He found a chronic inflammation of the stomach with 50% more acidity in the gastric juice than normal. The man prescribed a diet for me, the carrying out of which entails foods that are impossible to procure here. In total one would need roughly

10 pounds rice
5 " semolina
5 " macaroni

as well as the most substantial quantity possible of zwieback. The cure is supposed to take 4–6 weeks. The relevant supplies of my relatives[4] are already virtually depleted. If you believe that through your influence you could obtain the permit to export these things, I ask you sincerely to please do so. For I do not see how else I could get out of this misery. You will receive a medical attestation as soon as possible. Otherwise I am being excellently looked after; my cousin cooks everything herself with almost no salt and without anything else that excites the wrath of the evil spirits.–

You have, I hope, finally received the money for my family.[5] If there are bills for doctor, medications, etc., please always send them to me as soon as possible, because the bank has difficulties with money transfers to Switzerland,[6] so there is always a delay. Let's hope there will be no difficulties to my trip to Switzerland in the spring. Words cannot say how much I am looking forward to seeing my children, you, and Besso again.[7] The prospect of peace appears to have shrunk again very much, so this year bodes little good. The saying that humans can even get used to hangings is holding true, alas.

I have been regularly receiving very welcome messages from my Albert.[8] The boy is acting very well toward me under these difficult conditions. Scientifically there is nothing worthy of note. With age one becomes, bit by bit, stationary and short of ideas, as is fitting. For that there are young ones whose tracks in the brain are not yet so worn.

I occasionally think of Ratnowsky. Has nothing happened yet to make his position somewhat stable? If you hear anything about that, please do let me know.[9] Do you sometimes go to the physics colloquium? It must be very interesting (Meyer, Weyl, Abraham)[10]. Over here much is laying waste because of the long war.[11] The younger people are perishing on this evil treadmill. This year Sommerfeld found very fine things about spectra (in quantum theory). It is by far the most important thing that has appeared in our science these last few years.[12] But we have come no closer to an understanding of quanta.

In spring I shall be staying near Zurich, not in Zurich itself, in order not to arouse unpleasant feelings in my wife and the children, like last year, by not going to the apartment.[13]

Cordial greetings, yours,

Einstein.

# Vol. 8, 291a. To Heinrich Zangger

Berlin, 1 February 1917

Dear friend Zangger,

Your humorous letter amused me boundlessly. But it didn't even occur to me to regard your vivacious emotion and your vigorous efficiency in human affairs with any lack of respect. That would be the pinnacle of cynicism. So what's the latest chateau[1] of my Bessolet called? I don't know yet, for he is almost as lazy a correspondent as I am, just with the difference that he's copious once he's past the salutation, whereas with me the best you get are spot shots. I thank you heartily for the promised invaluable victuals, or chicken feed.[2] In the hands of a prominent specialist (Prof. Boas)[3] my erstwhile stomach ailment has transformed into a liver complaint, which anyone who looks at me gladly believes. He should finally be sending you the attestation in the coming days. The war madness has now reached its boiling point;[4] Switzerland seems to be on the verge of it as well, if it's true that it had to trade a lot of cows to Germany in exchange for coal.[5] In any event, I'll do my utmost to be able to travel in April or May to see you all. Scientifically, the little hen[6] duly laid an egg, albeit not a golden one. According to the general theory of relativity it is probable that the universe is not infinite but closed in upon itself, somewhat like the surface of a sphere.[7] I'll tell you about the details later. Isn't it inconceivable that *I*, of all people, should get a Kaiser Wilh[elm] Institute?[8] Fate sometimes has a sense of humor, even though its grimmer qualities work much more powerfully. Modern man spared God the task of showering ash and brimstone; he takes care of that himself.[9] My former tribal fellow really let himself be nailed up in vain; he could have ended his life in peace on a diet of gruel. Don't fail to read Ph. Frank's[10] excellent article on E. Mach in the *Naturwissenschaften* (latest issue); it's a small masterpiece.[11] I asked him (Frank) to have an offprint sent to you. Then show it to Besso as well, who has a great interest in such things.[12] Is it true that your inner life is also rebelling? I hope you faked this only as sign of friendship, but with a false estimate of my brainpower.[13] Your impish smile upon reading this last sentence confirms my view. Dear Zangger, why am I not receiving any bills for my wife's medical expenses?[14] The nurse? I saved up a full $2 \cdot 10^4$ marks here[15]—albeit not enough to secure the future of two growing boys. But if I live a few more years, it will suffice.

Cordial greetings from your

Einstein.

# Vol. 8, 297a. To Heinrich Zangger

[Berlin,] 13 February 1917

Dear friend Zangger,

A hale family, indeed! The wife is sick, one child sickly, the husband sick. It is touching how well you're attending to such a pitiful crew. Today I brought your question to Prof. Boas,[1] who also promised to send you word. The liver is sick, that's certain; he verified it today again by feeling in the front under the lower edge of the rib cage and on the back at a particular spot on the right, where a characteristic oversensitivity was detected. The nursing is excellent; my cousin cooks everything exactly according to the doctor's prescription.[2] Without your care packages, though, the stock of suitable chicken feed would soon be exhausted.[3] I get, in addition, Mergentheimer mineral water twice a day. I spend the afternoons in my relatives' home because I cannot prepare the right supper myself.[4] He, too, is very much for the Tarasp cure; but who knows what will happen by then.[5] Let's not make any plans so far in advance. I should just commit myself to the extent of determining the time of my trip to Switzerland. When do you think I should come? I also asked my Albert, but haven't received an answer yet. In any case, the spring break is an unfavorable time, because I can't go on any excursions with the boy then. How is your health; do you really have to be patched up as well?[6] I hope not. You have even less time than I. The analysis on the cosmic field of gravitation has become very interesting. If matter is as dense everywhere as in the fields of stars within our range of view, then the universe should (be finite and) have a radius of approximately 10 million light-years. Unfortunately a direct proof will probably never be possible, because we can only see up to a few thousand light-years. It is difficult for an intuitive thinker to consider the universe unbounded and yet finite, similar to the surface of a sphere but three-dimensional.[7] This subject reminds me of our friend Besso's castle in the air, which you withheld from me without regard for my not inconsiderable curiosity.[8] The war doesn't seem to want to end at all anymore before Europe has deteriorated completely.[9] But the people don't deserve otherwise. Or should one excuse it all due to suggestibility and unfortunate circumstances? Destiny does not ask about guilt but proceeds according to natural laws. But one thing constantly occupies me. Why don't we have an analogue to the medieval monastery, a place of refuge for people who want to withdraw from all worldly dealings, renouncing certain things commonly held to be worth striving for? Couldn't such an international establishment with cultural objectives be created? However, the accursed utility of modern science for so-called practical purposes would pose difficulties for such a thing. Recently I was horrified to read that America is also planning compulsory military service;[10] what a despicable world! Who knows whether hard times aren't approaching even for Switzerland, now that imports are being jeopardized everywhere.[11] Be this as it may, I am enormously looking forward to the time I can sit with you and Besso on your veranda again and peer out over the lake[12] and discuss all sorts of things.

Cordial greetings, yours,

Einstein.

# Vol. 8, 299a. To Heinrich Zangger

[Berlin,] 16 February 1917

Dear friend Zangger,

Your letter about the condition of my youngest scares me less than you might think. Well-deserved punishment for my having taken the most important step in life so rashly. I begot children with a physically and morally inferior person and cannot complain if they turn out accordingly. Only *they* will accuse me one day when they are old enough; they will be only too right, unfortunately. So send my poor boy wherever you and Bernstein see fit,[1] if you really think something of it. And even if you silently say to yourself that every [effort] is futile, send him anyway, so that my wife and my Albert think something is being done against this evil. I am going to try to send 500 marks to Zurich. I would be very unhappy if I believed that I could have begotten valuable progeny with another woman. But if I look around among my own family and see the banal people, tolerably healthy though they are, then it seems to me that my contribution to this pitiful business mustn't be valued that highly, either. I console myself with the fact that life still goes on through the fruits of labor. The happy consciousness of having really acted productively and liberatingly in this way, and lastingly so, is a consolation for me that nothing can destroy. With this thought I will know how to bear the experiences of my children, sad though they may be; if only the cursed drive to beget children didn't aim to extend the misery into infinity! This drive, in concert with the medical arts to keep alive something that is not viable beyond the years of fertility is undermining civilized humanity. So it would be urgently necessary that physicians conducted a kind of inquisition for us with the right and duty to castrate without leniency in order to sanitize the future.–

I asked Mr. Boas to report to you about my condition.[2] He has quite certainly established that the liver is sick. But he himself is not entirely sure about the nature of the sickness for the time being. In any case, I don't believe that this is a minor, temporary thing. I have been suffering from this pain for about 18 years and have always looked pretty rotten. But two months ago something snapped, expressing itself in rapid aging, frequent chills, and otherwise general discomfort. Since I have been on this strict diet, the pain is gone but the ailment is naturally still there. Because death is the worst thing that can happen in such a case, and I don't make much of that, I don't let the affair upset me at all. Added to this is the fact that I am being very carefully nursed by my cousin,[3] better than would be possible at a sanatorium.

Your hints about Besso's plans have raised my curiosity to the boiling point,[4] without my being able to get an idea about what it involves. So I assume that he will investigate the secret of natural life in the Wild West under the informed guidance of an Indian chief, or that he will think this through so precisely that he won't need to leave at all but will find the solution through purely theoretical means, with the usual marginal error of 2%.

Dear Zangger! Don't think that I turned into a hypochondriac. When we're together in Zurich or wherever else, we'll saunter around with light minds, relishing the sunshine, enjoying the hour of day and existence as we did before. Life should not be negated for the sake of death nor for the sake of war nor for any other madness. When should I come to Switzerland this year? I asked my Albert as well but he is hesitating with his reply.[5] Perhaps you can speak to him about it sometime.

Warm greetings to you, yours,

Einstein.

P. S. The bird feed, if it really arrives, will last for a long time.[6] We don't need to think beyond that. Who knows what will be by then.

## Vol. 8, 300a. To Emperor Franz Josef[1]

[between mid-February and 29 April 1917][2]

Your Majesty,

Under the pressure of an imperative obligation I am allowing myself the liberty, Your Majesty, to submit a request. ⟨Some time ago⟩ The political murder ⟨which⟩ of which Fritz Adler made himself guilty has profoundly shaken the spirit ⟨of all well⟩ of every right thinking person. I do not wish to varnish this horrible act as such with a single word. Considering the psychological state of the perpetrator, however, ⟨he⟩ ⟨the act⟩ it seems that one deals rather with a tragic accident than with a crime.

Few would know Mr. Adler as well as I. ⟨We were⟩

I know ⟨Mr. Adler⟩ him since we both studied theoretical physics 20 years ago in Zurich. A few years ago he was still my closest colleague as lecturer in this field at the University of Zurich; at that time we were also housemates. I have come to know Adler in these years as a ⟨example⟩ man of honorable character, of ⟨an⟩ almost unique self-sacrifice.

⟨I have⟩ Few are known to me who are as unconditionally reliable and honest as he, who devote their energy as much for impersonal matters and with such an overcoming of their own wishes.

⟨So as not to inexcusably omit⟩ Under these circumstances I fulfill an imperative duty if I now ask you with all my heart, Your Majesty, to make use of the right to pardon, in case Adler is sentenced to death. A valuable life could thus be preserved.

# Vol. 8, 308a. To Heinrich Zangger

[Berlin, before 10 March 1917][1]

Dear friend Zangger,

You are fretting too much on my behalf. One package arrived safely, the other did not.[2] I will try to file a complaint at the post office. But nowadays the mail is very unreliable. ⟨Much is being stolen and nothing is being refunded.⟩ Let's hope it still turns up. I will not fail to try everything. Boas seems not to have reported to you carefully.[3] Perhaps you just received the medical statement from him but not the personal report. He promised to send you one. I think that he is surely right insofar as the sick liver is concerned. But I don't believe it about the gallstones. I don't have cramplike attacks with sudden stabs of intolerable pain. Incidentally, the pain has been plaguing me on and off since I was an adult. It evidently involves an old complaint that is long-entrenched and got worse in the last few years (weight loss, bad appearance). Since the diet, the pain is gone. It used to come like this: light pressure in the liver region in the morning upon waking. Throughout the day, gradual increase in pain, eventually to such a degree that I had to lie down. Rapidly better after lying down in bed. My appearance is still bad but a little better since the diet. Hands always cold. I am being very well cared for. My relatives are nursing me excellently. My cousin cooks everything personally and—what is much more difficult—comes up with the food. Without her I could not manage here.[4] Don't worry yourself anymore for my sake, for the time being; if it really gets urgent, i.e., if nothing more can be scraped together for me, I'll ask you again, but not before. I know how precious your hours are. And ultimately, there remains the option of fleeing to Switzerland. I hope the conditions there won't become as severe as they are here![5] Nobody is upsetting me here; on the contrary, everyone is spoiling me, if not worshiping me, so at most I run the risk of losing my self-criticism and modesty.– Now to the main thing. My wife and Tete are tuberculars.[6] I'm afraid that my Albert could also get infected and this fear leaves me no peace. Children are never careful. I would like to have Albert live with me and teach him myself.[7] Imagine what valuable guidance I could give the boy. I would let tutors teach him foreign languages. I think that this would be splendid for the boy. The household in Zurich would then perhaps best be dissolved and my wife and Tete could be accommodated according to medical criteria. My poor youngest, whom I loved so inexpressibly much will now probably soon be taken completely away from me. Better dead than forever-suffering! Just punishment for my irresponsibility![8] Wouldn't you also find it right if I took in my Albert? He wouldn't go into any school here, of course, so Berlin could do him no harm. He would constantly be under my supervision. I'll read stories with him, do mathematics, physics and chemistry with him and many other things that I would first have to teach myself for the purpose. 3 hours every day. Add to that a walk. What do you think of this? About the treatment of my poor little one, please act on your own judgment in agreement with Bernstein.[9] 400 francs have gone out to you for this purpose.

Besso wrote me himself; I'll answer him soon.[10] In personal, human affairs I do not like to turn to him; because human matters make him so terribly miserable.

Cordial regards, yours,

Einstein.

When should I come to Switzerland? Should I take the oil when it arrives, despite what I wrote above? Should I inform Boas about it?[11] I often cannot read parts of your letters despite the great effort I devote to reading; don't be cross with me but I absolutely must tell you this, because bad consequences could arise from it.

Yesterday I spoke at length with a woman who is certainly suffering from gallstones. In the process I was convinced after all that Boas is entirely right with his diagnosis. My pain is genuine gallstone colic. Since the cure, the thing has not recurred (4 weeks). I am curious whether you consider my plans with Albert sensible; I am filled with the firm conviction that I would do it well.[12] Maybe my wife will also agree with the plan; she really must realize that I mean well with them despite the personal rupture with her.[13] Write me without restraint what you think about the future of my little boy; your prognosis cannot possibly sound worse than what

I think.[14] Was the infection already present at birth or was there only a disposition? ⟨The second package could still arrive; I hear that mailings are frequently in transit for weeks on end. There is, of course, also some theft. Things generally are astonishingly topsy-turvy.⟩ The second package also arrived. Hallelujah! Once again, most hearty thanks for your friendly assistance!

# Vol. 8, 319a. From Hans Albert Einstein

[Zurich, between 1 April and 22 April 1917][1]

Dear Papa,

I want to write you one more time because I just thought of it. I haven't written you until now because I didn't have much time ⟨at all⟩ and because I don't like to write at all. It's always such a major business for me, a letter like this.[2] At the beginning of the holidays[3] the weather was tolerably nice, but now it's raining when it's actually not snowing. In this way, a game we've been playing has been made impossible. Namely, we made little aerial ropeways from one house to another where a child happens to be living. So we always sent each other things; but the installation was really the nicest part! Over telephone cables, below telephone cables, and even above the tramway cables.[4]

At school we had quite a lot of homework toward the end but, this way or that, it's always the same thing! Vocabulary, and more vocabulary, etc. I think that it will be more interesting in the 2nd form [2nd year of Gymnasium], in which I am going to be now.[5]

I can't write anything about the summer holidays yet because it's still not clear *how circumstances will permit it then*.[6]

Now I'm not taking lessons with the carpenter,[7] nor taking piano lessons;[8] even so, I always have something to tinker away at, although "the mess" does not quite appeal to the others. I also play music & I'm playing a lot of Mendelssohn songs and Mozart sonatas, also a few pieces by Schumann and a few old ones.[9] Mama assigns them to me and corrects me when I make a mistake. Everyone says that I play nicely. Now I'm taking care of Mama, too, because we don't have a nurse; Mama was too annoyed with them.[10] We still have enough to eat, although it is a bit expensive; we are all fat and plump. Are you too?

Mama asks you please to send us the money directly, because otherwise it takes so long for us to get it.[11]

Many greetings from

*Adu and Tete.*

# Vol. 8, 326a. To Heinrich Zangger

Monday, 16 April [1917]

Dear friend Zangger,

I hope you recuperated well on vacation. I am also feeling well; every day I take some of your oil, ever since I haven't received any more milk rations.[1] It is a true [blessing]. With A[dler] I took other measures. I wrote him I should be summoned as a witness for the trial. I have a few things to present that won't fail to leave an impression on the judges.[2] If the Zurich Phys. Soc. approaches this matter with the little warmth it seems to have, *then it should dally off home.* (Sleepyheads!)[3] You received the money, I hope; if, as a consequence of the unfortunate circumstances, more is needed, I'll gladly send it.[4] How are my wife and my little boy doing?[5] Albert is enveloped again in deep silence.[6] When does he have summer holidays? I would like to plan my trip to visit you accordingly. Recently I had another strong attack; I now also believe that it's gallstones.[7] I suffered many such attacks already in Berne, even more frequently than now.[8] I am being very selflessly nursed by my relatives, despite the truly serious difficulties.[9] Ex oriente lux—who would have thought that![10] The future is less certain than ever but more full of hope, the influence on the minds unmistakable. Raised self-esteem among the multitude.–

In science I am slowly brooding on; the path is stony and the legs are weak. The gen. theory of relat. is finding enthusiastic supporters throughout the world. The *Naturwissenschaften* has an excellent exposition of the theory by the philosopher *Schlick* in Rostock.[11] I am sending you the paper with the request also to give it to Stodola and Besso.[12] That's a fellow one could appoint to a chair in Switzerland!

Cordial regards, yours,

A. Einstein.

# Vol. 8, 330a. From Hans Albert Einstein

[Zurich,] 28 April [1917][1]

Dear Papa,

Unfortunately I have to inform you that I still don't know when I have vacation. But I'll let you know as soon as I myself know.[2] Tete is doing well, and Mama and he are feeling very comfortable at "Bethanienheim" hospital.[3] They have a nice room with a balcony. My address is now, for the time being, 25 Berg St., Zurich 7.[4]

Many greetings from

Albert.

# Vol. 8, 330b. To Heinrich Zangger

[Berlin, before 29 April 1917][1]

Dear Zangger,

I am unsettled about your health because I know that you are not one to complain without reason. What's ailing you? Do inform me about it very soon.[2] Regarding A[dler] I hesitated whether the submission made any sense. But now I am in favor of it after all, thinking that it *can do no harm*.[3] At the same time I am writing to Besso[4] and sending a telegram to Beck.[5] Scientifically I don't think much of A., but came to know him as an unusually selfless and decent person of rare goodness. He is simultaneously almost pathologically stubborn. In his case, the denial of egotism veers into its pathological opposite (voluptuous suffering).[6] He sent me a scientific manuscript written just now: unproductive sophistry presented with prophetic pathos—poor fellow—what should I say to him without lying, without offending him, and without concealing what could be useful to him in his external circumstances! I'm constantly studying the bulky manuscript.[7]

Cordial greetings, yours,

Einstein.

# Vol. 8, 332a. To Heinrich Zangger

[Berlin,] 4 May 1917

Dear friend Zangger,

I thank you sincerely for your long letter, which unfortunately makes me distinctly aware again of how sad the circumstances of my family are. In the last few days I've been reconsidering things very much again and am quite worried. One thing seems quite certain, namely that the dissolution of the Zurich household is final. So the primary question is how my Albert should best be cared for. I have abandoned my original plan of taking him in again.[1] For one, the local environment seems unsuited for the boy; furthermore it would be too hard on my wife to let the boy be so far away. Then the housekeeping here would also pose considerable difficulties. So three possibilities come into consideration. (1) I send him to Lucerne to my sister, a childless, intelligent and sweet-tempered woman, who is married in Lucerne to a civil servant.[2] (2) I send him to Tanner, my former student, who is a teacher in Frauenfeld at the Cantonal School.[3] (3) Country boarding school in Glarisegg.[4] (1) would be most preferable to me but would encounter resistance from the suspicious wife. (2) would not be bad, (3) perhaps too expensive for my circumstances.

From now on, you see, I can count on 13,000 marks remaining after deduction of the high taxes, whereas in past years I had quite significant supplementary earnings.[5] If one takes the exchange rates into account along with the circumstance that I also have to support my mother, even though this is not very significant,[6] it is necessary to save quite a lot, especially since something definitely has to be put aside annually for my boys so that their schooling and higher education are secure.[7] I am not even thinking about the grim prospects for my little boy; one more reason for careful housekeeping. This year I sent my family over 8,000 marks;[8] I can't do that for the long term.

From this vantage point one must consider how my wife and the little boy should be supported over the long term. This is a mystery to me, but I think that she cannot under any condition start to manage an independent household again; she probably won't be able to recuperate enough for that anymore. Couldn't the household perhaps be completely dissolved and the furniture sold? But my wife won't want that, it cannot even be suggested to her. Besides, there is no hurry with that decision; we can resolve that in 2 months when I am there.

It is moving that you and your wife are taking Albert into your home.[9] But I would very much wish that he could be definitively accommodated very soon. I would most prefer it at my sister's, whom I *value very much* (Bramberg St. 16A). I would not take any objection by my wife into account. She is compulsively hostile and distrustful. Even opposition by the boy would not be important because he is being influenced. It would be very nice for him there. I'll also ask Besso, who knows all the parties involved.[10]

I was very glad that now the A[dler] affair has been taken up with such warmth. Even if the matter only tips the balance slightly, it really is most welcome that our Zurichers can be counted on when there is something decent for them to do. I am curious how the matter will fare.[11] A.'s father thinks quite optimistically about the outcome, as you probably noticed.[12] A. sent me a manuscript which, according to his father, he is supposedly keenly enthusiastic about; but it is really worthless sophistry, lawyerly methods in physics.[13] So you had a good nose with your harsh professional judgment, as always in your assessment of people. You would have the mettle to replace old Kappeler to a quite different degree than the former less enlightened despots of our Swiss schooling system.[14] It may well happen!

I am brooding over a manuscript on the quantum issue. It is a decisive step forward; the matter leads in particular to specific mathematical problems, i.e., away from endless pondering.[15]

Warm greetings from your

A. Einstein.

Don't seal your letters; they are always opened by the milit[ary]. Perhaps that's the reason why some letters did not arrive.

# Vol. 8, 333a. To Heinrich Zangger

[Berlin,] Saturday evening, [5 May 1917][1]

Dear friend Zangger,

This evening, after the letter to you had barely left,[2] I received news according to which the reduction in my earnings will *not* occur for the present.[3] My comments in this regard consequently do not apply. It is very good that among all the other evils *this one* is not added as well. For the present at least we do not need to think about that yet. Couldn't my sister take care of something that might relieve you or Besso of some effort?[4] She has a free ticket,[5] you know, and would gladly do anything. (I am thinking of dealing with the things in the apartment, putting in mothballs, etc.)

Cordial regards, yours,

Einstein.

Sunday morning, [6 May 1917]

Your and my Albert's report[6] delighted me, with the exception of the last remark, the content of which enraged me. If only people would finally leave you in peace! The tragedy of it lies in that the phlegmatic cannot grasp the agility of your intellect and interpret it wrongly.[7] Genuine ill-will would rarely play a role.–

It is touching how you always revert to a belief in the elevation of human beings! I am convinced that even ethical conventions are subject to fashion. They unfortunately change like the weather. They emerge epidemically in the herd. Albert's report delighted me.[8] It is impossible for me to leave before July (course; phys. society, etc.).

# Vol. 8, 343a. To Heinrich Zangger

[Berlin], Wednesday, [23 or 30 May 1917][1]

Dear friend Zangger,

In the past few days I have been thinking constantly about how my family could be properly provided for without causing an unsustainable financial situation, and I arrived at the following results, which I now present to you.

1) With this letter another is going out to my sister with the request that she travel to Arosa and make inquiries on location about how my boy can be accommodated there well but less expensively than by following your suggestion.[2] My sister will then immediately report the results of her research to you so that you can make your assessment appropriately, in accordance with the circumstances.[3]

2) It would undoubtedly be possible for me to intervene successfully from here so that my sister-in-law Zora Marić receive the travel permit to Switzerland.[4] If in your opinion, in that case, my wife could return to the apartment, I ask you please to refrain from renting it out.[5] Albert could then be with his mother again. My sister-in-law is impeccably reliable and loves Miza and the children tenderly. So the question is just whether my wife's condition allows this solution. As soon as you send me positive word I will immediately write my sister-in-law and take the necessary steps with the authorities. I beg you to reply very soon so that I can make the arrangements before my departure to Switzerland, which will take place in the first few days of July.

I have the strong, well-founded suspicion that my ailment has not yet been quite correctly diagnosed. I will tell you about this in Zurich. But please don't tell Prof. Boas anything about this remark.[6] For this very well-meaning man shouldn't be upset. My sister wrote my cousin worriedly about my condition, whereupon anxieties were generated here again. These things should stay between us, whose nerves are more resilient.

I am very glad that the small-minded mischief-making against you is not finding any support.[7] Cordial greetings from your ⟨crucifix⟩ unfortunately so troublesome

A. Einstein.

# Vol. 8, 344a. To Hans Albert Einstein

[Berlin,] 26 May 1917

My dear Albert,

I am always so pleased to read your good and cheerful news and especially also that Prof. Zangger and his family like you.[1] Now I will soon be coming to see you.[2] Together we'll go to meet Tete in Arosa and be up there together for a week or two.[3] That will be very nice. Do write me soon when your vacation begins so that I can make my plans accordingly. I'm holding twice as many lectures in June than before so that I'm finished by the 1st of July and can chug off.[4] I'm very much looking forward to our conversations; you're soon going to be a big son already, who can form his own opinions about everything he sees and hears. I am currently playing quite a lot of music, in a quartet, too.[5] Just be diligent with the piano so that you gradually also learn how to play. Having lessons is less necessary.[6] Bullheads like us prefer to learn on our own.

I have few obligations and still so much work.[7] The scientific correspondence alone keeps me quite busy, and then there are the calculations. ⟨The⟩ My health is better.[8]

Be sure always to be very nice with Mr. Zangger. He watches out for all of us so touchingly during this difficult time, even though he has so much to do.[9]

Just one more month left, then I'll have you back! Kisses from your

Papa.

Give the enclosed letter to Prof. Zangger.[10]

# Vol. 8, 346a. From Hans Albert Einstein

Zurich, 1 June 1917

Dear Papa,

I'd like to write you once again, but I don't quite know what. First of all, I'd like to report that Tete is probably traveling to Arosa this week already.[1] He's looking forward to it enormously & draws attention to himself in the entire hospital garden, because he jumps around everywhere.[2] Mrs. Zangger has offered to deliver Tete up there.[3] It was actually arranged that Tete be packed in a package & delivered up there by post. But now quite suddenly the undertaking has become serious; earlier it was all just a vague idea.

You asked me how school is getting on. To this I must answer that my memory has the following peculiarity: everything has great difficulty getting into my head, but once it's in there, it stays all right. This bad characteristic of my mind frequently gives me no end of trouble, because while the others learn everything at school already, I have to gnaw at it for a long time at home before I can do it.[4] This causes me a great deal of trouble now, especially in Latin. Now we are going through the deponents from the 3rd, you see, & they simply refuse to get into my "bird's brain," as our Latin teacher calls it.[5]

In mathematics we now have 3 subjects:

*1. Geometry.* We have the textbook by Spieker-Benke, or something like that.[6] We are now coming to the congruency of triangles, & I wonder very much how our dear O. Scherrer will explain it.[7]

2. Algebra: We are at the very beginning, & often it's so boring that everyone yawns through the entire lesson. He just has to go through everything very slowly in the beginning, because otherwise there will always be someone who is not able to follow.

*3. Arithmetic*: Here we're discussing a few individual cases of interest computation. Such as, for example: the interest rates; besides that, we're repeating what we did last year.

In French, Mr. Hardtmann is showing us more about how French can be derived from Latin.[8]

In German, we have to memorize something occasionally, otherwise it's quite boring. Once we came to talking about nonsensical phrases. He then gave the example of "a hairy bald-head" [*e harigi Glatze*].[9] Such a thing is often said without thinking about what it actually means. We all laughed then, but every one of us had certainly said it at least once before, even though we wouldn't want to think so now. When you think about it, you can find countless such examples, especially in dialect. When we had to write essays, he always assigned us topics, but usually such silly ones that we almost haven't a clue what to write & have to make something up.

I don't know what to say about the other subjects.

How are you doing with your illness?[10] In case you come here, I would like to inform you that, as far as I know, I have vacation *from mid-July* onward.[11]

I also tried to read that little book you had written & sent to Prof. Zangger.[12] I understood the first half fairly well; but the snag came where the equations started. You'll have to explain it to me when we're together. I'm very eager to hear about what it means. Now I'm getting piano lessons again, at the music school, that is.[13]I [like] it much better now than before, probably because the teacher I now have is a very nice lady.

Here at the Zanggers' there's a big garden[14] & Mrs. Zangger & I tend it. All the things that have to be done in it are fun. This is what has been planted: lettuce, spinach, beans, peas, rhubarb, carrots, etc. Then it also has gooseberries, currants, etc.

Many greetings from your sauerkraut-Latinist,

ALBERT.

# Vol. 8, 349a. To Heinrich Zangger

[Berlin,] 2 June 1917

Dear friend Zangger,

Your last letter makes me worry anew because I see that the upkeep of my sick family has acquired a ruinous quality. My net income (after deduction of taxes, etc.) has been reduced to 13,000 marks (this case has now in fact come to pass).[1] From that I need for myself, in order to make at least an appearance of maintaining the kind of lifestyle rightfully expected of me, 5,000 marks. If I don't want to save up a single penny, what's left is 8,000 marks = 6,150 francs. More I cannot and will not give; and ways and means have to be found to make do with that. If there is a small overrun, the savings could be used. But that should only happen in case of emergency. For I don't know whether these small savings are going to have to cover the children's education, in the event of my early death.[2] If I thought otherwise, I would simply be a heel.

So it cannot even come into consideration that 10 francs per day (= 4,000 francs per year) be paid for Tete.[3] Accommodations for him have to be sought that are suited to my financial situation, and the same applies for my wife. For this does not involve a transitory situation but a permanent one, as experience has unfortunately taught us.

It is not correct that I would have met my obligations better if I had stayed in Zurich.[4] There it would not have been possible for me to accumulate any kind of savings for the benefit of the children until now either, especially considering that my wife has certainly not been frugal. I also have to say, frankly, that the goodwill I encounter among my professional colleagues and the authorities here *obligates me to show the greatest gratitude.*[5] Everything that they are able to read in my eyes is simply being done. I am a good Swiss; but I make a distinction between political conviction and personal connection.[6] Without these local colleagues I would surely have remained an "unappreciated genius"; I must constantly bear this in mind.

It's not true that I disappointed my boy by coming over only in July.[7] When I suggested to him in spring last year that I might come and see him again in the fall, he acted quite negatively. Nor was a reappearance this spring either wished for or sensible, because we would not have known what to do with ourselves in the bad weather. Don't turn my boy into some saint! He is cheerful and happy-go-lucky and his letters happily reflect lightheartedness. He prefers to do things with his own age group rather than with a gloomy and venerable papa, thank God. Anyway, when I visit him nevertheless in July and go hiking with him, it will be more of a pleasure for *me* than for *him*; the love between parents and children is always somewhat one-sided, but nevertheless not unhappy! But when I get preached to by Mrs. Besso[8] and by you about love and conscience and am reminded of my gross infractions to paternal duty, then I can't suppress a smile.

I must confess that it weighs heavily on my mind that you are being burdened this way by me. This awareness haunts me all day long. I beg you sincerely to discuss with my sister[9] how she can relieve the burden which is not suited for your already so heavily burdened shoulders. Help Miza gain the necessary confidence in my sister so that she can take Albert, as would be the obvious thing under the prevailing circumstances.[10] Don't give Albert the booklet; he isn't mature enough for it yet.[11] His interest in such things is still playful, not actually intellectual, as suits his young age.

Cordial greetings, yours,

Einstein.

# Vol. 8, 350a. To Heinrich Zangger

[Berlin,] 12 June 1917

Dear Zangger!

Many thanks for the letter. The letter from my Albert was the greatest joy I've had for the past year.[1] I sense with bliss the intimate tie between us. I cannot describe how eager I am to see him again. I received a prize of 1,500 crowns from the Viennese Academy,[2] which we can use for Tete's cure in Arosa.[3] At the beginning of July I am coming to Zurich, then we can talk over everything. I'm not going to Tarasp; you yourself will become convinced that my illness is not situated in the liver.[4] It seems that what I need most is physical and mental peace and I plan to spend the vacation primarily at my sister's in Lucerne. But beforehand I am going with Albert to Arosa to see Tete.

Warm greetings from your

A. Einstein.

# Vol. 8, 352a. To Heinrich Zangger

[Berlin,] 17 June 1917

Dear friend Zangger,

Your letter pleased me very much, mainly because you aren't nearly as angry at me as I expected or feared.[1] But I did not believe I could do otherwise because the financial affairs were threatening to take a turn toward the impossible. If the apartment is now properly rented out,[2] and if Tete's stay in this sanatorium is not for more than 3 months, I see no immediate threat.[3] When I come to Zurich around 6–8 July, we can think about everything together. You'll then see that it's not bourgeois pettiness that makes me so obsessed. Affairs also need to be arranged so that *you* are no longer burdened as much as lately. You are one of those people who have *too* little egoism; this is a fine fault, but a fault nonetheless.

I am currently trying to obtain my travel permit, which is no small matter. But they won't be able to deprive me of this legitimate trip, if I set the right levers in motion. At the present time I'm working very little and lie outside very much, almost never walk. This prescribed lifestyle of mine sets my corpse into a visibly better condition. Unfortunately, I'm not going to be allowed to walk around much in Switzerland, either, but must save every step that can be avoided. I write you this so that no excursions are planned that I won't be able to execute. I feel sorry for my

Albert that I have to be such a lame companion for him. But I will try to compensate for the deficiency somehow. The money transfers are causing growing difficulties which, however, with the help of the Ministry of Culture can surely be overcome. I have already taken measures for the next transfer (1,400 francs on August first). If a further transfer should be arranged before my departure, which will take place on the 29th of June already, please inform me immediately so that I can attend to it right away. I am living a very quiet life, particularly considering that most of my professional colleagues have been deprived of their scientific work.[4] Who knows when people will revert to the pursuit of finer goals again.– In the *Naturwissenschaften* an exceedingly fine article appeared about Aristotle's and Hippocrates's theories on heredity, cast in the light of modern advancements in this area.[5] This article will perhaps also amuse an expert like you.

Couldn't we accommodate Tete elsewhere in Arosa for 2 years, when daily medical treatment is no longer necessary?[6] Let's discuss this then. In any event, I'm not going to leave anything untried, if in your opinion an actual recuperation is still within the range of possibility.

Cordial greetings, yours,

A. Einstein.

Many thanks to your wife for everything she has done for my boy, and a kiss for my Albert.

# Vol. 8, 357a. To Heinrich Zangger

[Berlin,] 24 June 1917

Dear friend Zangger,

Thank you for the two packages with the splendid comestibles, which arrived here safely. I've been shuttling around endlessly about the travel permit, but with God's help it will work. The day after tomorrow is my last lecture,[2] Friday night I am traveling to Frankfurt.[1] Saturday I am delivering a talk there.[3] Then I'll be spending 5–6 days at my mother's[4] and will travel, if all goes smoothly, between the 5th and 7th of July to Switzerland, a hard-earned but great reward; there seem to be some lively goings-on there, too.

Cordial greetings to you, your wife and my dear Albert,[5] yours,

A. Einstein.

I'm unspeakably pleased to be seeing my boys. I hope I'm not too shaken when I see my Tete; you do write that he looks all right.

# Vol. 8, 358a. To Werner Bloch[1]

[Berlin,] 27 June 1917

[Not selected for translation.]

# Vol. 8, 359a. To Elsa Einstein

[Frankfurt-am-Main,] Saturday, [30 June 1917,] 9:30 in the morning.

Dear Else,

The journey proceeded magnificently;[1] I looked out the open window from a prone position as long as the unusually fabulous sunset lasted. Later I slept for the most part. I discovered that my apartment keys were still in my pocket.[2] They are being sent out to you together with this postcard. The hotel really is one of those Christian old maids' homes, with a little black prayerbook on the night stand;[3] Wachsmuth is a c[lumsy] o[x] for having led me there.[4] Now I'm going back to the hotel and shall await the visits like a pasha.[5] Regarding edibles, I have obediently adhered to your orders.[6]

With affectionate g[reetings] & k[isses], yours,

Albert.

# Vol. 8, 359b. To Elsa Einstein

[Frankfurt-on-Main,] Sunday morning. [1 July 1917]

Dear Else,

It all went well. In the morning Mr. Oppenheim was here to see me, a very intelligent, kind man, as fine as his son.[1] Mr. Oppenheimer came shortly before 1 o'clock. I went home with him for lunch (cab).[2] A very interesting man, Nussbaum, a painter & graphologist, was also a guest with me there.[3] I enjoyed myself very much. Then Laue came to see me and it was very agreeable.[4] Then I drove together with him to the institute and held my lecture—moderately well.[5] Then a group meal at the Rathskeller, likewise very interesting. I was extremely careful about eating.[6] 11 o'clock in bed! State of health impeccable. No one paid anything for me (stingy). Today, ½ past 9 o'clock, trip onwards to Heilbronn.[7]

Lots of love to the children & parents,[8] a kiss to you, from your

Albert.

The journey has been very picturesque so far.

# Vol. 8, 359c. To Elsa Einstein

[Heilbronn,] Tuesday, [3 July 1917][1]

Dear Else,

Hallelujah! You came, were seen (and heard!), and conquered.[2] From now on I shall always send you in advance, like Josef Adler used to send his Rosa when he was afraid of our dog.[3] I would never have managed it. You received the passport, I hope. Send the thing very soon so that I can steam on. I am being cared for very attentively, am not making any visits, and lie down often.[4] While doing so, I've been reading Spinoza, which I bought yesterday.[5]

G[reetings] to our old folks,[6] g[reetings] & k[isses] to you and our girls,[7] yours,

Albert.

Please send my mother a copy of Rathenau's publication.[8] Several copies are lying around in my flat.[9]

# Vol. 8, 359d. To Elsa Einstein

[Heilbronn,] Wednesday. [4 July 1917]

Dear Else,

The appeal by Harnack, Fischer, etc., honestly pleased me,[1] likewise Rathenau's fine essay,[2] and the letter by my dear confidence woman, who is fighting so lovingly and selflessly for my trip. I hope the last adventure now lies behind you, so that I can travel Friday evening or Saturday morning.[3] My lifestyle is strictly medically vegetative, without visits, outings, etc.[4] Yesterday Paula Weil (Flamme) was here all day. It is to be hoped that this writing spell, which is so foreign to my nature, will last to your satisfaction; such a tranquil lifestyle makes it easier to carry out pious resolutions. I hope we can execute our Ensingen project;[5] prepare yourself for it. Mother's hopes of traveling to Switzerland have sunk, because her patron has to take a trip.[6] Milk and honey are trickling sparingly here, too, only I, lucky fellow, am being pampered everywhere!

Kisses from your

Albert.

Greetings to the little minxes and your parents. [. . .][7]

# Vol. 8, 360a. To Elsa Einstein

[Zurich,] Monday morning. [9 July 1917]

Dear Else,

Yesterday evening, 9 o'clock, I arrived here after a comfortable trip and very decent and lenient customs clearance at the border[1] and after staying a number of hours at the elder Habicht's, who has become the happy father of three sweet boys.[2] I telephoned Zangger immediately.[3] My boy was already in bed;[4] I am expecting him here right now. Zangger went to Winterthur today; he is coming to see me today at 4 o'clock at the hotel. I shall not allow myself to be invited anywhere under any circumstances, except to Besso's, if his wife is still away and he is alone here with the household staff.[5] There is no white bread here either, but excellent standard bread, twice as expensive as white bread in peacetime. Nutritional conditions similar to ours in the first year of the war.[6] Apart from the brief telephone conversation with Zangger, I haven't spoken with anyone yet;[7] the familiar locality alone has made me feel at home. My mother is still hoping keenly to come here.[8] It seems to be easier from there.

When my drama has advanced further, I shall report again.[9] In the meantime, kisses from your

Albert.

# Vol. 8, 360b. To Elsa Einstein

[Zurich,] Tuesday. [10 July 1917]

Dear Else,

Albert has developed splendidly, is in glowing health, cheerful, very curious, intelligent, and modest; I am elated.[1] On Saturday we're traveling to Arosa;[2] before that I'm also going to Lucerne.[3] I was resting for a good part of the day yesterday, today likewise. The diet is being followed very strictly.[4] My sister-in-law will be coming here in the next few weeks (she obtained the passport);[5] now it is a problem that the apartment has been rented out (until October).[6] Maybe the apartment will become available again sooner. At noon today Zangger will examine me—probably including an X-ray.[7] It's raining incessantly and is cold. Zangger has little time; I've not been able to talk to him much yet.[8] I moved to his house for the remaining days, after all, so that he can observe me and he does not waste more time than necessary. I also just saw Besso[9] for a moment.

Warm regards & kisses, yours,

Albert.

Greetings to the little minxes and the elders.[10]

# Vol. 8, 361a. To Elsa Einstein

Lucerne, Thursday. [12 July 1917][1]

Dear Else,

Here I am outside Maja's flat in a deck chair in the sun and admiring the view in wonderful weather.[2] We are enjoying ourselves very much together; this is the first real relaxing moment on this trip. On Saturday I am going with Albert to Arosa for a few days;[3] then he and I are ⟨going⟩ staying with Maja for about 4 weeks (!). When that's over, it's back again to Ensingen, where we shall then meet.[4] I have arranged that I stay with the Zanggers for a total of only two days (one still to come), because it is very uncomfortable for me there; his wife, in particular, is a real pain in the neck for me.[5] But my Albert is happy there.[6] At Maja and Paul's it's cozy beyond words; I immediately thought that we must try to make similar arrangements for ourselves.

Greetings & a kiss from your

Albert.

Warm regards also to our little minxes and the elders. [. . .][7]

# Vol. 8, 361b. To Elsa Einstein

[Zurich,] Friday. [13 July 1917]

Dear Else,

I have to stay here until Tuesday, after all, because Zangger still wants to keep me under close observation.[1] His diagnosis agrees quite precisely with Rosenheim's.[2] He also attaches great importance to treatment in a warm climate. Today he took a blood test and determined that there must be an internal inflammation in my body. After a brief Arosa stop, a rest cure of several weeks in Lucerne, then perhaps another couple of weeks in Tarasp, which I would certainly change to Mergentheim[3] if you were to come over there. Please go to the bank and make sure that the 1,400 francs finally arrive.[4] Also ask there whether De Haas has received the 1,500 crowns.[5] My Albert[6] gives me very much joy; I could not have wished for a finer fellow. But that doesn't mean I'm less fond of our little minxes![7] I am taking him along to Lucerne, so that he can spend his entire holiday with Maja and me.[8] If a water cure does not become necessary, we shall meet in Ensingen.[9] Paul's sister Rosa is remarrying next Tuesday after a long widowhood. If I'm still there, I'll be a witness at the wedding.[10] Today I was with Uncle Jakob.[11] He is as jovial as ever & lives in the Baur au Lac hotel.[12] He is advising me to transfer all my money from Berlin, despite the bad exchange rate.[13]

Greetings & a kiss, yours,

Albert.

# Vol. 8, 361c. To Elsa Einstein

[Zurich,] Monday morning. [16 July 1917]

Dear Else,

This postcard from the platform just before departure to Arosa, where I intend to stay with Albert for exactly one week.[1] Zangger's thorough analysis also yielded slightly high blood pressure (heart somewhat affected by it).[2] I did not visit anyone in Zurich. Albert has been a bit refractory again (clear influence of his mother).[3] The flat can probably be vacated on September 1. My sister-in-law is already here.[4] Zangger made some quite unfavorable comments about my wife's character. The Bessos are touchingly kind to me. Mrs. Besso keeps house without maids; that probably explains her somewhat hesitant manner regarding invitations. She is frail and does everything herself.[5] I don't know yet whether Tarasp is necessary.[6]

The train is about to leave, I have to hurry. Greetings also to the little minxes and the aunts,[7] and a separate kiss for you from your

Albert.

# Vol. 8, 361d. To Elsa Einstein

[Arosa,] Tuesday. [17 July 1917]

Dear Else,

We have been in Arosa for one day now already.[1] The area is wonderful, the care very good, but not quite up to spa standards. I am very careful about what I eat, though, and walk about little.[2] Tete looks splendid, not fat, but ruddy like a healthy farmer's boy. He is indescribably funny. The sanatorium is managed conscientiously but militaristically; I shall see if I can bring him to Mrs. S[tahel] without spoiling relations with Z[angger].[3] It's like in a family there, a friendly woman who made just as favorable an impression on me in person as her letter did on us. Besso does not believe in Tete's illness; his Vero[4] had supposedly had the same symptoms, without anything serious having been behind it at all. The doctor[5] claims the contrary but expects full recovery after a year's stay in the mountain air. In any case, I have never seen a better-looking or more cheerful little fellow. The absence of their mother[6] does both of the boys a lot of good, outwardly as well, in that their behavior has improved. I happened to travel up to Arosa with Mrs. Hurwitz.[7] She was very friendly; I told her the reasons for the separation, more or less.[8] Albert is diligent at school, and that without a trace of conceitedness. He is also not nearly as reserved and has a strong sense of orderliness, oddly enough! On Saturday or Sunday we are returning to Zurich, where Zangger wants to keep me under observation for a couple of days longer.[9] Then I'm going, with or without Albert, to Lucerne for a couple of weeks.[10] It is still uncertain whether I must then go to Tarasp as well (for bowels).[11] I hope we can meet in southern Germany instead.[12]

Kisses from your

Albert.

# Vol. 8, 361e. To Heinrich Zangger

Arosa ⟨Monda⟩ Tuesday. [17 July 1917]

Dear Zangger,

Yesterday after safe arrival and a splendid trip we visited Tete.[1] He looks completely healthy, is lively, but not nervous. Dr. P[edolin] says the glands are infected, a stay of a whole year is necessary.[2] Care evidently unobjectionable. But the institution seems to me too militaristic.[3] I visited Mrs. Stahel today with Albert; I liked it there *very* much. Just a few children, like a small family. Cost 6 ½ francs, as of the fall 7 francs.[4] If the boy should stay here for a year, which P. might be right about (only you can judge that), then we could consider bringing Tetel there later. Today we had fine weather and could enjoy the fabulous view from Arosa in brilliant sunshine.

Warm greetings to you and yours, from your

Einstein.

# Vol. 8, 361f. To Elsa Einstein

[Arosa,] Thursday. [19 July 1917]

Dear Else,

It's wonderful up here; I wish you could also have the pleasure.[1] Gorgeous nature and food on demand! Both my children give me much joy; fine, intelligent boys. I don't believe that there's anything wrong with the little one;[2] but all precautions should be taken. I'm also spending much time with the Hurwitz family; the ice melted as soon as they saw that I was the same as ever.[3] I'm feeling remarkably well, despite lying down very little and puttering about a lot. The mountain-air myth does seem to have some truth to it, after all. On Sunday I unfortunately have to go back down again, maybe already on Saturday. I'm often alone with my little boy, because Albert goes on longer walks with the young Hurwitz boy.[4] The little one is very mischievous and is already a bookworm, in that he devours whatever he gets hold of. I am sending Ilse and Margot a picture of the boys I arranged to have taken here, which has come out well.[5]

Greetings & a kiss from your

Albert.

# Vol. 8, 361g. To Heinrich Zangger

[Arosa,] Friday. [20 July 1917]

Dear Zangger,

We had a fine time here. Sunday evening we are returning. I'm happy that Tete looks so well; he lends an impression of complete health. Perhaps just the overactivity is abnormal.[1] I like your booklet very much. But in my slow way I only managed to read through a little over half of it.[2] My mode of life was quite in accordance with regulation and no suspicious signs of life inside made themselves noticeable, either. Albert is a healthy, alert, and independent fellow; I had many occasions to delight in him.[3] He hasn't yet been infected with the pallor of contemplation.

Cordial greetings to you and yours, from your

A. Einstein.

# Vol. 8, 364a. To Elsa Einstein

[Zurich,] Tuesday, before departure to Lucerne. Train station. [24 July 1917]

Dear Else,

Many thanks for letters. I'm writing continually, aren't I. Rebukes not justified. Albert is coming along to Lucerne.[1] Health impeccable. Pain not even *once*.[2] Go to the Dresd[ener] Bank and press for *immediate* execution of money transfers.[3] My ex-wife is in difficulties.[4] Soon I'll not have anything left, either. Please arrange ⟨right away⟩ to have the remittance from the L.V.G. aeronautical company deposited in my account at the bank.[5] Please adhere to the termination notice of my apartment;[6] I cannot approve of your steps toward extending the lease; I gladly take the opportunity to move close to you all.[7] Please inform Mr. Abbé of this.[8] I did not pay any visits in Zurich, but yesterday I went sailing with Albert for 1½ hours. He can already do the crabbing technique. If only you could be here as well; it's so lovely in Switzerland. The food-supply calamity is threatening here also, though.[9] I am leaving the handling of the taxes to my bank. I'm not filing a reimbursement claim because I also received unexpected payments subsequently.[10] My children are both attached to me now with genuine fondness. But your and the little minxes' places of honor in my black soul are not jeopardized because of it! I miss you all terribly much.

Heartfelt greetings to all of you (young & old)[11] and kisses to you from your

Albert

Among the "friends," feelings toward you are very amicable.[12] Time heals all wounds. Everywhere fine weather.

# Vol. 8, 364b. To Elsa Einstein

Lucerne, Wednesday. [25 July 1917]

Dear Else,

I read the newspaper articles in the *Frankfurter* with sincere joy; dawn does seem to be breaking.[1] Don't worry all the time about my health. All things, possible and impossible, are being done. Yesterday Zangger came here especially to instruct Maja.[2] I am feeling like a peach, at that; and I already look much better than at my departure, am getting *ideal* fare, rest for the greater part of the day, and warm my venerable belly three times daily with a hot-water bottle.[3] If only you all were here. Margot would be quickly back to normal as well. I have a lot to study, both professional and otherwise, and also plenty of correspondence. Do what you wish with the flat.[4] It is odd that the L.V.G. is still sending me anything.[5] Please check that the Dresdener Bank relays both cash remittances promptly to Switzerland![6] It is scandalous how they are making me wait. Please write a little more clearly; your letters are almost as hard to read as Zangger's.[7] Do you really seriously advise me ⟨not⟩ to return soon? I planned to be back in Berlin before the 1st of September. And Ensingen?[8] Do come over there! Be enterprising and happy-go-lucky! What has come over you to be so jittery all of a sudden? Now you'll be the mistress of the house for a while.[9] Is Blaschkina the Terrible still your only help? The weather is fabulous, and if I were to follow my heart rather than my reason, I should go hiking! My Albert is a great guy who brings me much joy; the ice has melted.[10]

Now, be cheerful, stop whining, and be kissed with the little minxes by your

Albert.

# Vol. 8, 364c. To Elsa Einstein

[Lucerne,] Thursday. [26 July 1917]

Dear Else,

I'm writing to you nonstop, yet you wail about neglect! The post is simply operating irregularly. Zangger has made *exactly* the same diagnosis as Rosenheim, so don't grumble (duodenum).[1] Tarasp would only come into consideration after a full recovery of the stomach, for fortification.[2] It is quite certain that I'll not be going there; so never fear. When I come across the border again, e.g., end of August, let's meet in Ensingen.[3] You must enjoy yourself once too. Don't worry constantly about the future! A bird in hand is worth two in the bush! Don't send letters. They were returned because the censors don't conduct such voluminous checks.[4] Bloch[5] is deluding himself in vain about the institute. I do not intend to engage anyone and, frankly speaking, also don't want to have him. I did know who was there at the meeting at the time, but forgot again.[6] I don't want Moszkowski to drag me before the public as a private person (a novel!); he can satisfy his muse on whatever else he wishes, of course.[7] That I'm not as attached to you as you to me is slander! At the end of August I'll be with you again, even if it means going hungry. Miza is still bedridden but supposedly looks quite all right, and it is believed that she will learn how to walk again if she makes an effort.[8] Tete is very comical, red cheeked, and looks robust. But the glandular infection has been established objectively by X ray. The doctor is sure that his health will be fully restored. The fever is gone, although his temperature is not quite as steady as for a healthy child. He is staying in Arosa for the time being *with my consent*.[9] Health goes before money. But later he should go to Mrs. Stahel's, which would still cost 7 francs.[10] I've become convinced that this isn't too much, at today's prices.

Kiss from your

Albert.

# Vol. 8, 364d. To Elsa Einstein

Lucerne, Saturday. Morning [28 July 1917][1]

Dear Else,

I'm spending my days pleasantly but adhering to the strictest medical regimen.[2] Nobody comes to disturb the idyll. Uncle Jakob is also here, but he never comes up here to us because it's too far for him.[3] I recline almost all day long and study. Today Albert is going away with Pauli[4] until Sunday to climb a mountain.[5] Mama is cleaning vegetables right now, Maja is doing the shopping, Pauli is at the office,[6] Albert is getting potatoes, and I am lying under the great linden tree next to the house and am letting myself be roasted by the sun. Since Berlin, I have gained 1.3 kg; the remainder is sure to come. This to reassure you. Albert is a clever, healthy boy, but is often cheeky with me; yesterday I gave him a good dressing down. My girls really are much better behaved; Margot's letter pleased me very much, especially since I know that she is not such an excessively avid writer. What a shame that all of you aren't here as well. The weather is splendid, and life is indescribably comfortable.

Greetings & kisses from your

Albert.

The money to Miza has also arrived.[7] Best regards to the little minxes and to Uncle & Aunt.[8]

# Vol. 8, 365a. To Elsa Einstein

[Lucerne,] Monday. [30 July 1917]

Dear Else,

I hope you are receiving my messages more regularly than I am yours. The day before yesterday we went on a cruise to Weggis with Uncle J[akob] in splendid weather.[1] He is brimming with health and mirth. I am feeling *very well*. My outward appearance is improving visibly.[2] At the end of August, let's meet in Ensingen; I say this in every postcard so that you get used to the idea.[3] Mama is going to Weggis at the end of the week with Uncle.[4] I feel very much at ease with Maja. Albert is going back to Zurich tomorrow;[5] yesterday he climbed a mountain near Vierwaldstätter Lake (2300 m) with Pauli.[6] Now you shall wield the scepter of the house,[7] certainly a difficult reign. I hope our little Margot is completely well again. I miss you very much already along with the little minxes, as nice as it is for me here. From this I see, better than ever, how much I belong among you!

Greetings & kisses from your

Albert.

# Vol. 8, 367a. To Elsa Einstein

[Lucerne,] Wednesday. [1 August 1917]

Dear Else,

I have been without news for almost an entire week and am close to sending you a telegram. Is everything all right at your end? I am feeling remarkably well. I have been here a week now already.[1] ⟨Maja⟩ Albert traveled back to Zurich yesterday evening.[2] The harmony between him and me still leaves quite a lot to be desired despite his quick mind and interest in everything.[3] I don't know whether there is any prospect of it ever improving. There are deep differences, I believe. Life flows by calmly and very pleasantly. Maja and Pauli are very relaxed and charming toward me.[4] Mama is going to Weggis with Jakob at the end of the week.[5] Ogden was also here;[6] I did not see him, though. Are you coming to Ensingen?[7] I invite you there. We don't want to hurry onward to Berlin. Greetings & kisses to you and Margot, and greetings to Uncle, as well as to both the excursionists.[8]

Do write finally sometime to your

Albert.

I write virtually every day, at the latest every 2nd day. If a longer break occurs, the post is to blame.

# Vol. 8, 367b. To Heinrich Zangger

[Lucerne], Wednesday. [1 August 1917][1]

Dear friend Zangger,

Your letter just arrived. Retaining such good humor with a throbbing pain in the head is rarely achieved, not to mention writing a letter then as well. I hope the matter gets resolved so that everything is soon over.[2] What atrocious colleagues you must have, if they're constantly busy plaguing you! My physicists are better people after all; they are generally well disposed toward one another. Is this because physics doesn't offer any substantial opportunities for making money? It almost seems that way.[3]

I have nothing against a conversation with my wife, provided someone else is present, although I don't understand what good it would do.[4] If I go there once, it would have the consequence that next time it would be expected that I go again. Thus, constantly growing diplomatic difficulties would be engendered. An additional complicating factor is that I am no match for my opponent . . . In any case it should be agreed in advance that this be a one-time event that should produce no consequences in the above sense.

I'm not for the Nobel society, because I earnestly fear that it—could be rotten [beschissen]. As long as I have something in Berlin, I will exchange it at the going rate, however unfavorable. Only when this has become impossible for some reason or the money simply isn't there, will a radical change have to occur, but we don't need to rack our brains about that yet. I don't try to anticipate such things. Time will tell. If in this regard I have a wish, it is just that we accommodate Tete for the rest of the envisaged year with Mrs. Stahel. I was there and came away with a very favorable impression of the lady and her small institution, housing just a few children. The price now is 6 ½ francs, as of the fall 7 francs.[5] Perhaps you could make more inquiries about the lady. It would have to happen early because she has no room right now and so advance registration is necessary.

The stay here is excellent. I would wish you the same peace. Albert left here yesterday evening because his mother wanted it.[6] He was enjoying himself. Mentally he has developed well but is somewhat rough emotionally, particularly in his behavior toward me; this is grounded in the circumstances.

Now I wish you a good recovery and an enjoyable vacation![7] Please send me your vacation address. Cordial regards, yours,

Einstein

Puzzle: How did I happen to write on two sheets? Hint: The sheets come from a single pad.

# Vol. 8, 369a. To Elsa Einstein

Lucerne, Monday. [6 August 1917]

Dear Else,

Communication is so sluggish, owing to the miserable postal connections, that we're almost losing the thread.[1] I had a visit the other day from Michele, Anna, and Vero,[2] which was very enjoyable and entertaining, despite the miserable weather. Michele is still here. Zangger is in Tarasp on holiday; he also has trouble with his innards.[3] Maja and Paul[4] are leading a very happy life and are incomparably good to me. What a pity that you can't come to Ensingen.[5] We shall simply meet in Sigmaringen,[6] then or somewhere else around the first of September. By then Aunt and Ilse will have returned and then you must also have some freedom.[7] Have you also spoken with the Meissners about the apartment, so that I don't end up falling between two stools?[8] I relay my thanks to Ilse for the kind postcard. I have become convinced that postcards & letters have gotten lost. So don't make such an exact reckoning with me. My health is steadily good.[9] I shall prepare myself as well as possible for the winter. If only I could share some of this with you as well!

Kisses also to Margot from your

Albert.

I am already eager to share house again with my three women.[10] [. . .][11]

# Vol. 8, 369b. To Elsa Einstein

[Lucerne,] Tuesday. [7 August 1917][1]

Dear Else,

So, no express delivery, after all. That does not befit Asia's sons and daughters.[2] Have no qualms about the apartment. I am not going to have to camp on the street, never fear.[3] Rather, think about where we should meet.[4] Maybe Thuringia would be suitable in September. Michele has left.[5] No one is ill-disposed toward you, not even the "friends."[6] Health good; I hope all of yours as well.[7] Kisses also to Margot from your

Albert.

# Vol. 8, 370a. To Heinrich Zangger

Lucerne, Wednesday. [8 August 1917][1]

Dear friend Zangger,

Today I received your inquiry. Saturday to Monday I had a quite hefty attack. First, discomfort with a barely localizable sensation of too much volume, then gradually rising painful pressure, also difficult to localize but roughly in the area you determined, accompanied by acidic belching. Then quite rapid but certainly not sudden recovery. Now certainly not *more* sensitivity toward pressure than before the affair. Stool regular. Fever none. Mood always good, mind clear. As the source I regard quite definitely a large amount of quite tart stewed apples, which I ate around Friday and Saturday.[2]

Who knows, Nicolai may well have been right, after all, to regard the gastric acid as the *primary* cause.[3] It is possible, isn't it, that this caused this mess, if more gets into the intestine than it is able to neutralize. Would you consider this out of the question?

Michele happened to be here during those days.[4] We were enjoying ourselves very much together. The fact that it mostly rained and that on Sunday I couldn't leave my bed altered nothing. He is an eminently fine and good-willed person who would be happy if he were capable of regarding his introspective, scarcely active disposition as a kind of gift of nature instead of reproaching himself, and if he weren't harassed by his wife.[5] By upbringing she is a good person but she completely lost her equanimity through an overdose of kindness and consideration, perhaps also weakness, on his part. She ought to be separated from him at least for a while so that she could shake her habit of regarding him as the cause of all her bad moods and the lack of fulfillment of all her unclear wishes.–

The planned visit with my wife gives me quite a knot in my stomach. I have far too little confidence in her decency and honesty.[6] After being in Albert's company for a longer while, I noticed again that she must speak about me with much ill-will and lack of respect.[7] It was as if the beneficial but less deeply entrenched influence you had on the boy in this respect had gradually receded into the background again, against the earlier influences of greater mass effect.[8] I know that I must accept this as a natural consequence of the now prevailing conditions, but do feel a deep abhorrence toward enlarging the contact surface beyond the absolutely essential, especially considering that this could hardly be of any use to the woman anyway. But I will make the visit to see her nonetheless, if you as a discerning and good-willed bystander deem it desirable, despite all this, that I do so.

I hope you recuperate very well up there and also learn to be a little lazy, despite your lack of talent on these points.[9] Picture for yourself very vividly that in two years the world will suffer such a severe earthquake that no living being will be spared; that way you might become acquainted with the attitude that makes life sweet for us indolent orientals.

Cordial regards; also to your wife, yours,

A. Einstein.

# Vol. 8, 370b. To Elsa Einstein

Lucerne, Thursday. [9 August 1917]

Dear Else,

I am living as nicely and peacefully as a cow on alpine pasture, except that instead of flowers and grass, I'm devouring fantastic amounts of zwieback, butter, honey, and milk.[1] And it's even working. There's very little diversion here, thank God, just my correspondence with De Sitter and also with Levi-Civita belongs in this chapter.[2] Both are extremely interesting. I am again easing myself into working on scientific matters as well. Moszkowski, my friend, don't bite off more than you can chew! That would be a tall tale for people to relish; write to the editors in this sense so that they know that it is not *I* who recommend him for the review. I am firmly convinced that he cannot meet this task.[3] Maja and Pauli[4] are living an extraordinarily pleasant life. You can hardly imagine such harmony, peace, and sense of security. That little sacrifice of tidiness and cleanliness is well worth it. I wish we could arrange it similarly for ourselves as well. Then I too could join the liberation-from-Rome movement with full conviction![5] I procured an adequate amount of underwear to pull me through. Incidentally, I suspect that during the winter we'll all be in bed quite a lot (heating substitute).[6] Here it will be likewise.[7] I am considering traveling to Ensingen around September first and staying there for about a week.[8] Where shall we meet then? I'll still have plenty of money left for the two of us, as it seems. Find out about a suitable spot; it suffices, of course, if you send me word about it in Ensingen. Margot is also heartily welcome, in case you would like to take her along.

Kisses from your

Albert.

Greetings to Margot and Uncle, as well as to the two excursionists.[9]

# Vol. 8, 370c. To Elsa Einstein

[Lucerne,] Saturday. [11 August 1917]

Dear Else,

I just wrote to turn down Zangger, who wants me in Tarasp.[1] I want to go to Ensingen at the end of August, stay there one week, and then meet with you somewhere. Think about where this could be. In Weimar, perhaps? Or somewhere else in Thuringia, in a quiet place, if possible.[2] I am in very good health, which is certainly no wonder, with such abundance of food. Yesterday I received a dear letter from Ilse, which delighted me. She has probably steamed off at last with Aunt[3] and has entrusted Samson and Delila[4] to Margot's care. I congratulate Margot on the fecundity of her muse;[5] I am curious! You seem to have a vehement friendship with Mrs. Fechheimer.[6] Be careful; strict masters do not rule for long; that's what one says when it's raining down hard.

Greetings & kisses from your

Albert.

# Vol. 8, 370d. To Heinrich Zangger

[Lucerne,] Saturday [11 August 1917]

Dear friend Zangger,

I can't make myself go up to Tarasp,[1] especially since I am determined to cross the border on August 31st in order to travel to Ensingen to see my priest.[2] I am very comfortable at my sister and brother-in-law's[3] and it is the first time that I have been visiting with them for any length of time; so you should shut a medical eye to it. They want to talk me into staying even longer in Switzerland. But I would like to go somewhere with Elsa in Germany in the 1st–2nd week of September and I have to be back in Berlin on October 1st because I am moving.[4] My health is beyond reproach. Sensitivity to pressure is not definitely establishable by me.[5]

Please don't be cross at my obstinacy. My own belly is at stake, and Elsa, who has not got away from Berlin for 2 years now, should also be able to enjoy herself a bit.[6]

I compared the theory of relativity with thermodynamics not with reference to its *content,* but with reference to the *method*. Both are based on a very general principle:

1) There is no such thing as a perpetuum mobile.

2) No state of motion has preference over any other.

Both draw their consequences from the general principle without needing a more detailed model-like theory. Therein lies their certainty but also their limitation.[7]

Cordial greetings, yours,

Einstein.

# Vol. 8, 370e. To Elsa Einstein

[Lucerne,] Monday. [13 August 1917][1]

Dear Else,

You are carrying on our correspondence with such fine devotion that I can't constantly keep up. I'm sorry that the story with the apartment came to nothing, but it serves us right because we did go in for Abbé's offer and cannot go back on it anymore against his will. But maybe it's good like this after all; for since an uncertain destiny awaits Meissner's flat, I could be forced to move around like the Wandering Jew, which is not uninteresting, in itself, in these great times, but is a bit wearing.[2] I would have liked to attend the little diplomatic congress at Moszk[owski]'s[3] and in doing so see the charming younger generation; I actually do quite like that young man, and he seems to me to be considerably less nebbish than all of you think. But women are more authoritative and clear-sighted than we men in such things. I am very sorry that our priest will not be in Bensingen anymore on September 1st;[4] but I can't go away from here quite yet. That would be too short a stay for me here.[5] I have already written to him. Now it depends on when Aunt and Ilse return home.[6] Then we can meet wherever you think is suitable. I won't abandon this plan now; you must also have a little freedom for once, my dear Cinderella![7] This afternoon I'm going to visit Uncle and Mama in Weggis for the second time.[8] They're a very merry lot over there. Mama walks about vigorously again, which is surely attributable to a curative lack of fat. Uncle swims and rows like a youth and is enjoying life. He unfortunately seems to have gone off the idea of founding a household again:[9] the rascal. Where should I get the courage from to cast a stone at him for that? You can afford to do so more than I; but the goal will attract you less, you black soul.[10] The volatility of the exchange rate, connected with the necessity of leaving my little boy in Arosa for the time being, make my financial situation seem increasingly precarious.[11] Sooner or later it will probably become necessary that my wife go to her parents.[12] I don't see any other way. I spoke with Michele about that yesterday, who also agrees with me. She would take the little one with her.[13] Then Albert[14] would be accommodated at Maja's. Miza[15] is doing all she can to have me visit her, and is trying it with sweetness, even. I am not going there, though. Let her stir Albert up against me even more, by God, if she finds satisfaction in that. Yesterday Aunt Ida and Edith were here;[16] their tireless chatter sweetened their speedy departure. Maja and Pauli[17] are taking care of me so touchingly; the stay agrees with me splendidly as well. Nothing surpasses a simple, peaceful life. Pauli has a very active mind and is a very pleasant companion. He paints attractively and enthusiastically and reads a great deal in his free time, takes care of the garden, helps Maja in the house, is friendly, and in cheerful spirits. Every day at noon they drink jetblack coffee with or without brandy. I congratulate them on their gem of a housekeeper but with reservations; may she also persevere!

Kisses also to Margot from your

Albert

Best regards to Uncle,[18] furthermore also to Mr. Sparrow and to Samson and Delila.[19]

I shall obtain the agar[20] ⟨when I go to Zurich⟩. The news from Mrs. H[ochberger] about her son[21] pleased me exceedingly. Give them both my best regards.

# Vol. 8, 371a. To Elsa Einstein

[Lucerne,] Wednesday. [15 August 1917]

Dear Else,

Your touching telegram against hunger arrived yesterday comically truncated: "He is coming . . ." rather than "you receive. . . ." I have written Brandhuber to cancel.[1] From September 1 onward, I am at your disposal anywhere except in Berlin. Don't scold me then for being here in Lucerne for so long: in exchange, I am eating with all the conscientiousness a schoolmaster can muster.[2] Otherwise, I am reading *Auch Einer* [Another One of Those], a very droll book.[3] There are hefty storms daily over here; I always think how charmingly afraid you all would be. Today Pauli[4] is on a mountain; he must have gotten thoroughly soaked. I am very eager to see you all. Are you bringing Margot along, or are you coming alone? (Where to?) Don't allow our venture to fall flat![5] Maja[6] plays the piano splendidly; it is amazing what she has learned in her solitude. Today I helped her hang up the laundry. Around the 25th I am traveling to Zurich and shall stay those few days in the care of the Besso Sanatorium;[7] address until departure: 33 Universitäts St., Zurich.

Kisses also to Margot from your

Albert.

Greetings to Uncle, the decampers, Kraft, I[lse]'s parents-in-law in non spe, your new bosom friend,[8] as well as my flames of the past, etc.

# Vol. 8, 371b. To Elsa Einstein

[Lucerne,] Friday. [17 August 1917]

Dear Else,

The letter from Aunt and Margot just arrived.[1] I am traveling to Germany on the 31st. I'm going to spend the night in Heilbronn[2] & drive onward on Sept. 1st. Now there's the problem that you probably can't leave yet on September 1, because Aunt and Ilse are away.[3] Or does it work anyway? I would be glad to wait for you somewhere in Thuringia[4] but can't get anything to eat there, because I obviously have no ration coupons and also have no change of address notice, which would entitle me to obtain coupons. In the last days of August I'm going to be at Besso's:[5] 33 Universitäts St. in Zurich. If I receive news from you by letter with meeting arrangements or a telegram from [you] to Besso's address, I shall act accordingly. If not, no choice remains for me but to travel up to Berlin. But if you can arrange it somehow, we'll meet on August[6] 1 at a place indicated by you. I would like it very much. Health very satisfactory. Weight 72.1 kg (compared to 70).[7]

Margot's little letter pleased me very much, the pretty little picture also. What a pity that the soc[ial] event at Moszk[owski]'s[8] disagreed with her so. What's done is done![9] Kisses also to Margot from your

Albert.

Best regards to Uncle and the sight-seers.[10] The "express delivery by messenger" is absolutely worthless, as I have gathered from your postcards. On the other hand, postcards work better than letters.[11]

# Vol. 8, 372a. To Heinrich Zangger

[Lucerne,] Tuesday, [21 August 1917][1]

Dear friend Zangger,

Seippel's letter unfortunately arrived in my hands late, because his letter was incorrectly addressed ("Zurich" instead of "Lucerne" on the address).[2] Seippel learned about my view on the issue of peace from Weiss, whom I had been speaking to in great detail;[3] it satisfies me very much and confirms to me that my impression agrees with Foerster's.[4] I have been thinking much about the political situation recently and hit upon a more hopeful conception. I now see a way and offer it for your comments:

Now, right away during the war, a pacifist union ought to be founded of as many of the Entente and perhaps neutral states as possible, according to the following principles:[5]

1) A court of arbitration to settle cases of conflict between these treaty states.

2) A common institution which should decide to what extent these states apply and may apply the principle of universal conscription. Common foreign deployment of troops. Reduction of the standing army according to the possibilities afforded by the foreign relations of the treaty states. ⟨Mutual military assistance⟩

3) Principle of most-favored customs policy among the treaty states, connected with the inclination to gradually eliminate the customs barriers among them.

4) Any state can become a member of the union that fulfils the following conditions:

a) A parliament elected according to democratic principles.

b) Ministers dependent upon a parliamentary majority (which ministers must naturally have the executive fully under control).

5) Military pacts with states not belonging to the union are not allowed, on pain of loss of membership in the union.

6) The union guarantees each treaty state its land holdings against outside aggression.

The advantage I see in this proposal is essentially that even a federation comprising by no means *all* states can be very valuable in that, at the price of forfeiting expansion, it secures the existing property of its members and draws with it a greater reduction in the military burden as more states join the union. If the Entente could bring about a union of this type, which had the United States, England, France, and Russia as members, it could make pacts with Germany without concern, which would eventually be forced economically to seek to join the union without anyone being able to say that the nation's "dignity" had been injured.—If you yourself find this matter reasonable, I ask you please to give it your support. I only know Weiss personally on the Entente side; I intend to speak to him as soon as I get to Zurich, which should be soon.[6]

I'm going to write to my wife myself; I recently wrote her and also received letters from her.[7] Feel free to tell her that I had written you and that we had been in direct touch with each other. I'd like to propose to you that we have Tete brought to Mrs. Stahel as of the 1st of September, provided he has to continue to stay in Arosa for months (for reasons of economy).[8] Your silence shows me that you really did pay for Tete for the month of July out of your own pocket. I'll send you the money as soon as I'm back in Berlin.

I have failed again in my resolve to return already, after all. Too much speaks for a longer stay; even the reports coming from Berlin make me hesitate.

Cordial greetings! Yours,

Einstein.

# Vol. 8, 373a. To Elsa Einstein

[Lucerne,] Wednesday. [22 August 1917]

Dear Else,

Thank God that Ilse is on the wing again. I hope the two will have finally left the nest.[1] So I am going to allow my passport to expire and shall stay here in September, welcoming your courageous resolve with joy.[2] Around the 1st of September we shall try it Mama's way.[3] The peculiar letter by the doctor from Stuttgart amused me very much. The facts he provides are naturally genuine doctor's gibberish.[4] Recently I caught a cold in Weggis,[5] with accompanying stomach complaints;[6] all is right again, though. My writing laziness is growing again; but I do think that, on the whole, I have earned a good mark. Your malicious interpretation of the motives behind my letter is a devilish invention.[6]

Kisses from your

Albert

How disconsolate the little chick Margot will be when Mama hen flies off.[7] How she will sing her song of sorrow with Samson and Delila![8] Send her my best regards, also to Uncle.[9] On September 11, I'll be in Zurich at the mathematicians' conference where Hilbert is giving a talk.[10]

# Vol. 8, 374a. To Elsa Einstein

[Lucerne,] Thursday. [23 August 1917]

Dear Else,

You've received my second telegram, I hope. All of the travel plans were upset again by a letter from Brandhuber that he is expecting me on September 1.[1]

Evidently I am going to have to travel there now. In Benzingen I expect to hear news from you about when and where we should meet.[2] I intend to stay there (at Brandhuber's) for about 10 days (at least). This is superbly convenient for you as well.[3] This is a curious to and fro, actually quite to my taste. From Sunday onwards I shall still be in Zurich under Anna's protective wing.[4] Tomorrow Mama is traveling homewards to Heilbronn.[5]

I hope the honorable office of housewife agrees with you well and that Uncle is very proper; otherwise he must be banned to the . . .[6]

Greetings & a kiss from your

Albert.

[Greetings] also for Uncle, [a kiss] also for Margot.

# Vol. 8, 376a. To Heinrich Zangger

[Lucerne,] Sunday. [26 August 1917]

Dear friend Zangger,

There is no other way, I have to leave, telegram notwithstanding. My priest in Benzingen writes me that he is staying home for me and is expecting me on September 1, the condition I had set for my promised visit.[1] I recovered satisfactorily after following the prescriptions you gave me *precisely*, although the local sensitivity still hasn't disappeared completely.[2] Today I am going to Zurich, where I'm going to stay under Mrs. Besso's care until Sunday.[3] I pulled myself together and gave my wife notice that I am coming to see her.[4] Then I'll discuss everything with her. One really has to regard the option of accommodating her with Tete in the Black Forest in Württemberg or in the Bavarian mountains for the remainder of the war, which would be feasible together with the sister.[5] That way the boy could stay in good mountain air without our becoming financially ruined.[6]

I hope you have recovered well from the painful operation and overwork so that you can resume your burden with refreshed energy.[7] A gathering of forces is less necessary in my case, with my tranquil lifestyle and my light load of obligations.

Cordial greetings in parting! Yours,

Einstein

Romain Rolland wrote me twice;[8] he really is an otherworldly mind, too soft to look rough reality honestly in the eye.

Please also thank your wife heartily again on my behalf for everything she has done for me and my family.

# Vol. 8, 376b. To Elsa Einstein

Zurich, Tuesday. [28 August 1917]

Dear Else,

You fidgety filly! Another telegram! Everything is going well. Thursday at noon I am traveling to Schaffhausen.[1] On Friday it's onwards past the posts to Benzingen.[2] Over there I'll expect your message of when and where we plan to meet. No scolding, just enjoy setting off for Thuringia, for example.[3] Maja is here today; tomorrow I am also going to be with Albert.[4]

Regards and a kiss from your

Albert,

also to Margot. Greetings to Uncle.[5] [. . .][6]

# Vol. 8, 376c. To Elsa Einstein

[Gottmadingen, Württemberg,] 31 August [1917]

Dear Else,

Today at half past 10 I set off from Schaffhausen and am still here in Gottmadingen at ⟨7⟩4 o'clock.[1] I hope it goes more quickly now. I'm spending the night today in Sigmaringen and am going to the priest tomorrow.[2] At the border the Swiss are very punctilious; but I had very little with me, just enough for the trip. I just saved 1 piece of fine soap and a tube of toothpaste for you. Write soon where we are to meet. I am thinking of Thuringia. Find out about a charming, quiet little place.[3]

Greetings and kisses also to Margot from your

Albert.

# Vol. 8, 377a. To Elsa Einstein

[Benzingen, Württemberg,] Monday. [3 September 1917]

Dear Else,

Surely you have calmed down again and are in good spirits.[1] So you're coming *with Margot* to meet me on the way to Jena, for example, or wherever else you think around the 10th, and we shall then pick out a suitable place for ourselves.[2] I am delighted to have become a Haberland-man now after all.[3] I fear a lawsuit, though. As soon as you have somehow settled with him, it's dangerous.[4] Aren't you afraid? In any case, don't worry about it; I know that you meant well. It is very nice here. There is full agreement between my and the priest's views.[5] Now I am expecting your message, not about *whether*, that is, but only *where* and *when*. I very much am looking forward to our little excursion.[6]

Kisses also to Margot from *your*

Albert.

I am in very good health.

# Vol. 8, 378a. To Elsa Einstein

[Benzingen, Württemberg,] Thursday. [6 September 1917]

My Dear,

Your letter delighted me, as I see that everything is back in order now.[1] Postcards must have gotten lost. I never waited longer than 2–3 days. I would very much like you both to come and meet me. I still have enough money to take care of the two of you as well.[2] If you do decide to come and you have not already reported it to me, I request a telegram. This postcard is the last before my ⟨return home⟩ departure. Even though it is so beautiful here[3] (weather includ[ed]), I am very impatient, to the point that I am counting the days on my fingers. Now it will be very nice in Berlin; we can make ourselves a bit independent and have meals at my place more often.[4]

I fear I forgot to greet Aunt in my postcard to Ilse. In any case, apologize for me when the two come back.[5] Now Ilse has also had bad experiences with me.[6] What's done is done.[7] Who believes the remorse of a poor old sinner?? He must simply be taken as he is. He has many a big vice but is also considerate regarding those of others.[8]

See you soon (preferably in Thuringia).[9] Greetings & kisses to all of you from your

Albert.

# Vol. 8, 380a. To Heinrich Zangger

Berlin, 15 September [1917][1]

Dear friend Zangger,

Now I am sitting here again, to be more precise, in another apartment (5 Haberland St.). In the interim I visited my priest in Benzingen, who, despite his black robe, is my political friend through thick and thin.[2] Here I can take better care of myself now than before insofar as I have an apartment next to my ⟨uncle's⟩ cousin's and don't have to go out anymore for meals. I haven't had another attack and am quite considerably fortified.[3] But a certain local sensitivity to touch on my abdomen still seems to be there as before.

Arriving home, I found colossal tax obligations waiting and the bank informs me that my store of cash is at an end. If it continues like this for a while, my finances will become unsustainable. So I resolve that Tete should be back in Zurich on October 1st and am giving my family 2,000 marks per quarter. Sending the money for Tete's sanatorium without documentation would not be possible.[4] I therefore request you pay the months ⟨July and⟩ August and September for me as well, as you have done earlier for June. (July is paid for.) Thereupon please send me a bill for the months of June, August, and September, which I can present to the Reichsbank as proof of my debt. Then I will sell a part of my securities and immediately send you the money. I ask you please to do all this exactly as I have said, so that I can get rid of this worry.

I know that the 2,000 marks quarterly is a meager support for my family, owing to the unfavorable exchange rate. But more cannot be given without definite bankruptcy developing. If it becomes necessary, I would have to send my wife and Tete to Bavaria or Baden, to a small town where one can get twice as much for one's money and where the food situation is hardly worse than in Switzerland.[5] This option is open to me despite the recent travel restrictions. But [Hans] Albert may under all circumstances stay in Zurich, as this is necessary for his schooling.[6]

I already discussed all these things with Besso and his wife,[7] although, at the time, I considered it possible to keep Tete up there for another half year. Anna Besso took excellent care of me in Zurich and appears, in general, to have much talent and an unlimited preference and goodwill for this type of work. I noticed that she likes it if I tell you so. Perhaps she is considering taking up nursing as a profession. Speak to her about it, but don't be put off by her silly talk. Her deeds are better than her words.

In Benzingen I read a very fine book by a Viennese (*Neue Staatslehre* [New theory of political science]) by Menger, from which I learned many things.[8] Take the time to dip into it when you have some leisure (?) again. Over here, a few of my best Academy brethren (Brauer, a zoologist, and Frobenius, a mathematician) died.[9] What is even more distressing is that the exceptional and still young *Smoluchowski* died of dysentery, one of the best persons and minds I have ever known; another war casualty, since he certainly would not have contracted the disease in normal times.[10]

Please do follow my wishes regarding the settling of my family's affairs, because it really is necessary. In the first place, my wife must return home again, of course, especially considering that I am convinced that her stay in the sanatorium is entirely purposeless.[11]

I hope you have recovered completely and are energetically back to work.[12] *But take care of yourself.* We've simply reached that age when one substitutes dignity for working energy. . . .

Cordial regards, yours,

Einstein.

## Vol. 8, 385a. To Michele Besso

[Berlin,] 6 October 1917

Dear Michele,

On September 11 I already wrote to Miza and Zangger that Tete *must be brought back immediately to Zurich*, because I cannot take the great expense upon myself anymore.[1] *See that this happens immediately.* I cannot see any other way because of the imposition of heavy new tax burdens, and I am firmly convinced that this precaution is exaggerated as well.[2] I'm not taking this step without having spoken with specialists.–

I am feeling very well. Where is your institute supposed to be located?[3] Best of luck to Vero on his studies.[4] I'll try to send you the statutes if possible.[5] But articles, etc., don't make it abroad now anymore without a special permit,[6] so the mailing might cause more trouble than running it down at libraries there. If you can't find the thing in Zurich, I'll get a shipment permit.

Warm greetings, yours,

Albert.

See that Tete is taken home right away. I'm going to write Zangger as well.

## Vol. 8, 390a. To Michele Besso

[Berlin,] 15 October 1917

Dear Michele,

Many thanks for your detailed news. Healthwise I've been very well; I haven't had another attack.[1] I feel very sorry for my wife. She seems to have a genuine canker on the spinal cord, without hope of recovery.[2] It was right, in the end, that you left Tete up there.[3] I'll try to send Miza money for him. If it doesn't work, I'll have to ask you all for bills that I can present to the Reichsbank as proof. I simply have to rely on your dispositions. But remember that the most important thing besides the exigencies of the moment is that my boys are somehow provided for in case of my unexpected death.[4] If it continues on like this, there'll soon be nothing left. Additionally, I'm convinced that anything spent on Miza is in vain. And as far as Tete is concerned, I don't believe that *such a long* sojourn is useful because what essentially seems to be effective is the contrast of the climate.[5] Preferably shorter and more often.[6] But since I now see that my instructions are not being followed at all, I don't feel responsible anymore. On the contrary, I have *great trust* in Anna[7] and am glad that she is taking care of Miza and has some influence on her. She'll arrange everything, satisfactorily. When are you all going to Rome?[8] I've been working properly again and with a clear mind. Bold plans but what will be the outcome? You know how [sticky] a business this is. Cordial regards to Zangger[9] from your

Albert

Your institute pleases me immensely! When is the launch?

What a pity that we have to be even farther apart.

# Vol. 8, 391a. To Heinrich Zangger

[Berlin,] 15 October [1917]

Dear Friend,

Now something (abscess?) seems to have been localized on my wife's spinal cord, after all. If so, the poor thing will always need to remain lying down. I'll send money to my wife (1,000 marks, for now) for the sanatorium. If it doesn't work this way, you will have to send me official proof that I truly owe this money.[1] I'll have to leave the decision to all of you about how long Tete must stay up there.[2] I cannot judge. But think in what dire need my children are of all my modest savings in the case of my unexpected death![3] If this continues on . . .

I am extremely pleased for Besso. He is surely very happy to find such a fine work milieu.[4] But how much traveling that will mean for me! He's so hard to coax onto a train, quite the opposite of both of us. I am feeling very well. No more attacks, though I've been following a strict diet and not smoking, reclining a lot.[5] I haven't been to the doctor at all. I'm working hard again and with pleasure.

It is very fine of Dr. Pedolin[6] that he is so accommodating toward me.

Cordial greetings, yours,

A. Einstein.

# Vol. 8, 424a. To Werner Bloch

[Berlin, 3 January 1918]

Dear Mr. Bloch,

I cordially return your kind greetings. On doctor's orders I have to stay in bed for the rest of the semester because of my stomach ulcer.[1] I will make good on this course in the coming semester, however.[2]

With kind regards, yours,

A. Einstein.

# Vol. 8, 435a. From Hans Albert Einstein

[Zurich, after 14 January 1918][1]

Dear Papa,

Yesterday I received your postcard.[2] Your books were here on time,[3] but the postcard took more than 3 weeks, and that's why I am writing you now "registered." Tete hasn't come home yet and we sent him the storybook.[4] If you feel unwell again, you must definitely try taking "Dr. Hommel's Haematogen." It served us very well when Tete was sick last year. He had a local inflammation too. "Haematogen" is very good for improving overall well-being. You see, since he started taking it, he has never been so completely weak and miserable, despite having a high fever.[5] You start with a very small spoonful and, if it agrees with you, increase to a small soup spoon.

We have constantly had a lot of snow this winter, so I could almost always go sledding or skiing; only coal was not particularly abundant, but we haven't ever had to shiver yet.[6] Groceries are so expensive, and as if by plan, now, of all times, I always have such a huge appetite.[7] I still have holidays from school at the moment.[8] Next week we are starting music school[9] again. Now I've played 2 Beethoven sonatas, the "Moonlight Sonata" among them.[10] I'm managing quite well, I just can't assume the correct tempo yet. Tete is continuing to stay in Arosa for a while, until it has thawed up there. There is too much fog here, and that's not good for him. I am sending you here the last 2 bills for Tete and ask you to pay them and to send the money to us because Mama still needs some more for clothes and lessons.[11] He has to prepare himself for the 2nd gr[ade], you see. Uninstructed, I would also like to ask you whether it wouldn't be better for you to send the money for the next quarter *now*, because the German exchange rate is so high at the moment.[12] I hope that you are better soon. Do you have bleeding in your stomach?[13] Write me about it. I hope you're eating the right food as well.[14]

Friendly regards from your

Adu!

# Vol. 8, 442a. From Hans Albert Einstein

[Zurich, after 25 January 1918][1]

Dear Papa,

I received your letter[2] and hope you are feeling better already so you can get up.[3] Write me sometime in detail about when you felt worse again and what may be the probable cause. I believe you did not do well to stay here for such a short time in the summer. You'll surely remember how much better you were in Arosa, and that's why I would like to ask you to go to such a mountain spa for a longer time, if possible, this summer, because this seems to be the only means of improving your health at least a little.[4] What use is a little bit more or less money to you if you must always be in bed, or at least can hardly move?[5] It's better for us, too, when you're in good shape again, because this way we obviously don't gain much at all from you. Besides, these health resorts still receive just about the same good food as before because deliveries are always made there first.

I find that you are being very unfair about Tete. We asked various doctors, and all have found this to be the only thing that can be done. You write that he is being "pampered."[6] But that's absolutely not true. He's outdoors all day. (One moment he lies down, the next, they go sledding and ice skating), and everyone who sees him says he looks a lot healthier now. Anyway, he is coming down in the spring when the snow has melted up there. Here it's constantly foggy, and the sun hasn't shined for over 2 weeks.

You can't judge at all what ought to be done about Tete, for you haven't a clue about all the things he's been through. Mr. Zangger can judge best because he has been examining Tete constantly, you know.[7] We can count ourselves lucky that he takes such pains with us. What would have happened to us if he had not been there when Mama became ill?[8] Also consider that! Something must be done for his health this once. We know absolutely nothing about each other; you have no idea what we need and require; I know nothing about you. Now I know Mr. Zangger or anyone else here much better than I know you, and that's where the misfortune lies.[9]

Now I am playing the 5th Beethoven sonata (C minor)[10] and I enjoy it very much.

Presently I am trying to make the train electric and Mama is helping me with it as much as she can. I am doing this because of the coal shortage[11] and would like to finish it for when Tete comes. If you were here, you would surely amuse yourself with it as well. Mr. Zürcher[12] finds the greatest pleasure in such things, and I think when mine is finished, theirs will also become electric.

Write me soon about how you are. Many greetings from your

*Adu.*

# Vol. 8, 461a. From Mileva Einstein-Marić

Zurich, 9 February 1918

Dear Albert,

I received your letter.[1] Exactly two years ago, you pushed me over the brink into this misery with such letters,[2] which I still can't rid myself of; and now you evidently think it just too nice that I have, as you say, neither fever nor attacks and write me the same again. I can reply to you neither yes nor no today, since I absolutely must get advice from a lawyer about the matter; I could not justify it to my children to act otherwise. I wanted to do the same back then as well, but since I never felt well I constantly postponed the matter until I became ill.[3] Also, Besso told me that you were not thinking about that anymore, and had given up the idea.[4] D[ear] Albert, why do you torment me so endlessly? I would never have thought it possible that anyone, to whom a woman who had devoted her love and her youth, and to whom she had given the gift of children, could do such painful things as you have done to me. You can't possibly imagine what I suffered through these last two years, not even to mention what was before.[5] And now I'm not even recovered and the business is supposed to start all over again. I really don't deserve this from you. You ought not to have subjected me to the relentless, insensitive, and callous protests about Tete either; you know that he was sick, and if we weren't going to allow him to deteriorate, something had to be done for him, especially since I myself am ill and could not take care of him and provide for him. And how do you manage to imagine Tete's treatment as pampering?[6] If only for Zangger's sake, you ought not to have said this. He has expended so much effort on us, has never spared his time, or even cost, when he could be of help to us; he has done many a thing that would actually have been your responsibility, because he is fond of you, and you ought to acknowledge him and not hurt him.[7] The whole misfortune is that more money was spent this year than had been projected. And if you had let us succumb to sickness, would this couple of thousand francs have made you happy? Do consider also that everything is so expensive nowadays, many things now are 3 times their price;[8] how am I supposed to afford everything? I don't understand you. Do you really want to place yourself on the same level as Elsa's husband?[9] From her you will know how hard it is to survive, and do you really want to have the same bestowed on your wife and children as well? Consider that the matter is even more serious for me insofar as I have no inheritance, or nothing to speak of,[10] and that after this illness I am not going to be able to consider earning anything soon. And what if your health doesn't improve? This has not been the case since the summer, Albert,[11] and you don't know how much this worries me! And what if your illness should take an unfortunate course? This must also be considered. In this case do you, after all, also want to take away from me the only secure thing that we really have left, namely the pension, by divorcing, and to leave me at the utter mercy of your 2nd wife[12] about whether she will pay out the pension sum to me, so that I have to fight with these people as well for the rest of my life?[13] Please consider all of this; it's impossible not to admit that anything is more appropriate than closing your suggestions with a threat.[14]

I wish also that your health may improve soon and beg you to do everything toward this as well; that is surely the most important thing for the time being.

Miza.

# Vol. 8, 471a. To Heinrich Zangger

[Berlin, 27 February 1918]

Dear friend Zangger,

Endless is the chain of illnesses afflicting our families.[1] How could your little girl possibly have contracted such a serious sickness in the Engadin?[2] But I confidently hope that things will improve, now that you are back in Zurich.

The affair with my wife's sister is a new misfortune,[3] which however makes some things more explicable to me. A few weeks ago I received a letter from her that stumped me. My wife had withheld from me the existence of this sister throughout the 20 years of our relations, so I was beside myself with surprise at being addressed as "brother-in-law" by a woman I had never heard of.[4] The frightful mistrust with which my wife has been plaguing me and all others is also of a pathological nature. I was of the impression that the other sister of my wife, known to me as *Zora*, was in Zurich.–

Now it is absolutely imperative that we emerge from the provisional measures for my family. I have now decided upon the following solution.

1) Albert goes immediately to my sister's in Lucerne, permanently.[5]

2) My wife is put permanently in good care either in Zurich or in Lucerne. The latter would have the advantage that she could then see Albert daily.

3) Tete stays up there at Pedolin's until June.[6] Then I'll collect him myself and take him with me to Berlin. I won't permit that the boy should grow up so unattached to his own family. I'll then take him along whenever I go to Switzerland.

4) The household is dissolved in the summer. I shall come with Elsa to Switzerland and we'll take care of everything ourselves.[7]

5) What will become of my poor sister-in-law, i.e., when can she be given back to her family, only God knows.

---

The main thing is right now that Albert should go to Lucerne. I'll ask my brother-in-law to pick him up immediately.[8] I beg you to support me in this matter. My sister is a highly intelligent and excellent person, in whom I have unlimited trust. It will be wonderful for Albert there.

I want to keep Tete here until he has become somewhat stronger and gets to the age when he must step into broader surroundings, i.e., until he is 12 or 13 years old. Then Albert will be a student and independent, and Tete can come to Lucerne; in any case, he will come back to Switzerland later. For I won't send him to school here. He will be in excellent care here physically. If his ailment returns, he'll be sent from time to time to the lake or into the mountains with Elsa or with her daughters.[9] I would never have taken him away from my wife. But since she cannot keep him with herself now, anyway, it is certainly healthy and natural that I'd want to have him with me. I am doing this not because I am yielding to an emotional weakness of mine but because I consider it the right thing to do.

These constant interim solutions and seesawing are becoming more and more unbearable. If I could travel now, I would come right away. But that is out of the question with my still unstable health, particularly in the current traveling conditions.

Would that your little girl were back to normal by the time you receive this letter. Warm greetings from your

Einstein.

## Vol. 8, 471b. To Heinrich Zangger

[Berlin, after 27 February 1918]

Dear friend Zangger,

An express letter has just gone out to you,[1] one major point of which I have to immediately retract again, however, after speaking with Prof. Rosenheim.[2] I now understand that we have to accommodate Tete permanently at high altitude.[3] Now I submit to all your prescriptions without the least contradiction as concerns Tete, and will pay for everything without objection. *Please forgive me for all my unjustified meddling.*[4] You have been advocating the one and only correct view.

| Father, I have sinned!! |
|---|

Again I urgently beg you please to give Albert to my sister.[5] He can, of course, visit his mother as often as he wants. We shall have to accommodate her permanently in a sanatorium.[6] The main thing is that we arrive at a permanent state of affairs so that the ⟨endless⟩ terrible irritations finally come to an end, as far as they can be avoided.

Warm greetings from your penitent

A. Einstein.

# Vol. 8, 475a. From Mileva Einstein-Marić

Zurich, 5 March 1918

Dear Albert,

In answer to your last letter of 31 January 1918[1] I inform you of the following:

Your proposals regarding my and the children's financial security in the case of a declared divorce I accept, with reservations with respect to the 3rd point concerning the provision of the widow's and orphan's pension for our benefit. This point is too vague and unclear and must still be clarified before the filing of the claim.[2] If my right to this pension can be maintained through a directive by you, then I ask you to provide it. This will probably not be possible, though; do you then intend to secure the matter by depositing a sum (which, in case you are granted the Nobel Prize, you can then remunerate out of it), a bond, or make true otherwise?[3] Please inform me about this. My dispensing with this pension just like that would be an injustice on my part, and I couldn't justify it toward my children. Your and my conditions of health are such that we must think of everything;[4] it could happen that upon losing this pension in the case of your death I am unable to work and am left without any means whatsoever with the children, which you surely don't want, either. That is why I request that you settle this point. As concerns the filing of a claim, I shall comply with your wish if you agree that I do it here in Zurich; for you it is of no import where it happens, of course.[5] Dr. Zürcher was so kind as to prepare an opinion on our case,[6] which I am enclosing with my letter. I ask you to read through it and to inform me about whether you consent to the content of point (1) in the same, upon which the claim should be based,[7] and won't cause any difficulties about it.

With kd. regards,

Mileva.

# Vol. 8, 475b. From Maja Winteler-Einstein

Lucerne, 6 March 1918

My dear Albert,

Last Sunday your letter of alarm arrived.[1] It is really no trifling matter for you to still have such agitations, in your ill state. You can be sure that we are doing whatever is in our power to put your mind at ease.[2]

Yesterday, for instance, on Tuesday (Monday I had ironing), I drove with Pauli[3] to Zurich to find out the essentials, and if possible to take little Albert with us right away.[4] That was a tragicomical odyssey and I want to tell you about it in extenso so that you have something to smile about. Immediately upon our arrival we telephoned Zangger and received the information that he would not be home until the evening; he was away on a trip. With the present horrendous connections, the last train from Zurich to Lucerne leaves at 6 o'clock in the evening; so looking at each other with dumb-founded, very long faces we proceeded to draft our plan of attack under the changed circumstances. Pauli was to return to Lucerne and I wanted to spend the night at Anna's,[5] in order to orient myself and, if possible, to speak with Zangger the next day, also to go to Miza, etc., etc.[6] I had to pay another visit and Pauli went first to Anna's to procure night quarters for me. He did not want to talk to her at all about our mission, since, as you know, one can let her talk, of course, but not talk with her. But she guessed the purpose of our coming, and when Pauli confirmed her suspicions, such a torrent of accusations, scoldings, and threats poured forth from the mouth of this injured Justitia, who in her blindness never strikes wide of the mark (or always must?), that Pauli took his hat to escape and fled. I was just coming through the front door as he was halfway down the stairs and heard Anna's excited voice still giving him all sorts of gracious exhortations from above. Paul wanted to drag me out onto the street, but Anna now pounced on the new prey, who naturally had not the faintest idea about the preceding events and whose very perplexed gaze was alternating between her screaming sister-in-law and her enraged husband. Pauli finally did pull me away, and then we were left standing out on the street. Pauli then recounted in unvarnished prose what had happened, whereupon I resolved to return to the lion's den to find out something about the facts after all. Also, I did not want a new family feud. As soon as I came up the stairs, I was deluged again with the absurdest of accusations while I sat there mutely. After a quarter of an hour of accusations against you, Elsa,[7] and me all mixed together and a valiant defense of Zangger (whom I had not attacked at all) whom she protected with her own person, she felt like the embodiment of justice and we parted peacefully.[8] Then I wrote to Zangger about a rendezvous and we steamed off together again back to our peace-loving household penates;[9] you can also say: Let Providence protect me from my friends, for I can protect myself from my enemies.[10] It certainly is remarkable how we all are at once prey and hunter. Or is it only that everyone thinks he is being pursued? I cannot quite sort it out. So we are awaiting the things to come. Would you consent to having little Albert go to Azzolini's[11] if it were completely impossible to abduct him to here? I spoke with her yesterday about it. If little Albert could come to us, I would consider it best that you gave us little Tete as well.[12] Since both the boys are so attached to each other, it would be a great pity to separate them without good reason.

What you say about your state of health is not encouraging. It would plainly have been better if you had not returned to Berlin so quickly.[13] I am very much looking forward to your both coming to Switzerland. Then you will certainly soon regain your former weight.[14] Although our years of plenty are over as well and we also are living off foods which for the most part can only be had for a little slip of colored paper, and quite a small packet on top of it.[15] But for the time being, it's not necessary to tighten our belts yet. For me neither, unfortunately.– More soon! In the meantime, an affectionate kiss,

Maja.

Heart[felt] gr[eetings] to Elsa, the children, and the others.[16]

# Vol. 8, 482a. From Mileva Einstein-Marić

[Zurich, before 17 March 1918][1]

Dear Albert,

I would like to ask you sincerely to send the money you have intended for us directly to me;[2] likewise, when you wish any information regarding me or the children, turn to me directly.[3] Otherwise misunderstandings are unavoidable, which always involve agitation for me. In view of my so fragile state of health, I hope that you will not reject this request of mine.[4]

Miza.

# Vol. 8, 482b. From Mileva Einstein-Marić

[Zurich, before 17 March 1918][1]

Dear Albert,

Dr. Zürcher is going to draw up a contract as soon as he has time,[2] and I'll send it to you then.

As far as Albert is concerned, there is absolutely no reason why he should have to be cared for somewhere else.[3] My state of health is now such that I can lie down quite well at home; although I can't get up, I can very well occupy myself quite a considerable amount with the children, and this makes me very happy and contributes much to my well-being.[4] My doctor, in whom I have great confidence and who has much experience specifically in conditions such as mine, gives his full approval. I sunbathe at home and do a few other things as well, and more cannot be done in hospital, either; forcing such things accomplishes nothing, as I have already seen from personal experience; every attempt to force anything had severe consequences. At home like this with the children, I have the joy of living together with them, and it is better for the children as well; they have a home and grow up together under my supervision. I beg you, therefore, to regard this matter as settled; it is surely best this way.[5]

I would like to request once again that you directly forward the money and communications intended for me.[6] I think you will understand without my dwelling on it and not refuse me this request; it would spare me many an upset.

As concerns Albert, it would surely be good if you would answer his letter in at least a few words; if he must constantly appear as a culprit in your eyes, he will soon lose all openness toward you; it was certainly not his intention to hurt you.[7] What are your thoughts on this?

With kd. regards,

Mileva.

# Vol. 8, 494a. From Vero Besso

Zurich, 28 March 1918

Dear Mr. Einstein,

The postcard you sent to my mother was really not nice;[1] for she had taken such great pains, as though she were doing it for her own brother. And she certainly did not have any intention to dictate or dominate, as far as this is at all possible for a concerned housewife.

She had started to take charge of the matter because she had been asked to do so, because Papa went away on a trip—taking over, so to speak, as an inheritance—and finally because she does not drop her hands in her lap with pity when she thinks she can help, or when what is due seems to her to have been denied.[2] And having once started, she carried it out to the best of her abilities.

Just remember the insular conversations, and although the content was a quite different one, her character came to light just as clearly then as now through her actions; with the difference that now matters have been touched upon that lie within her experience, from which a certain instinct has formed, leading (relatively and provisionally) toward what is right, whereas in that other case her weak logic fell short.

That is why it seems to me that more attention ought to be paid to her actions than to her words, especially when they were said heatedly to someone with whom she *had been* on bad terms for a long time previously.[3] It is easy to grasp that she was excited, also because beforehand she had gotten into an argument with Paul about a somewhat unfriendly comment about Maja.[4] Besides, her words would not have offended you in any way, if you had *heard* them yourself; you would just have laughed and would have toned down their sense a little. In an hour of reasonable conversation you would certainly have come to agree with each other on the entire affair, whereas in writing misunderstandings always arise.

Your sister is temperate and reasonable. Every word of hers also has an intended sense, and so it is easy for her to voice an opinion of someone without much fanfare, without saying it directly; that is why it is also much more impressive and credible. You know well that your sister is not well disposed toward my mother (just as vice versa);[5] she may say what [she] wishes, yet precisely because of this, her view will surely be one-sided, even though it does not seem so.

Now I beg your pardon for the liberty I took, for three reasons:

1. because my mother was unfairly hurt (not that she had broken out into lamentations, as is often heard said),

2. because I like you personally,

3. because Papa is surely essentially in agreement with my mother (and not the contrary, which the "humorous report"[6] insinuates), and both are anything but unshakable.

Very cordial regards, yours,

Vero

Read Dostoevsky one day, when you have time.

# Vol. 8, 494b. To Vero Besso

[after 28 March 1918][1]

Dear Vero,

Your letter, with the fine and conciliatory manner in which you deal with touchy subjects, pleased me very much.[2] His father's worthy son,[3] who already views things with circumspection. My postcard was not meant to be malicious. It was a kind of defense against further abuse; it was as if someone opens his umbrella when he's surprised by a hail storm. My sister's reports don't have that effect. She illustrated and described the reception with humor.[4] What really put me off was the insufferably impertinent and presumptuous tone that Anna assumed toward me.[5] The help was well meant and gladly granted but so painful and disagreeable to me that I prefer the worst mix-up. It was *my* mistake to have asked her to do something inappropriate for her. She doesn't have the conciliatory manner necessary for smoothing the waves of passion, nor the ability to understand her dear neighbors. She responds to virtually everything simply with moral indignation. But it was *my* mistake to ask her to do such a thing, and I thank her again for fulfilling my request.

I am trying always to settle everything directly with my wife now.[6] She is currently acting remarkably well and reasonably with me and she will certainly not suffer by it this way.– But enough of the misery of private life. From Dostoevsky I read the memoirs *The House of Death* with fondness and admiration and intend to read everything else by him.[7] He reconciles, without hoodwinking life's tribulations. I hope you don't let the Poly spoil your pleasure in science.[8] Do what you find interesting and don't waste away out of a bleak sense of duty. Don't let yourself be impressed, either, by the masses of things that are being preached at you. Almost all of it is bound to be forgotten. But delight in science is among the most precious things there are; it also lasts one's whole life long. It's a crying shame that Weyl is leaving Zurich.[9] He is a great master.

I am looking forward to seeing you very much at the beginning of July. I'll be coming then to strengthen my insides a little in the high mountains. Calm Anna down, and cordial greetings also to your two elders from your

Albert

Zangger's tragedy affected me deeply.[10]

## Vol. 8, 496a. To Mileva Einstein-Marić

[Berlin, 3 April 1918]

Dear Miza,

Yesterday a local jurist[1] friend of mine assured me that divorces in Zurich are very long-winded whereas here one can do it in 3 weeks with little expense and effort in a single session! There is also the consideration that the divorce on the grounds of "adultery with Elsa" wouldn't help me because of the interdiction to remarriage attached to the judgment.[2] But here we could execute the business easily with a stand-in who takes the adultery upon himself for convenience. The main thing for you is that the divorce occurs on the basis of the contract agreed with Zürcher; where the formality takes place is of no consequence, however. I am very willing to submit to you the claim in advance in order to assure your approval. Please talk this wish of mine over with Dr. Zürcher; he will surely also find it justified.[3]

This summer I am going into the high mountains for 2–3 months.[4] At the beginning of July I'll come and visit you in order to take Albert and Tete[5] along with me for a while.

Best regards, yours,

Albert.

Please send the acknowledgment of receipt for the money promptly.[6]

---

It is extremely troublesome and protracted to achieve a lifting of the interdiction to remarry. Such a petition goes all the way up to the minister and then takes a year!

Still no answer from the ministry.[7] However, the money can be immediately deposited.[8]

Einstein's emendation: "By special delivery."

# Vol. 8, 496b. From Mileva Einstein-Marić

[Zurich, 4 April 1918]

D[ear] A[lbert],

I received the check yesterday for 2,000 francs[1] and thank you for it.—My complaint at the bank was unfounded; I had mixed something up, the remittances were all in order.[2] A contract is being prepared and will be sent to you one of these days.[3]

With kd. regards,

Mileva.

# Vol. 8, 513a. From Hans Albert Einstein

[Zurich, before 22 April 1918][1]

*Dear Papa,*

I have holidays now. After these holidays, everything will be different: I'm going to be in the 3rd form and therefore in upper school.[2] But I've gotten tall too, I'm a good 1 meter 65 cm by now, and have grown out of all of my things in the process. Tete will be going to school as well after the holidays, into the 2nd grade.[3] He doesn't quite know what face to pull when people talk about it.

During this vacation I drafted and made a model for an electrical locomotive.[4] It came out quite well. I made it out of thick cardboard, but what with the wheels, the motor, etc., it's all a very complicated affair, because everything is so hard to come by, and what is available is expensive and bad, and so I've had to do away with finishing the carriage for the time being and wanted to change over now to making a flying machine. But then I was made aware of the fact that everything has its hitches, namely, the carpenter,[5] who was supposed to make me the little slats, never had time or was never in the workshop; in short, I just have to wait until Mr. Master Carpenter deigns to make me the little planks.

I had vacation now at music school.[6] Even so, I played almost every day, because it's already lots of fun for me. I'm already playing many things by Schubert, Mozart, Mendelssohn, and Beethoven. The newest ones now are Chopin waltzes. Although they are a bit hard, if you keep at it a bit, it goes all right.

At home I always have lots to do as well until I have all the necessities in life together. Everything's so inconvenient now. Write me how you are feeling and how your stomach is doing! Can you get up again, or do you have to lie down like Mama? She's a little better now, on the whole, but she always has to be lying down.[7]

Warm greetings from

Adu.

# Vol. 8, 514a. From Mileva Einstein-Marić

Zurich, 22 April 1918

Dear Albert,

I have received the contract[1] with your alterations and inform you of the following:

1. Regarding the sum to be deposited, it's the same whether it happens here or at a bank in Germany; but it should take place in the manner indicated by Dr. Zürcher in the contract.[2] The suggestion of making the deposit in Switzerland was a consequence of the assumption that the interest accumulated from this sum would already form part of your contribution, and thus be a part of our livelihood; having this part in Switzerland for safety's sake, just in case, would be to our advantage; however, I do not insist on it.

2. I approve of your changes regarding the annual payment of 8,000 francs and the pot[ential] Nobel Prize.[3]

3. As concerns the formulation of the point regarding the children,[4] I would like to add the following explanation: This formulation should not demonstrate the intention of making it in any respect more difficult for you to meet with the children but rather it has an entirely different reason. We have considered this point very thoroughly and have taken the view that it is necessary that I *not be obligated* to send or deliver the children at your request to you in Germany, e.g., even if for no other reason than for the children's care. It's anyway impossible to formulate the

contract to cover all eventualities, nor do we have any way of knowing what shape all of our lives will take. It may well be possible that our circumstances would be such that I could not send the children to you that distance away without financial loss. If in such a case you wanted to insist on your right, I could fall on hard times. I just want to remind you with what resolve and persistence you demanded that Tete be taken away from Arosa last year,[5] and yet now you know how necessary this stay was for him. I absolutely do not believe that you did this out of a lack of kindness toward the child but rather because you didn't really know his condition and also didn't know what was necessary for him. Naturally, it is not even possible for you to know from there every detail about what is happening here, and this provision in the contract should be a precautionary measure, so that no difficulties arise from which the children could possibly suffer. For the time being, a visit by the children in Germany is out of the question anyway as long as the war is going on. Later we shall surely be able to agree on each individual case. It is merely the thought of the children having to visit you in Berlin itself that I would find disagreeable; but I hope that you'll not demand that of me at all, and will respect my feelings about this one thing, at least. I make this request of you and hope that you'll not reject it.– With these explanations, the last point of our agreement will not seem so hard, and I ask you please to leave it as it stands.

Dr. Zürcher tells me that he or H. Zangger wrote you about whether the matter should be carried out here or in B[erlin].[6] Mr. Zürcher considers it very unlikely that the affair would be settled there so quickly; he also thinks that the matter would work here without complications.[7]

With amic. regards,

Mileva.

# Vol. 8, 532a. From Mileva Einstein-Marić

[Zurich, before 8 May 1918][1]

Dear Albert,

Your last letter with the new proposals for the contract[2] arrived with some delay; also, Dr. Zürcher[3] was away for 3 days; that is why I can only answer you today. He points out to us that securities nowadays are a very unreliable thing.[4] Since I believe I can assume that you really do have the good intention of assuring us the annual sum of 3,000 marks (I believe that is how much it is) in the case of your death,[5] I beg you to consider properly what the situation is with securities. I cannot and do not wish to demand more, but this amount should be genuinely secure. Wouldn't it be more sensible to sell off these securities available to you and to deposit the cash so that one has some form of pure and clear account of it? Prof. Hurwitz's wife[6] already told me on various occasions how many losses they've already had from securities and how much of their money they must regard as altogether lost; please do take this into consideration and spare me these worries. Please reconsider this matter and write me how you conceive it.

I hope that you received Albert's letter in the meantime;[7] if not, please tell me.

With best regards,

Mileva.

# Vol. 8, 533a. To Heinrich Zangger

[Berlin, before 8 May 1918][1]

Dear friend Zangger,

Here some scattered remarks on your comments about probability.[2] Let's talk about it when I come to Zurich, which will be around the 1st of July. I am now very satisfied with my health. A large part of it is probably due to my lying out on the balcony very often and letting myself be roasted by the sun.[3] A Director Huguenin from Zurich has arranged that condensed milk be delivered to me through a local company.[4] This is probably owing to your supplication. In any case, I thank you very much. Lately I have been living a loafer's existence and exerting myself very little, reading with delight Rousseau's memoirs and Anatole France, a wonderful, mild, perceptive fellow.[5] On the other hand, I was not able to warm to Balzac. What ever did I write that you allude to as so terrible and having juridically insulted Dr. Zürcher?[6] The written word certainly is a deceptive means of transmission for thoughts and feelings, when exact science is not involved, when one isn't a linguistic expert. But we'll soon be seeing each other and can talk through everything. I'm also indescribably eager to see my bosom friend Besso; I just want to avoid his wife; she was too insolent with me in her letters.[7] I truly am not sensitive; but one doesn't like to sit on an anthill, even if one is endowed with a resilient behind. The meeting on probability in Lugano will resemble the building of the Tower of Babel a little (I'm also reading the Bible); but it may well be amusing. Just please don't oblige me to give a sermon there.[8] For I can talk only when I really have something to say, that a general audience does not know yet. But that is not so here. You will object that one can preach about any old thing so long as one preaches *well*; but I am lacking precisely this skill, so my halting speech can only be tolerated when the novelty of the ideas seems to excuse it. Weyl wrote an ingenious paper in which he strove to unify gravitation and electromagnetism along the lines of the general theory of relativity. I consider the thing physically incorrect, however.[9]

Tell Besso he should ask for the story. I'm curious if he finds the same fly in the ointment as I do. At the Academy there were a couple of droll scenes about the paper, about which I'll tell both of you later.[10]

A satisfactory relationship has formed between me and my wife, namely through the correspondence I am conducting with her about the divorce![11] A funny opportunity indeed for reconciliation. I am currently trying to obtain the permission to give her my savings and send them to Switzerland. Up to now I have not been able to carry it through at the Reichsbank.[12] Together with two large prizes I recently received, it is 40,000 marks.[13] You are certainly laughing at my telling you this with a certain pride; but for my children it does mean something not to be quite so dependent on friends in the event of my death.[14] I experienced that myself!

Cordial greetings from your old

Einstein.

I[15] wasn't able to attach any meaning to the first two and a half lines.

Any judgment about probability *W* of complex events is based on certain set premises about probability *w* of elementary ⟨primitive⟩ events, ⟨From these premises probabilities⟩ out of which the more complex ones are composed. The link between the *W*'s and *w*'s is made according to certain rules that probability calculus chooses in order for it to be advantageously applicable to as many series of observations as possible.

The same probability for the 6 throws of a die has logically nothing to do with the impossibility of preferring one such throw over the others.

The statement about equal probability of two events has nothing to do with the statement about the independence of events from each another. One can be valid without the other.

Probability considerations and causal considerations may indeed be mutually exclusive, but in nature the causal connection, even if it is unbroken, allows room for probability considerations. This comes from the fact that it is often not possible nor intended to completely deduce the causal relations.

We don't know whether the demand for strict ordering of observational facts according to cause and effect is satisfiable, and we shall never know for certain. In practice we have to dispense with it in almost all areas, because a certain and full drawing of the causal chain almost always exceeds our abilities.

---

Hitch: Rationally defined probability intrinsically does not permit application. Structure of the theory, setting out from the empirical definition of probability, runs up against major obstacles.

# Vol. 8, 539a. To Max Jakob

[Berlin, 17 May 1918]

[Not selected for translation.]

# Vol. 8, 545a. From Mileva Einstein-Marić

[Zurich, before 23 May 1918][1]

Dear Albert,

I don't want you to misunderstand me. If in my last letter I raised objections to securities,[2] it was certainly not in order to satisfy my greed for money, rather only out of concern for the children.[3] Their youth is already not exactly nice, without a father and with a sick mother; I'd like them at least not to come to know direct privation. I'd like to raise a small point now as well; wouldn't it perhaps be sensible to have that portion of the deposit which is made in currency, placed under my name at a German bank, in order to circumvent the loss of interest?[4] For if this loss goes to our account, it would have very bitter consequences for our budget; groceries, clothes, and everything is so expensive and is constantly getting dearer that we hardly have anything left over.[5] Please consider this.– I don't want to discuss a thing with Michele.[6] His wife behaved so badly toward me this spring that, for me, friendship there is out of the question. She tried to meddle in my affairs in a way that reveals potential human malice.[7] I showed letters of hers, in which she informed me of how she would like to "help" us, to Dr. Zürcher and to Zangger;[8] they almost fell over backwards. As far as I know, you've been informed about this, and I hope you'll understand that I want nothing more to do with her. I receive her when she visits and that is as far as it goes.– It would be good if you wrote me about how long the children are supposed to be with you.[9] They have holidays in mid-July; if you should come to Switzerland earlier, they would meet up with you later at the place agreed upon. They really should get used to taking their responsibilities seriously, you will surely agree with that. That you are considering Engstlenalp surprises me; that place is only accessible by foot, a walking distance of 4–5 hours according to Baedecker; with little Tete, you would have to calculate 8 hrs.[10] He mustn't become exhausted in any way, and also, if something should happen there, how do you think you can get help? At these remote spots, amenities aren't particularly good either, nowadays. Regarding Tete, I would at all events still have to ask the doctor whether he could come along. It's surely not the right thing for you either; since you ought not to walk and need special food,[11] a more civilized region where things are accessible would certainly be better. Why don't you consider Rigi?[12] There are smaller villages also in the Zuoz[13] region, where good, reasonably priced private lodging is available. If you like, I can easily make some inquiries about something from here. It is very generous of you to contemplate visiting me on this occasion; thank you for the thought.[14] It is out of the question, of course. I saw far too well in the last few years, last summer as well,[15] how reluctantly you do so, for me to be able to accept it.

I hope that you are feeling well and remain with kd. regards,

Miza.

# Vol. 8, 557a. From Mileva Einstein-Marić

[Zurich, after 4 June 1918][1]

Dear Albert,

I received the report from the local Swiss Bank Association about the arrival of the securities you sent.[2] The deduction of the interest income from the sum you are sending is a bit involved, since the coupons become due at different times; would you agree to the following mode of action, which in my opinion is the simplest: You send 2,000 francs quarterly, as you have been doing up to now.[3] The bank takes charge of the collection of the coupons and enters all deposits in a savings booklet; at the end of the year it provides me with a written account of all that has been deposited, which I send to you so that you can subtract the total interest accumulated in the course of the year from your January remittance; thus on 1 January you would send me 2,000 francs less the interest of the previous year.– Do you agree with this? I find that it would be the simplest this way; Mr. Zürcher[4] also finds this the best way. Please write me about this.

The children were very disappointed that you aren't coming; it's a great pity, as well; this one time they could still have had you for themselves, and now nothing is to come of it.[5] But one request you'll surely not refuse me. Don't promise them anything before you are certain that you can fulfill it as well; why always disappoint them, why always tantalize them with something that they then can't have after all? If you cannot find any other way, they should become used to thinking that they have nothing to expect from you and nothing to hope for; if they then get the pleasure of a visit, it would be a gift to them; this way, with the many cancellations, naturally a bitter feeling develops, from which the children really ought to be spared. I'm enclosing a letter from Tete;[6] he always thinks of you with great affection and asked me, yesterday even, why you don't work in the physics building, which is located so conveniently close to us.[7] Albert is sending his letter separately, because he can't write today[8] and I want to send the letter off. Tete is an extremely gifted child; his favorite subject for now is geography; he knows more in this field than the lot of us, and his greatest joy is a fine atlas; aside from that, he is an absolute daredevil and very childish.– But enough of this chatting for now; I don't even know if it interests you.

With amic. regards,

Miza.

# Vol. 8, 557b. From Hans Albert Einstein

[Zurich, after 4 June 1918][1]

Dear Papa,

Today there was talk again at our school[2] about the holidays. Then all the plans that we had made popped up, which you had upset completely in your second letter. Write me please why you aren't coming, at least.[3] I and all of us are very sorry that now we must stay at home like this for the whole vacation. Wasn't it nice last year, when we were together in Arosa?[4] I really would not have thought that you would not be coming again now. How is your health?[5] It had also benefited so much from the stay in Arosa, you know, and you wrote yourself that we were planning to go into the mountains, since it was good for you.

I had quite a lot to do recently with school and piano. Not long ago I was the accompanist for Dr. von Gonzenbach, who lives in our building and who plays the violin very well, in playing the 5th Corelli sonata,[6] and otherwise I often play the piano accompaniment for a young lady from our building who also plays violin.[7] As we both don't have so much time for practicing, we only play single movements, but we've already ventured into bigger things as well. Now and then I also accompany Mrs. Wohlwend,[8] who sings songs in Italian, e.g., and by Schubert, etc.

I've been very busy with that, and when I have any time left over, I have to go shopping and anyway make sure that we get something.[9] Consequently, please don't hold it against me that I haven't written you for so long. You must accept Tete[10] writing sometimes when I'm not able to at the moment.

Since we cannot go anywhere these summer holidays, I have planned for this time that we, Tete and I, go on walks whenever the weather is fine, and that in bad weather I build either a ship or an aerial tramway, which I'm going to design myself, of course.

I hope, though, that the weather will be very good and that we can go on walks very often. Please write me very soon how you are doing and why you have disappointed us so greatly.

Lots of greetings from

*Adu.*

# Vol. 8, 557c. From Eduard Einstein

[Zurich, after 4 June 1918][1]

Dear Papa,

How are you? I go to school now,[2] I like it very much. I have friends, a boy and a girl, to play with. I am feeling well. Can you send me a book? I like to read books of travels. Why aren't you coming to us during the holidays? I had already been looking forward to it.[3]

Lots of greetings and kisses from your

Teddi.

# Vol. 8, 561a. From Maja Winteler-Einstein

Lucerne, 10 June 1918

My dear Albert,

It's already been ever so long since I wrote you, and when your last letter arrived I wanted to reply immediately. But now a fortnight has passed after all, because I've been away. I was in Filzbach for 9 days, you see, and held classes at school for Peter, who had to have surgery in his nose.[1] Being schoolmistress appealed to me quite a bit. The children enjoyed themselves very much with me, and I with the children. We laughed a great deal and still learned something in the process. I had to teach the 3 lowest grades. Shall I tell you some funny episodes from work?—One pupil from the 1st gr[ade], somewhat dumb, but all the more self-assured, goes to the door without asking. *I*: Heineli, if you want to go to the toilet, you must tell me first. *He*: Oh, but I can go piss!—And he struts proudly out of sight.

The same pupil is put in the corner by me because he is always disturbing the others. *I* (after he has been standing for some time): It's not nice to have to stand there, is it, Heineli? *He*: I've stood in more unpleasant places at home!

It is really funny being with the children but a big strain. When I didn't know what else to do, I told them a story. That always perked the children up.

It was a great disappointment for me to hear that you don't want to come this year.[2] Above all, I'm firmly convinced that the mountain air agrees with you particularly well.[3] You do still remember how much good Arosa did you. You also enjoyed yourself there exceedingly with your children. Here, too, you were very content with Albert[4] for a time, and in Zurich you even praised him very much. I don't know either whether you are doing well to neglect your own children for the sake of Elsa's.[5] You can go sailing in Switzerland as well, you know.[6] Does

Margot really have to be operated just at that time? Why couldn't the children go along to Switzerland? It would certainly do everyone some good. What's the news on your divorce, by the way?[7] Paul and I are personally still very sorry, of course, that you don't want to come.[8] Think it over once more, both of you! There are mountain spas where you could also go sailing.

Dear Albert, don't worry in the least about our health. We had just as much influenza[9] in our first apartment; there were no mosquitoes there, and it wasn't damp. The Lucerne climate, with all that foehn, does not agree with either of us particularly. Paul didn't have the flu this winter, incidentally. He suffered from insomnia, and that is why he had to take a leave of absence for a longer period. He's in excellent health again now, though. He looks like an Indian again. Mosquitoes don't come into consideration anyway as the carriers of our influenza, since we come down with it only in the winter when there aren't any mosquito bites.

I'm enclosing for you the bill for the packets. I hope they have all arrived intact. At the end of June I'm going to order another one.[10]

Alice and Ogden are delighted with Freiburg—up to now.[11] Ogden just wants to earn his doctorate as fast as possible—vanitatis causa—he doesn't care about anything else.[12] What was it that you read by Anatole France?[13] We also like him very much. He's an extremely fine and witty skeptic.

Mama will have departed in the meantime.[14] I hope she can come. We are living in great isolation at the moment and are particularly receptive to visits. You should take pity on us!

Now I've written you a good amount. Give my regards to all the Haberlanders along with dependents, especially to Elsa and the children.[15] Continue to stay so well, and beware of newspapers, which are more discouraging the longer they are.

Fond kisses, yours,

Maja.

## Vol. 8, 561b. From Paul Winteler

[Lucerne, 10 June 1918][1]

Dear Albert,

What hopes you've dashed for us by imagining that your sickbed would be softer anywhere else than with us and that, for various reasons, you would have to pitch camp at Müggelsee instead of along the shores of Lake Silvaplana or Oeschinensee.[2] Even on the summit of Frutt, 1,800 m up, a quite acceptable little lake was available where one can trim one's sails to the wind.[3] Tell Elsa that I send my heartfelt greetings[4] and that there are excellent pastry shops in Lucerne with many fine sweet things and I would pick out the best ones for her when you all come. We can even get real butter fruit-loaf [*Butterstollen*] here sometimes,[5] although others usually skim the cream off the milk.– Don't let yourself be irritated by the newspapers; it will go on for a couple more years, no doubt, and set off a bang here and there, and yet we aren't abandoning hope that eventually everything will turn out well.

We read your last postcard of March 17th to Anna, but she misunderstood it, as it seems, and Michele as well. She has constructed a link as though it were not *her* letters to you, but our last one to you, in which we humorously described to you our escapades in Zurich,[6] that takes the blame for why you're asking her not to write to you anymore.[7] Hence, in her eyes, Maja is the patsy who is supposed to have obstructed the affectionate correspondence between you and Anna. Oh, blessed foolishness! You don't need to come back to this misunderstanding with Anna and Michele, since I've taken pains to set the matter straight. We naturally want to be on good terms with Michele, and are, too; Anna is Hecuba to us,[8] she naturally has quite some influence on Michele, though, as all wives do on mainly intellectually minded husbands.

The two of us are doing splendidly, I am frequently on mountain peaks and therefore tanned like an Indian. Come and imitate me! Cord[ial] greetings,

Pauli

# Vol. 8, 563a. To Hugo A. Krüss

Berlin, 13 June [1918][1]

Highly esteemed Professor,

Many thanks for forwarding Eötvös's letter as well as for the enclosed information. I only needed the Eötvös letter to keep a man from the Academy from being nominated who would become Krüger's and Schweydar's superior *without any other arguments possibly put forward in his favor besides purely professional ones.*[2] I also would consider Mr. Runge's election the most welcome solution,[3] to which my colleague Planck[4] drew my attention.

In great respect, yours very sincerely,

A. Einstein.

# Vol. 8, 588a. From Mileva Einstein-Marić

[Zurich, ca. 17 July 1918][1]

Dear Albert,

I must have written down the wrong figure; 1,600 francs were sent to me.[2] As concerns the fall vacation, it's in October.[3] But I couldn't decide to send the children across the border under the current conditions.[4] We had agreed, you know, that they stay in the country for the present; this certainly is not in order to make a meeting with the children more difficult for you, rather I consider it really irresponsible. Tete has had this influenza that is going around now,[5] and the matter is so insidious that I can't ⟨or want to⟩ think of what would happen if they got something of the sort during such a trip. If I could at least get to them in such a case, matters would be different, but that's impossible in all respects, of course.[6] I hope that you understand me and share my opinion.—Maybe you'll have recovered enough by the fall so that, if you are already coming that far south,[7] you can make a stop in Switzerland and spend a few days with the children; they certainly would be pleased! They are both writing to you at the same time.[8]

Cordial regards,

Miza.

# Vol. 8, 588b. From Hans Albert Einstein

[Zurich, ca. 17 July 1918] [1]

Dear Papa,

How are you? Did you get my last letter?[2]

I recently read a description of a monorail. That gave me the following idea:[3] You install a pendulum (*f*) into the carriage and then let electric current flow through from *g* to *f*. Now, if the pendulum is positioned exactly perpendicularly to the carriage floor, the circuit is interrupted. But if the carriage tilts to the left (or right, resp.) then, seen from the carriage, the pendulum would swing to the left (or right) and therefore the current through *e* to *d* to the right (or left) and in this way *b*, by rotating around *h*, would be drawn toward the right and the carriage would be tilted toward the right (or left), and then it goes immediately to the other side again. But the gadget does have one disadvantage: it will always wobble a little; the better it works, though, the less it will be noticeable, but there is still the advantage that the carriage always leans into the curves to the point where a pendulum would be vertical and in this way would result in a comfortable ride and less wear on the wheels than the gyroscope system. Would you maybe be so kind as to study this a little and write me what you think of it?

We are enjoying the holidays now,[4] that is, we go on walks and laze around. Also, it's so terribly hot here now (up to 28° in the shade) and therefore we prefer the latter to the former. The lake water has warmed up to 25° as well,[5] so that almost all who are not away on vacation are enjoying the lake.

My coming to Germany would almost be more impossible than your coming here, because, in the end, I am the only one in the family who can shop for anything.[6]

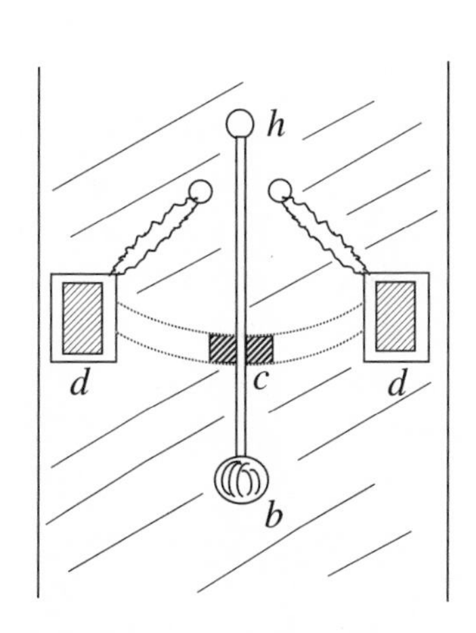


carriage floor from below

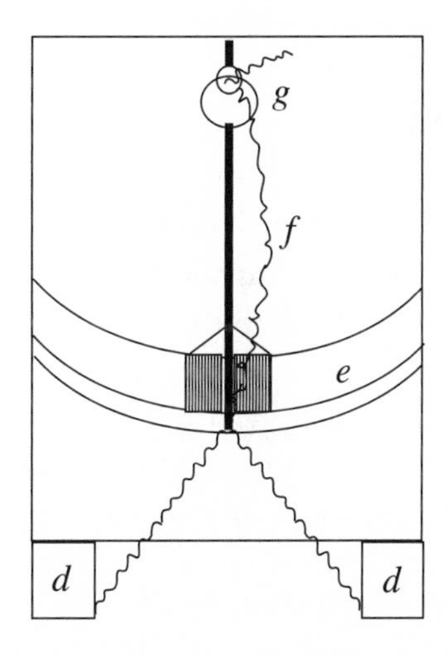


view through the internal pendulum

*b*) weight
*c*) iron block
*d*) electromagnets
*e*) internal pendulum tracks
*f*) internal pendulum
*g*) hook
*h*) weight axis
Greetings, yours,

*Adu*

# Vol. 8, 588c. From Eduard Einstein

[Zurich, ca. 17 July 1918] [1]

Dear Papa,

How are you? I hope very well. Thank you very much for the book, which made me very happy.[2] It's my birthday on the 28th of July, I'm looking forward to it very much.[3] I'm feeling well. Now we are on holiday.[4]

I send you many greetings, yours,

Teddi

# Vol. 8, 607a. From Michele Besso

Zurich, 28 August 1918

Dear Albert,

I just received your letter of the 20th, which was not in transit as long as my last one, but long enough as it is.[1] I'd also still like to discuss your suggestion for Zurich with our friends here.[2]—This isn't an embarrassment of riches in Heine's sense[3] but, as courteous as the people here are, it's rather home-baked loaf being proffered against fancy cakes, on the one hand, and a very fine sphere of intellectual life, on the other, which expects enormously much of you and is offering itself to you now, after a number of options that will perhaps soon be exhausted. There wouldn't be a lack of piety toward your Berlin friends attached to it, if I understand it correctly, which would reflect the exceedingly obliging conduct on the part of his Exc. Naumann.[4] But don't rack your brains over this anymore right now:

The worst thing is that I can't see from your letter *when* you are coming. (My time here is soon coming to an end already. Am I going to be able to see you at all?)[5]–

You will surely best be able to make your final decision here with a completely fresh mind.

I'm beginning to get a scare now, after all.[6] If it's really possible to show that the energy law is no longer capable of plausible formulation once the gravitation equations become fourth order, then this seems to me to be a very weighty argument against the theory of contact action driven to its extreme.[7] You had, by the way, overestimated the meaningfulness of my observations again: I was not aware that they had the meaning that an energy tensor for gravitation was dispensable. If I understand it correctly, my inadvertent statement now implies that planetary motion would satisfy conservation laws just by chance, as it were. What is certain is that I was not aware of this consequence of my comments and cannot grasp the argument even now.

[Likewise, although I comprehend well that a consistent theory of contact action still seems impermissible to you in physics for the time being, because it only suggests *one* consequence of experience and this one evidently is *not* correct,[8] I nevertheless don't understand at all why you think that it goes beyond implying to actually requiring this consequence of experience. It wouldn't surprise me, as I've already repeatedly said, if the actual constancy of the elements didn't belong among those inexplicable things that might only become comprehensible with a solution to the problem of quanta—:—so then the logically inconsistent theory based on the constancy of the [elements?] would, in the end, be a wrong track in the deeper sense. I also repeatedly admitted to you that a true physicist cannot let himself be impressed as much as I have been by such shaky "hypotheses." Shaky though Mach's confidence was in your theory of gravitation, my confidence in it is likewise—if possible—even shakier.

Whether I'm ever going to be able to retain this confidence, once I've understood your consideration (from your existing letter) about the energy question, I'll have to wait and see, of course.][9]

– On the question of the difference between past and future, I'd like to ask you: would you perhaps dare to regard the "fact" that we find ourselves in—or seem even to be promoting—a period of increasing entropy as based on the definition of the progression of time? This interpretation seems to me to be an unavoidable consequence of your view that *all temporal bias* is based on *order.* The initially striking reference to the experimental observation of thermal equilibrium seems to me ⟨now⟩, for some time now, merely an illegitimate interpolation in a deeper sense; this understanding of radioactive phenomena[10] [The fact that living beings use the entropy of the sun's radioactivity and not their own is simply a matter of course. We mustn't forget that entropy corresponds somewhat to exertion.] still seems to me very valuable as a working hypothesis, but both of us *don't* actually believe it.– –

– Incidentally: what influence does your $\lambda$-term have on the radiation? Is there perhaps also an absorption of the radiative energy parallel to the absorption of the gravitational lines of force?

– Your present description of the intuitive preference for the closed world with a $\lambda$-term doesn't appeal to me, by the way: What do I care about the constant irradiation of thermal energy into the infinitely empty surrounding space? (On the contrary, it guards me from thermal death in that it represents the reservoir of infinite capacity and absol. temp. 0°.– – – –

– ⟨Once again⟩ And back to the question of past and future. For me it is identical to the question whether world events can be understood as events between given physical objects entirely governed by natural law; ⟨on one hand, and as opposed to an infinite⟩ or as *in principle* unbounded, infinite creation. I repeat myself: Determinism, as a product of experimental constraints, as *subjugation* of the *mind* to an imposing limiting law as long as physical experience cannot afford any reason against this limit, . . . . . is, according to my current view, [still] a death-defying leap of reason (but a "forgivable" one). [It's wonderful, you know, how Spinoza anticipates scientific progress—defines something as "freedom" that *shouldn't* be a freedom at all in the sense of causal detachment: how well its structure entirely suits the word in the end, even in this sense, even if it only refers to the *cognitive* subject!][11] Now, however, after the transgression of the boundary has been exposed as illegitimate, adherence to determinism, *different from a working hypothesis* in natural science, therefore strikes me as a *belief*– – or as a very mystical faith. We must *assume* certain *primordial elements* of the mental "replica" of the universe: whence the primordial facts of consciousness: admission of the past ↔ possibility of the future; existence of ⟨at least one⟩ *constructive* principles within the framework of this *possibility*. Excluding precisely these primordial facts is a process that the history of science has justified as a temporary process which I, however, regard as infinitely "improbable," at least for the *final* mental state.

– The ultimate thing is the irreducibility of existence. A creative person creates for himself pictures of the whole, in which the creator's personal introspective focus, his "navel," *must* inhere—Spinoza's "navel" in this sense is *amor intellectualis*. Another navel is in modern-day physics of *processes* [*Geschehensphysik*], which essentially just describes a 4-dimensional *existence*, the cone of consciousness describing existence :: better still, what matters is what comes [out] the other end.

I do think, incidentally, that I'm battering down a wide-open door. You're surely *only* saying: "in *physics* this view doesn't have any rights of citizenship yet. As soon as anyone gets anywhere with it (and not before!), I may concern myself with it as a physicist." Against which I, of course, can't counter anything, nor do I want to.

Warm greetings, yours,

Michele.

# Vol. 8, 620a. To Heinrich Zangger

[Berlin,] 21 September [1918]

Dear Zangger,

You don't write a thing about my proposal that I'm willing to offer a lecture for a month each semester at Zurich.[1] Meyer *for his part* is agreed but hasn't spoken to anyone yet. According to his view, if all goes smoothly, I should appear for the first time on February 1st.[2] That's why I want to drop the trip for now, especially considering that the semester is already starting here and I have announced a lecture.

What's the matter with my wife? She's supposedly in hospital again, according to a comment by Dr. Zürcher.[3] My two rascals don't answer, no matter how often I inquire. Everything's hereditary, even writing laziness and confounded equanimity![4]

I really am a lucky person. Having virtually no wishes and then seeing the very few big, apparently unfulfillable ones fulfilled!

It's unnecessary that I go to Zurich for negotiations, since my proposal is entirely clear. If all of you consider it acceptable, it would only need to be approved by the people here.[5] They are always so obliging here that I have no serious doubts about it.

Cordial greetings, yours,

Einstein.

# Vol. 8, 630a. To Heinrich Zangger

[Berlin,] 5 October 1918

Dear friend Zangger,

I didn't mean to say that all was perfectly right with me; you could imagine that yourself! I am very sorry that my wife is feeling worse again.[1] I couldn't read the name of the illness on your postcard. Do you think she'll soon be healthy again? Albert wrote *very cheerfully*.[2] According to his postcard she seemed to have recovered again. I think he's right; as long as one's alive, one should keep one's chin up, and not complain. That's best also for the people around one. If I were there now, I couldn't do much for my boys either, because I couldn't live together with them.[3] But what good is the constant lamentation? My boys don't hang their heads, because they inherited their lightheartedness from me. My wife is apparently also quite satisfied, Besso writes, even though her health is very unstable.[4] All in all, better times are beginning now, and we should step into them in a happy mood.[5] So away with the furrowed brow! Quiet labor and humor should be the solution!

Cordial greetings to you from your

Einstein.

# Vol. 8, 639a. From Mileva Einstein-Marić

[Zurich, after 24 October 1918]

I received this notification from the bank and send it to you so that you can decide or give your approval on how the money should be reinvested;[1] perhaps you can write me about this.

# Vol. 8, 646a. From Mileva Einstein-Marić

[Zurich, before 9 November 1918][1]

Dear Albert,

I have received the money sent to me, 1,800 francs,[2] and thank you for it. I'd like to urge you again to accept the money transfer procedure I had proposed. Dr. Zürcher also finds this way the best—he had also suggested it, by the way; it really would be more convenient for both sides if the matter ran more or less automatically and we didn't have to balance accounts all that often. It was this: You send me 2,000 M quarterly, and once a year (1 Jan. perhaps) the bank draws up a statement of all the interest that has been deposited up to then, and at that time you send me only the remainder up to Fr 2,000.[3] I hope that you don't see behind this request anything other than my wish to simplify our business relations and that you will not hold it against me.

I hope that you are in good health and that you avoid catching the flu. Tete unfortunately got this nasty business again[4] but withstood it well; he hasn't had any fever for 10 days now, which rose to 40+°C during the illness. The doctor is advising me urgently to send him back to Arosa in the winter for about 2 months to get his strength back.[5]– Schools are closed[6] and the children are at home and are busying themselves with all sorts of things. Albert is practicing Fren[ch] conversation a little with an old acquaintance and builds machines tirelessly;[7] if he doesn't change, he's bound to become an industrious mechanical engineer.[8]

With cordial regards,

Miza.

# Vol. 8, 659a. From Paul Winteler

Lucerne, 22 November 1918

Dear Albert,

Many thanks for your card;[1] well, that really does it! Maja's birthday[2] was celebrated very nicely, she was *very* pleased with your note. Meanwhile, the birthday of the German "republic" has also taken place.[3] I don't know how the affair looks from close range; the child still has to get thoroughly washed before we can tell which parent he resembles, the eastern Bolshevik dad or the western revolutionary mom. It could also be a bastard, you know, who after being allowed to scream for a while is shown the door and smuggled via Switzerland preferably into France and Italy as an anonymous Bolshevik, as soon as the occasion arises. Clear evidence of this has been available here since conception. So we had the general strike,[4] which in the land of universal voting rights and political equality was supposed to mean a mini-revolution with the ultimate goal of moving political and economic power into the hands of the working *class*. The affair was handled with extreme skill. The posting of ca. 9 demands whose legitimacy, arising out of the Federal Council's stupid governmental measures during the war, the legitimacy of which was quite commonly acknowledged or at least could not seriously be denied: recall of the Federal Council, replacement of the Federal Assembly, 8-hour workday, old-age pension and disability insurance, voting rights for women, monopolization of imports and exports (here an unsuspecting little disadvantage: this demand, to be sanctioned as a wartime measure, means de facto nationalization of total production (work) and consumption, as a permanent institution of peace, but it is presented as if only imports and exports were at issue. Future development in this direction is possible, though, and arguable.) These demands, regarded by all as good or worthy of discussion, were not introduced by referendum and the exercise of voting rights but are being forcibly imposed by a general strike.[5] Upon the resignation of the Federal Council and the Federal Assembly, the Workers Committee would then simply have taken their place, a committee that would not have been accountable to anyone, nor would it have had to keep to any constitutional or legal regulations in carrying out the demands.[6] Thus *absolutism* of the first water, which would nominally have used the suspicious demand "nationalization of imports and exports" as a tool for an unrestrained reorganization of the economic conditions (I am not defending the present ones!), without any public opinion poll, while degrading the voter to a subject. Here you will immediately recognize the parallel with Bolshevism and Liebknechtism (Spartacist group?).[7] Otherwise as well: one of the members of the Workers Committee is . . . . . Grimm of all people, minion of the ex-Federal Council and pan-G[erman] time-server of Hoffmann, Grimm who traveled to Russia on a quasi-German mission and bolshevized there.[8] The almost *universal* impression here is that the general strike does not involve a suggestive and contagious wave of European ideology, perhaps in the sense of the revolutionary ideas of 1789, but is a means of using Switzerland as a medium for the transplantation of Bolshevism following the Entente, in order to obtain a more favorable peace that way. Then, since G[ermany] evidently has nothing more to lose, following the conclusion of a favorable peace, the extremists (a minority) in Germany could be allowed to disqualify themselves through *mismanagement* in the meantime so that the field is all the more favorably prepared for political legitimacy. Over the years, G[ermany] in particular has understood admirably well how to play off foreign action and reaction against one another (Russia!);[9] the orthodox will continue to deem this policy domestically advantageous. The extremes meet.

Our general strike petered out after a few days, despite participation by the railway men for a week.[10] It was indicative that right in the middle of it a *Berlin* communication arrived under the new government: "Revolution in France, Fren[ch] government has been overthrown!" This time the clumsy maneuver didn't work anymore like the Nuremberg bombshells. Propaganda against the Entente is in full flourish here again, from G[ermany]! False reports as well, as before. The people said quite unanimously: our Federal Council can go to the devil, but *first* and foremost the semi-*foreign* general strike—soviet "made in Germany."

I went into some detail because from there you've probably hardly had an opportunity to closely acquaint yourself with the conditions here. We hope that you find the matter amusing from your safe lookout onto the partisan goings-on over there: the jump from the 18th century into communism, without adopting the democratic attainments of the Fren[ch] revolution just leads from absolutism to absolutism. You can imagine who profits from it in the end. Incidentally: one isn't a republican and a democrat merely by waving a little red flag; this higher mentality needs the longest and most persistent education and self-education.

Do write back sometime very soon, we would like to hear your opinion. Take good care of yourself in the hullabaloo, and warm regards, yours,

Paul.

# Vol. 8, 659b. From Hans Albert Einstein

[Zurich, ca. 25 November 1918][1]

Dear Papa,

How are you? I am doing very well, but school has started again.[2] Please write me what's going on in Berlin.[3] You've probably heard that there was a national strike here last week.[4] It happened roughly like this:

On Saturday morning it was suddenly reported that the decision to strike had been made. So the first thing was, as always, that the "streetcar men" abandoned their duties and went home. No one was working at train stations and all workers left their bosses in the lurch. That was called Zurich's protest strike but it left the federal works unaffected. A lot of military were immediately brought to Zurich, probably around 10,000 men. On Monday evening it was suddenly made known that a general strike had been decided for Tuesday under the direction of the so-c[alled] Olten Action Committee.[5] Now telephone was the only means of communicating. But that wasn't supposed to last long: a portion of the Zurich student body offered their help to the government. Then suddenly on Wednesday at 11 o'clock in the evening a paper was circulated, as there had been no newspapers before, either. They were sold by students and printed by employers. Students also helped support postal operations. A few trains were running as well, operated by engineers and the military. Apart from twice when the military had to fire shots in order to disperse the crowds, the strike in Zurich went very calmly. And on Friday everything was back to normal. One amusing thing to happen during the strike was that by the 3rd day the tram rails were already completely rusty.[6]

We're all fine. How did these revolutionary days affect your stomach, though?[7] Write me a little about the state of your health and of Berlin. Teddi has to go to Arosa again in the winter because of his flu, but we haven't found anything for him yet.[8] Everything is filled up.

Lots of love from your

Albert.

Mileva's postscript: "On your last letter there was a comment by the censor that only letters indicating the sender are delivered; please bear this in mind.'

# Vol. 8, 659c. From Eduard Einstein

[Zurich, ca. 25 November 1918][1]

Dear Papa,

How are you? Although I did make an effort to write, we have already had 7 weeks of vacation.[2] Now I'm reading the history by Oechsli, I'm at the 14th century.[3] I usually play with lead soldiers or in the park with other children. Then I bring my weapons along.

Lots of love from

Teddi

# Vol. 8, 661a. From Maja Winteler-Einstein

Lucerne, 29 November 1918

My dear Albert,

I was in Filzbach[1] for a couple of days, that's why this letter is being sent off late. Many thanks for your birthday wishes, which delighted me.[2] But I was much more pleased about your optimistic view regarding the state of affairs.[3] Do keep us better up to date. You can imagine how valuable to us disclosures from you are.

Our Swiss people impressed me very much on the occasion of the general strike.[4] Even though only extreme Socialist papers appeared, practically all parties and levels of society were of one mind: the demands by the strikers are right and debatable, but the use of force in our purely democratic state was thoroughly reprehensible. It would surely be difficult to find such political maturity among the masses in any other European country besides England. This certainty of judgment really astonished me. The effect of the strike was a very salutary one politically: the Federal Council and the Federal Assembly, in a word, the almighty Liberals, got a good shock, and the pace of introducing proportional elections and other legitimate reforms was accelerated significantly;[5] on the other hand, the strike committee was discredited and had to go, so its power was broken. Unfortunately, the flu has taken very many victims again as a consequence of the large crowds. It's uncanny how this sickness is ravaging our little land.[6] How is it where you are?

It was only today that I was able to order the November package in Berne.[7] Sorry if it arrives a little late. The funds you had transferred to us for the packets are used up now. Forgive me for reminding you of it, but we are—as usual—on the rocks and would be very grateful if—provided you aren't too badly off as a result—you could send us another little sum.

Mama is coming to Lucerne again one of these days with Uncle Jakob.[8] I have to look for a room for her nearby, since our guest room is too cold and damp for her in the winter. She is suffering a lot from rheumatism.

Affectionate regards to Elsa, the children, Uncle, and Aunt.[9] Please write me soon, at least a little card. Fond kisses from your

Sis'.

Paul just showed me that the German exchange rate is very bad.[10] So don't send anything. Sooner or later it won't be possible to send any more little packages anyway.

# Vol. 8, 661b. From Paul Winteler

[Lucerne, 29 November 1918][1]

Dear Albert,

Enjoy the foodstuffs![2] It's unlikely that you'll gain back your former little paunch as a result, and even less likely because envious spongers will try to win your good graces. There are probably few sausage Croesuses like you anymore in your surroundings, of course. I could almost wish that our Federal Council had decided that Swiss citizens entitled to 5 kg be obliged to present themselves periodically for a weighing-in, to check whether they use it all for their own fortification as well. Instead of that, it exports cattle, unfortunately not the largest, out of sheer modesty. Otherwise, it's quite tolerable here, practically everything can still be had for money and we still do feel comparatively in clover in this land of political shepherd boys.

Michele is private lecturer at Zurich again and,[3] as we go there more often now, he, little by little, is becoming the provider for our intellectual appetites. He is really cut out for private lecturing, but I would like to wish him a larger professional circle; his efforts mediating between the disciplines is something more suited to mature people (and of those there are so few!) than for students and doesn't draw these aspiring examinees within the range of his spell. But I hope that gradually he will feel more and more that he was made to be a teacher and expand his material accordingly.

Don't become too opulent in your plenty, but always keep the ancient Swiss virtue in mind: "Each to his own, and for me a little bit more!" May you prosper, as a jewel of your fatherland.[4]

Pauli

# Vol. 8, 661c. From Max Jakob

Charlottenburg, 27 Kastanien Avenue, 3 December 1918

[Not selected for translation.]

# Vol. 8, 663a. To Max Jakob

[Berlin, 5 December 1918]

[Not selected for translation.]

# Vol. 9, 7a. From Heinrich Zangger

[around the end of February 1919][1]

Dear friend Einstein,

So you had a nice time with your sons. Right? [Hans] Albert is developing well and steadily.—The little boy is becoming healthy and strong.[2] I proved to be somewhat right with my optimistic prognosis. The world needs 1% optimists.

Your wife's sickness was surely partly the consequence of overexcitement.[3] Fame is a pernicious evil for the living. Once you are playing violin with Newton, it's no longer dangerous for one's circle. Fame is like a lobster: whoever sees a lobster feast through a big window cannot do otherwise but want it—even from one's dearest—but whoever knows the nauseous indigestions attendant on the object of envy, envies the Einstein from the patent office. The small relativity indigestions of 1905 had all the creative joy of Einsteinian visionary certainty, even if no one believed it, for a full understanding of your powerful feat of logic. Perhaps Besso understood you.[4]

I told your wife how a violent instinct impelled me to react toward an obtrusive journalist: he wanted to make my institute famous, he knew about the accomplishments, etc., that the university should enlighten its supporting public—I also knew the worth of the man from before and asked him: do you think I would allow myself to be handed out praise and censure by you here in private? – – – do you think I would allow you, among 20,000 people, to judge me irresponsibly?—So please.

Furthermore I explained to your wife: Descartes was famous even before Einstein—but he succeeded in living as expressed in the motto: *bene vixit qui bene latuit* [he lives well who has hidden himself well].[5]

You cannot protect yourself from curiosity—this is the curse, which counts on vanity, etc.—Go on a sailing boat—Wannsee—have a big sign made: Prof. Einstein is away—no one should give any other information, for that is the greatest risk of legitimate annoyance: becoming the source of misunderstanding. I am speaking from much experience of [indirect] relativities.

# Vol. 9, 25a. From Hans Albert Einstein

[Zurich, around 20 April 1919][1]

Dear Papa,

I've been anxiously waiting a long time already for any sign from you, particularly since very bad news arrived about the conditions in Germany.[2] That's why it was a huge relief when I got your letter of the 15th of March and heard that you're doing well. You're probably finding it very interesting in Berlin now, but your stomach that much the less, which is very certainly going to revolt again shortly.[3] But it is a consolation that you're coming here again for a while in the summer, to our great joy as well as to that of your revolting little stomach.[4] I was also very glad that you've already seen about my scores.[5] It's questionable whether they can be sent without any trouble. Please let me ask you also when you come again to bring along the photographic plates "From my Childhood." I've already made various plans for the summer, but we can't choose until you're here and we can sort of see what you can handle. I'm enclosing a postcard that just arrived. For the last few days I've been going to Mr. Holder, the mechanic at Prof. Maier's institute, and have been filing away for dear life.[6] I've got vacation, you see, and am making the most of every last bit of it. To the point that I hardly found the time to write you. Sometimes I also play music with Dr. von Ganzenbach[7] now and otherwise do things with Tete, who's been back here again for sometime already.[8] We can't go out on walks now, though, because it's wintertime again. There have been about 40 cm of snow for a couple of days now, but it's very wet so it's no fun to tramp around in it.[9] By the way, we need money very badly, because everything's so expensive and Mama needed the deposits at the bank for Teddi.[10] Please send us some as soon as possible.

Many greetings from

Adu and Teddy

P. S. Just now the money (2,000 francs) arrived and Mama sends her thanks.

## Vol. 9, 59a. From Hans Albert Einstein

[Zurich, before 13 June 1919][1]

Dear Papa,

Thank you very much for the music scores.[2] I already played some of the pieces out of them and last week I had an examination at music school.[3] It went very well. I played a few preludes from Heller (Op. 81).[4] The pieces are quite nice but the headings are even more wonderful, like, e.g., "While expressing the bitterest of pain." Heller thinks if the notes don't produce it, then the player must. Otherwise I haven't been playing anything interesting recently.

At school we started the Crusades in history and it's still about as boring as before.[5] We also started English, where the pronunciation seems quite ludicrous to us at first. In natural history, botany, we have a jumble of cell theory, plant taxonomy, tissue theory and all the rest. It's most interesting when there are experiments. We've also started physics now but aren't past the introduction yet. We're going to go through mechanics. Our teacher is Prof. Stierlin. For mathematics we still have young Mr. Folkhart,[6] whose lessons are very nice and one learns quite a lot. We've begun trigonometry and exponential theory and are using logarithmic tables with great relish when we want to write out the roots. It's quite a lot of fun, particularly that one can apply it so often. In geometry we just started in the last lesson the algebraic solving of geometric problems. Nothing difficult is taken into consideration there yet, of course.

Something just occurred to me that I wanted to ask you. If you have a sail of size $f$ on a sailboat, how can you calculate the sideward pressure of an average wind and how large does the weight hanging at distance $l$ from the rotational axis have to be in order to withstand the strength of the sail? One probably couldn't discuss this in just a few words; then we could save this up for the summer vacation. For I attached a weight of 200 gr just to see what happens and the result was that it immediately fell over. It's the boat that you saw in the beginning stage.[7]

I read at the university that you want to come for July.[8] How is your health, by the way? Could you perhaps go on a smaller walking tour, or could we even perhaps go on a larger excursion?[9] Please do give me more specifics about this so that I can draw up a few plans. We are unlikely to be able to leave Switzerland. What are the living conditions like in Berlin, by the way? How's your food, how's your stomach?[10] Please write me sometime soon and I hope you, too, follow soon, because I'm always worried about you.

Many greetings from

Adu

Please bring the photographic plates with you![11]

# Vol. 9, 59b. From Eduard Einstein

[Zurich, before 13 June 1919][1]

Dear Papa,

How are you? I'm glad that you're coming to see us.[2] In the summer I'm going to Rheinfelden to the saltwater baths.[3] I'm now in the 3rd form.[4] I like it at school very much. It's raining now in Zurich.

Many greetings and kisses to you from your

Teddi

# Vol. 9, 66a. To Elsa Einstein

Lucerne, Monday. [30 June 1919]

Dear Else,

Mama is suffering terribly, but is very glad to be with me. She feels quite unhappy here.[1] I'll try to get her accommodated elsewhere, cost what it may. She's hoping to be able to go to Berlin.[2] Maja is nursing her devotedly. She only has help for a few hours. Mama is being tormented by mosquitoes in the apartment; it makes airing out difficult.

In case the other postcard didn't arrive, I repeat that I'm settled in the "Sternwarte" [guesthouse] at a satisfactorily agreed price.[3] Tomorrow I have to return to Zurich.[4]

Fond greetings to you, the children, Mrs. Hellberg, and the parents.[5] Yours,

Albert.

I'm feeling extraordinarily well.

# Vol. 9, 68a. To Elsa Einstein

Tuesday, in the train on the return trip.[1] [1 July 1919]

Dear Else,

Mama senses that she is seriously ill but is still hoping for a recovery. Her only very strong wish is to be brought back to Berlin with me. The doctor doesn't exclude the possibility of fulfilling this wish.[2] Blame for this wish lies in her very strong antipathy toward Pauli and the housekeeping[3] and her yearning to be with Aunt Fanny. I'm for having this last wish of hers fulfilled, if possible; she wants to live with Aunt Fanny and would, of course, have to have a full-time nurse.[4] I'd bear all the costs. My presence is a great relief for her. The pain is horrible but not always quite so severe. Maja nurses her devotedly but cannot continue to do so in the long run. Accommodating a nurse in this snug apartment would be impossible. The possibility of an embolism persists, because the thrombosis persists—albeit to a lesser degree.[5] The swelling seems to deaden the pain somewhat. Appearance not bad, however, memory has suffered from the constant use of painkillers. According to the doctor, it could still take a year until the release!

Best wishes to you and yours,

Albert.

# Vol. 9, 69a. To Elsa Einstein

[Zurich,] Wednesday. [2 July 1919]

Dear Else,

This morning I finished off the expert opinion.[1] Toward midday I visited Zangger & discussed the illness and other matters with him.[2] After lunch I lay down until four o'clock. Then Edith came to see me about her dissertation.[3] She'll get sick and tired of the subject. Such a thing is unsuitable for women. At a quarter to 6 Tete[4] came on a visit until now, before dinner. My courses take place on Friday, Saturday, and Monday (in the evening). Today it's cold and frosty. I haven't gone out visiting anyone yet, hardly even went out for a walk. My boys are very nice and give me much joy. Albert[5] spent the whole evening with me. Miza is leaving with Tete around the middle of the month for Rheinfelden.[6] Then I'll be living with Albert in the apartment.[7] Until that time I'll stay at the "Sternwarte."[8] Mayer's[9] other boy also had an accident now. He was hit by a car and had a head wound. The other one's broken leg was not properly reset so he limps. Others also have much to endure!

Best regards from your

Albert.

Greetings to the children, the parents, and Mrs. Hellberg.[10]

## Vol. 9, 70a. To Elsa Einstein

[Zurich,] Thursday. [3 July 1919]

Dear Else,

I still haven't had a sign of life from all of you. So you won't have any from me either. So you're going to fret.[1] But it can't be helped! This morning, police [registry], etc. Tomorrow before noon again. This afternoon, at home in my room (studied). Thought up the method for Edith's thesis.[2] This evening, Phys. Soc.; Meissner is presenting a talk.[3] Tomorrow the course begins.[4] Tomorrow evening *alone* at Meyer's.[5] Met Mrs. Weyl and—didn't recognize her.[6] It's almost always raining. Haven't seen any acquaintances until now—happy solitude. Always rest after the meal until 4 o'clock. Your training was successful. Always go to bed early. There can't be anything more solid than living alone in Zurich (rhyme). Therefore venture out alone, leave the good wife well at home.

Heartfelt greetings from your

Albert.

Greetings to the children, parents, and Mrs. Hellberg.[7]

## Vol. 9, 70b. To Elsa Einstein

[Zurich,] Friday evening. [4 July 1919]

Dear Else,

Still no sign of life from all of you & I'm writing every day. This evening I was at the Meyers',[1] which was very nice; I was all alone there. The course is being very well attended. Before that I was alone throughout the day, quite contrary to my earlier habit here.[2] Weyl is suffering of asthma;[3] today he was in bed. I went quickly to see him with Meyer after the lecture. Along the way I met Mrs. Adler,[4] with whom we conversed a little. I am feeling well; just the thought of my mother's ordeals plague me.[5]

Send me a telegram so that I can at least see that you are receiving my messages.

Heartfelt greetings, yours,

Albert.

# Vol. 9, 70c. To Elsa Einstein

[Zurich,] Sunday. [6 July 1919]

Dear Else,

Yesterday your first postcard arrived at last. Everything's going well here. I'm healthy, but am also being very careful.[1] The course is very well attended; I'm also making an effort with it. Weyl is suffering very much from his asthma.[2] Yesterday evening I ate at Koppel's; he was very nice with me but is very worried indeed.[3] Today I want to spend the whole day with my boys;[4] I'm waiting for them now in my room. Always indicate the sender on your correspondence. I fear that my other postcards, which I wrote you, one each day, didn't arrive because of the omission of this requirement! Mama is now accommodated in a sanatorium, to my great relief. It could not have continued *that* way. It will be more tolerable for all parties.[5] The day after tomorrow I'm traveling to Lucerne again.[6] My courses are always Friday, Saturday, Monday, so I have time to be in Lucerne a lot.[7]

Heartfelt greetings, yours,

Albert

(No ink left today!)

# Vol. 9, 70d. To Elsa Einstein

[Zurich,] Tuesday morning [8 July 1919]

Dear Else,

If only you received my many postcards, which I wrote one a day! Up to now, all in all, I received from you *one* postcard! Write the sender's address on it! Yesterday I had the 3rd lecture.[1] Visited Stodola in the afternoon;[2] conversation about the social problem,[3] Hungary.[4] In the evening Edith was here to see me in order to be instructed about her doctoral thesis. The thing is very fine—but not her work.[5] The day before yesterday I spent the whole day with my boys. We went sailing for 1 hour in the afternoon. Albert is a splendid boy. Healthy, intelligent, and a well-rounded person, young though he still is. He's sure to make his own way. He fits into this world. It was certainly good and wholesome that he grew up here.[6] The little one also looks good but certainly doesn't exude such a finished personality. This morning I'm traveling to Lucerne, to return again on Friday.[7] Miza will then be away with Tete, and I'll be living there with Albert.[8] It's surely going to be very nice. Albert plays music *quite* well already. There was disgusting checking going on again at the lecture. But I tell myself quietly: It's the last time![9] Mama is now settled in a sanatorium, to my great relief. I just hope I don't have to take her along with me![10] I'm dreadfully afraid of that. If she absolutely wants to, it will just have to be.

Fond regards, yours,

Albert.

I'm probably spending Friday evening at the Karr's.[11]

# Vol. 9, 70e. To Elsa Einstein

[Lucerne,] Wednesday evening. [9 July 1919]

Dear Else,

Was with Mama at the private clinic the whole day.[1] She is doing reasonably well. Still thrombosis. Right leg always bandaged up and raised. Diagnosis isn't quite certain yet because thrombosis is also known to be a consequence of the flu.[2] The clinic and the nurses are *very good*. Transfer to Berlin would be foolishness.[3] From Friday onwards I'm going to be living with Albert at 59 Gloria St.[4] (Record achievement.) Healthwise I'm feeling very well. But am also being very cautious. Uncle J[akob], that sensitive man, is abandoning Mama here and going to Zurich, where it's supposedly more entertaining.[5]

Heartfelt greetings, yours,

Albert

Greetings to the children, parents, and Mrs. Hellberg.[6] Greetings from Maja & Pauli.[7]

# Vol. 9, 72a. To Elsa Einstein

[Zurich,] Saturday, 12 July [1919]

Dear Else,

Half of my lectures are almost over (Monday).[1] The audience has crumbled away a little as well. But Meyer[2] and the mathematicians are ⟨still⟩ present. I've been living with Albert since yesterday and am very well looked after and content.[3] The boy gives me indescribable joy. He is very diligent and persistent in everything he does. He also plays piano very nicely. Tomorrow at noon I'll be with Edith at the Karr's.[4] Zangger is out of town; I would have liked to have talked to him again about the diagnosis.[5] I'm continuing to feel very well. I think I can return home according to schedule.[6] Today I received a report from the Oppenheims.[7] They are still at Thun Lake.

I send Ilse many thanks for the forwarded letters.[8] Everything will be dealt with punctually.

Fond greetings, yours,

Albert.

# Vol. 9, 72b. To Elsa Einstein

[Zurich,] Monday. [14 July 1919]

Dear Else,

Living with Albert is very fine. *We* are building a small flying machine together.[1] This morning we went sailing in a stiff wind. Living here is entirely harmless.[2] Tomorrow I'm going to be traveling to Lucerne again.[3] Yesterday at midday I was at the Karrs', who are very kind, as always.[4] Their home is sumptuous and in the best of locations. Afterwards there's teaching again. In the evening I'll be with Weyl.[5] A bit much—but don't frown.[6] I sleep at noon every day and I'm feeling very well. This morning I bought you an umbrella—one that, at worst, one is allowed to lose. Today my coat is being mended and the laundry washed. Mrs. Fleischmann has died—of cancer.[7] I'm not convinced about the diagnosis of Mama's illness; the coming months will tell.[8] I still haven't seen Grossmann and Hurwitz,[9] I have so little time! Zangger is extremely busy, but he likes it that way.[10] Next year Albert will come to visit us during the holidays.[11] We want to go to a North German lake and rent a boat. That will give you another problem for your managerial genius.

Send many greetings to the children; I want to write them soon.[12] I'm considering setting off again after my lecture, without staying long. I'm not going to travel with Albert, but will visit Brandhuber.[13] Then I'm coming to see you in order to go somewhere else with you as well.

Greetings & a kiss from your

Albert.

# Vol. 9, 72c. To Elsa Einstein

Lucerne, Tuesday evening [15 July 1919]

Dear Else,

Mama had a decent time during my absence.[1] But today it's very painful again. Lying for months with a bent leg, without being able to move and in constant pain really must be terrible.[2] Last night I also visited Weyl.[3] This morning I drove back here. Albert[4] is a true joy, he cheers me up after the sadness and the exertions. But it really is a great relief that Mama at least has the right kind of nursing and good care. I'm not considering taking her along anymore.[5] I enjoy your many and good reports. I hope you all continue to feel well when I come back. Let's go on a small outing then as well, particularly since I canceled my trip with Albert. He gets more out of me in Zurich. When he wants to go out, he prefers to do so with his own age group.[6]

Greetings & a kiss from your

Albert

Best regards to the children, parents, & Mrs. Hellberg.[7]

# Vol. 9, 72d. To Elsa Einstein

[Lucerne,] Thursday. [17 July 1919]

Dear Else,

Things are going quite tolerably again for Mama and she feels relatively well in the neutral setting.[1] Today I'm traveling back to Zurich, because the lecture-filled half of the week is beginning again.[2] Camping in the lioness's den is proving very worthwhile. And there's no fear of any incident happening.[3] Around the 5th of August I shall leave Switzerland to return home to you via Brandhuber,[4] and completely safe and sound at that, if things continue as they have been going. The day after tomorrow there is a Gottfried Keller festivity at the university, which I'm also going to.[5] One lecture will thus be canceled, but I'll manage to finish off, anyway.

What do you say to my letter-writing diligence? One shouldn't praise a trip before it's over, but I really am quite proud of myself.

Greetings & a kiss from your

Albert.

I'm looking forward to Margot's new songs.[6]

# Vol. 9, 72e. To Elsa Einstein

[Zurich,] Saturday. [19 July 1919]

Dear Else,

Today was the Keller festivity and much bustling about.[1] But now I'm going home to have a rest. Habicht from Schaffhausen was also there.[2] Now I have 4 lectures left.[3] Then I'll go to Lucerne[4] and then soon afterward home again to you. I'm very eager to be back with you and in our quiet sanctum. Grossmann approached me again about escaping.[5] But that has time until the spring. Make inquiries about your assets in case we do leave before the end of this year.[6] Hilbert is quite certainly going to Berne.[7] But I believe that we had better follow the same old track.[8]

Fond greetings from your

Albert

# Vol. 9, 74a. To Elsa Einstein

[Zurich,] Monday. [21 July 1919]

Dear Else,

Everything continues to go well. Tomorrow I can't go to Lucerne[1] because I have to go to the Sterns', who are insistent.[2] It's not certain at all whether I can go this week. My life with Albert is very charming.[3] He has developed fabulously as a person and as a son. At the same time, a happy temperament and ever busying himself with something. Tomorrow he's going on a tour with other boys for about 5 days. I talked him into it myself.[4] I had the opinion transcribed here, tell Ilse, whom I thank heartily for forwarding the correspondence and for her letter.[5] Now I'm thinking of staying here a week longer after the conclusion of my course, after all, because I'd like to spend a few more days in Lucerne as well as with Albert (and Tete, if he's back here).[6] I just don't know where I should live for such a short time, since the hotels are so far away. Yesterday afternoon I played music at Mr. Greinacher's (physicist);[7] was at Hurwitz's (that morning); the latter is now seriously ill.[8] They are still complaining about their politicized daughter.[9] I try to talk them out of it. I haven't had any news from Brandhuber yet.[10] If none arrives, I'll travel directly to Berlin.

Greetings & a kiss from your

Albert.

# Vol. 9, 74b. To Elsa Einstein

Zurich. Tuesday [22 July 1919]

Dear Else,

Albert went on a mountain tour with his friends.[1] This afternoon I'll be at the Sterns', where Julie Ansbacher will also be on hand.[2] Next Monday I'll be lecturing for the last time.[3] Then I'll spend a few more days with Mama in Lucerne,[4] then probably a couple of days here again, so I'm probably going to be at Brandhuber's before the 10th.[5] I'm looking forward to being home again with you, where we can comfortably live all on our own, as is proper for such a young married couple.[6] The flow of correspondence has had a slight constipation, or is it the post office? But that's all right. I probably won't be able to make it to Lucerne this week; I'll be going next week instead. Mama will be very sad when I leave: she actually feels very abandoned there, which is partly her own fault.[7] I've noticed that my presence offers her much comfort. Uncle Jakob is nothing but a fatso.[8]

Heartfelt greetings from your

Albert.

# Vol. 9, 74c. To Elsa Einstein

[Zurich,] Wednesday [23 July 1919]

Dear Else,

At last I spoke with Zangger, who confirmed to me that it's not certain it's cancer—in any case, protracted and painful.[1] This week I'm not going to Lucerne.[2] Mrs. Ansbacher was very nice: Mama seems to have broken all ties with her.[3] Today Weiss (Strasbourg), who knew about a lot of interesting things, was here to see me; he unfortunately lost his wife.[4] He's coming again tomorrow. Weyl[5] will come along on a short walk. Albert is on a tour, unfortunately in the rain.[6] The Sterns were hospitable again. Dora Stern (from Berlin) is also here.[7] This evening I'll be visiting the Meyers.[8] I am being very well taken care of by the maid, who cooks well—I'd wish the same for you.[9] I received a friendly report from Brandhuber; I'll go there but will certainly stay less than a week.[10] On the 15th I hope to be with you again.

Fond greetings from your

Albert

also to the children and to Mrs. Hellberg.[11]

# Vol. 9, 74d. To Elsa Einstein

[Zurich,] Friday. [25 July 1919]

Dear Else,

Now it will soon be over. Today, tomorrow, and Monday still lecturing.[1] Then I'll go to Lucerne for a few more days, then to Zurich for a few, so I can travel to Brandhuber's around the 8th of August and will be with you at the latest on the 15th.[2] Zangger also isn't convinced yet that Mama has cancer. At least there isn't any proof.[3] Albert has returned again.[4] He is a dear and intelligent fellow. Give my warm greetings to the Moszkowskis and look after them.[5]

Last night I visited the Grossmanns.[6] It was very nice and interesting. We also talked about a certain possibility. But meanwhile I'm not so enthusiastic about being a schoolmaster, year after year.[7] Have you already inquired about your assets?[8] It won't make that much of a difference anyway! If I don't come, it'll probably be Debye.[9] He is young and completely healthy—it's actually the right thing.

Heartfelt greetings, yours,

Albert.

# Vol. 9, 74e. To Elsa Einstein

[Zurich,] Saturday. [26 July 1919]

Dear Else,

Today two letters from you. I call that loyalty! I think it *abhorrent* to stow Margot away in a sanatorium. Aunt Fanny really could manage alone there; so this mustn't happen.[1] Margot has to have her rest. Just say I "ordered" it. I also just received Mühsam's letter.[2] Zangger already suggested a blood analysis as well, which to my knowledge has not yet been done. The result wouldn't be *certain* either. For Mama it would be a new upset. So I'm not particularly in favor of it.[3] Monday or Tuesday I'll be traveling to Lucerne until Thursday or Friday.[4] After returning, I'll stay here a few more days, will spend the night at the Habichts' in Schaffhausen & then travel to Brandhuber.[5] I hope to be able to finish off the lectures today already.[6] Then I'll be glad to have some peace. The last lectures were still quite fine. From Planck I received a moving letter that I not be disloyal to Berlin.[7] *I won't be, either.* Come what may, we will hold through.[8] You shouldn't send any money. I have my reasons.

Greetings & a kiss, yours,

Albert.

# Vol. 9, 77a. To Elsa Einstein

[Zurich,] Monday. [28 July 1919]

Dear Else,

The lectures are now finished.[1] Today I also visited Mrs. Ansbacher and went sailing with Edith, unfortunately with virtually no wind.[2] This evening the Hurwitz children are also coming to see me.[3] Now I'm going to Lucerne until Friday.[4] Then I'll stay a couple of more days (2 or 3) in Zurich, despite your two telegrams.[5] Then departure for Benzingen,[6] a quite complicated itinerary. I certainly did not write anything about wanting to move here. I just put forth the possibility.[7] Send no money; I have my reasons. We're going to stay in Berlin, all right; my decision is firmly made now.[8] So calm down and never fear!

Greetings & a kiss from your

Albert

Albert left for Rheinfelden today.[9] What a fine chap. We'll see if he really does come to see us. I'll bring the stockings along.

# Vol. 9, 78a. To Elsa Einstein

[Lucerne,] Tuesday. [29 July 1919]

Dear Else,

I'm staying here until Sunday, Aug. 3rd, and will then go to Zurich for a couple of days in order to be with Tete a bit more.[1] Miza is still staying in Rheinfelden until I leave[2] (about 6th–8th August). Write Lange[3] that I dropped the plan to travel to Benzingen.[4] How impudent!

There can scarcely be any doubt left about Mama's illness, unfortunately.[5] Be careful in your letters; she is very suspicious! Why did you reveal Benzingen to Lange?

I'm very sorry for Mrs. Hellberg.[6] The new solution for Aunt Fanny appeals to me, also that Margot will now keep her little bit of freedom.[7]

Warm greetings, yours,

Albert.

# Vol. 9, 79a. To Elsa Einstein

Lucerne. Thursday [31 July 1919]

Dear Else,

Today I received yet another telegram from you, and a quite truncated one at that ('have' instead of 'Haber,'[1] etc.). But I managed to understand it anyway, along with the avowal of your innocence.[2] I would have stayed until Sunday anyway. Also, I already spoke with Meyer and Grossmann to the effect that my coming was very unlikely.[3] So all of that would not have been necessary. I also received a scared letter from Planck about rumors circulating about me.[4] But I already replied to him in a placating manner a few days ago. We're staying there in any case.

Lorentz's letter[5] shouldn't be forwarded to me but to *Miss Rotten*.[6] It's about commission business.[7] Mama is doing reasonably well. She's in excellent care here.[8] I spend almost all day with her and read the papers to her, write letters, etc. Erzberger's and Müller's revelations must be doing some good[9] (Moszkowski, etc.). I send my cordial regards to the latter.[10] Now I'm looking forward to our being together again, even if it takes a little longer.[11]

Heartfelt greetings from your

Albert.

# Vol. 9, 84a. To Elsa Einstein

[Zurich,] Monday morning. Train [4 August 1919]

Dear Else,

I'm on the return trip to Zurich.[1] Yesterday evening I saw Mrs. Haber; *he* hadn't arrived yet. I left him a letter. The matter isn't urgent.[2] I'm not ⟨going⟩ accepting at Zurich, if there's no pressing necessity to do so.[3] Mama is entering a quieter stage.[4] Her passionate will to live has diminished noticeably. Injections and suffering are numbing. Existence doesn't seem to be agonizing anymore. So I can leave more heartened. She doesn't complain anymore about her inability to move. The ability to suffer likewise diminishes with the progressing illness.

I'm staying 3–4 more days in Zurich, also in order to make all the preparations for the journey. My health continues to be good. I'm very eager to see you again. I didn't see Oppenh[eim] again but wrote him once more.[5] I'm bringing provisions and stockings along with me. We'll see if Albert[6] comes to see us next year.

Greetings & a kiss, yours,

Albert.

You will be, by now, quite alone. I think that we will both also leave Berlin for a little while. Until now I had everything except relaxation. The mirage of a sailboat is coming up to us again. You must also train for it.

# Vol. 9, 86a. To Elsa Einstein

Benzingen, Saturday. [9 August 1919]

Dear Else,

How unsettled you are! Miza was in Rheinfelden the whole time.[1] I left the day before yesterday at midday, she arrived Thursday (the same) evening. I'm firmly resolved to stay in Berlin. I didn't manage to speak with Haber. There's time for that in Berlin.[2] Yesterday evening the priest and Inge collected me in Veringendorf by carriage.[3] The trip was exhausting, thanks to the border business.[4] Umbrella and stockings are safe. For them, I left my comb and my hair-brush in Schaffhausen at the Habichts'.[5] It's beautiful here. On Thursday I'll be departing with Fidelia[6] and Inge and by human calculation should be with you in Berlin on Friday.

Warm regards, yours,

Albert.

I didn't write the children but they'll have left by now.[7] I wrote Mrs. Hellberg[8] from Zurich as soon as I received your postcard.

# Vol. 9, 87a. From Hans Albert Einstein

[Zurich, after 15 August 1919][1]

Dear Papa,

I am very sorry that this nice holiday[2] should have ended with an annoyance. But I don't believe that you left your things here. I inquired at the boarding house,[3] and there was nothing there; it cannot be the laundry, because Bertha did the washing, and Bertha[4] also did not take it, since nothing else has ever been missing yet. Also, Bertha thinks she saw the tie while packing, and it is quite probable that everything disappeared together. I would much rather believe that it must have gotten lost during your trip, e.g., at the customs office.[5]

Mama[6] makes the suggestion that as a replacement you perhaps take a few pieces of laundry from earlier, which Mama intends to have mended. Have you already asked the southern German priest[7] whether it isn't there?

The little plane[8] is finished now with a simple motor; the double-motor is too heavy aft. The wheels are also not strong enough, and I still have to change them; and then it still has to go on a test flight. I think that the motor is not suitable for actual operation, since it already runs out just as the little plane attains its full speed. I won't see until the test flight (which I'm terrified of) exactly how things stand.

Lots of love from

Adu!

P.S. Please don't be upset about it; maybe you also left something in *Lucerne*. In any case, we would like to provide you with what is here.

# Vol. 9, 96a. From Maja Winteler-Einstein and Paul Winteler

Lucerne, 29 August 1919

My Dears,

Almost a month has already passed since Albert left, and I still haven't written you.[1] I'm just always on the go. Mama isn't feeling well at all, she's in much and severe pain.[2] Thank heavens the nights are much better than ever because of the injections, but the longer it goes on, the more Mama loses courage, because she sees how it's getting worse and worse. She never has an appetite, either, so she doesn't even have the one pleasure she could still have, the pleasure of eating. The doctor[3] thought of cutting through the nerve to eliminate the pain in the leg. But after thorough reflection, she considers the disadvantages greater than the advantages after all. The danger of bedsores is supposedly significantly increased, which cause just as great pain, and the operation at Mama's age could easily be accompanied by serious consequences; i.e., it's not life-threatening but could have a very unfavorable effect on the general state of health.

She would like Zangger to come again.[4]– Guste is having great trouble getting the entry permit.[5] But I do believe she can come now, after we made another submission with an accompanying medical certification over here. Now, however, a new problem has posed itself. The conversion rate is currently at 25!! She's going to be living with us and won't need much, but if she doesn't get a credit account in Switzerland, she won't be able to come after all. She thought Uncle Jakob should open one for her, but he is no lover of such things.[6] Could you do it, dear Albert? If yes, get in touch with Guste directly.

The Koch/Steinhardt[7] family were here in full force for an entire week. You would have gotten stomach pains again. But they acted very nicely with regard to Mama; even Alfred and Robert are each giving 50 francs per month,[8] so Mama is now abundantly provided for. But it's necessary, too, because Dr. T[obler] thought yesterday that in the winter Mama should have a private nurse. I asked her to tell me what that would cost but haven't been informed yet.

Now you know how things stand. Sadly, I can't report anything more cheerful.

Fond greetings and kisses to all of you, yours,

Maja.

Dear Albert

I hope things [arrange] themselves again bit by bit so all of you are more comfortable.

Have you already seen Steinh[ard] perhaps? To remind you, there would be 2 things to discuss with him.

1. Whether a minimum of at least 2,000 francs hadn't been planned for the position on the governing board and whether perhaps more wasn't in prospect. Raising it to equal ranking with the others was approved, and it would therefore be a question of whether it had been negotiated. (Received amount per ½ year just 850 francs.)[9]

2. You were guaranteed to receive not 10,000 but *explicitly* 11,000, but received about 5,000 per ½ yr. I complained but haven't received a reply to this day. So it will be up to you to discuss this.[10] The 1,000 francs more are [now] to be ceded to Elsa in the intended sense.[11]

We are doing very well; in this glorious weather I am, of course, going hiking, and we're often on the lake as well. On the side I'm dabbling in geology and botany, but also have many interesting things in hand in the area of jur[isprudence].

Warm greetings to all of you from your,

Paul.

# Vol. 9, 101a. From Mileva Einstein-Marić

Zurich, 10 September 1919

Dear Albert,

I'd like to ask you for a bit of information: the Bankverein informed me last winter that a certificate you had deposited here in my name—it was "Deutsche Kriegsanleihe," I believe—had reached maturity and I was asked whether they should purchase another certificate for that amount.[1] I sent you this notification at the time to ask your opinion on what should happen with it[2] but then never heard anything more. Recently I received the notification that these 2,000 marks in question here have been deposited into your account. It's certainly not pettiness that impels me to write you about this but, since this is a matter that will likely recur frequently, I would like to ask you whether some error hasn't been made here. I also discussed this with Dr. Zürcher[3] and he finds the matter not right. If this matter repeats itself often enough, the entire security deposit would come into your hands again, which would go against our agreement.[4] So I'm awaiting an explanation from you.

In addition, I would like to make a suggestion that this deposit be put into a different bank because already the accrued interest was not recorded; also, this present notification is as unclear as can be, so I always have to watch like a hawk that nothing is wrong. The Kantonalbank is generally recommended as a very reliable and decent institute, and I would like to recommend it to you as well. Dr. Z. is also of this opinion and advises me to suggest it to you. I'd like to ask you, in any case, to send me an answer as soon as possible.

Kind regards,

Miza

# Vol. 9, 128a. From Maja Winteler-Einstein

Lucerne, 9 October 1919

Dear Else, dear Albert,

Today you are receiving a letter that should prompt all of you to consider whether it wouldn't be better, after all, if Mama were transported to Berlin. Since Guste has been here, Mama feels far better, as far as her morale is concerned.[1] This comes primarily from Guste's exceedingly pleasing personality, of course; but surely also from the fact that Mama isn't so alone and can't dwell on her dismal thoughts so much. Here I'm the only one who gives her any attention (the nurse has much else to do) and I'm no cheerful lady companion. Now Miss Tobler thinks it might perhaps be best, after all, if Mama were brought to Berlin.[2] She'll write you, dear Albert, herself after she has spoken with Zangger.[3] Mama doesn't talk so much about a trip to Berlin anymore; she's hoping for next spring. But being alone so much depresses her spirit terribly. I can't possibly spend more than 2–3 hours daily with her; this already means constant hurrying and commotion for me.—So please consider whether it's possible to accommodate her with a nurse at your home or whether she couldn't be cared for in a sanatorium not far from you, whether the coal shortage isn't grave enough to guarantee Mama, who chills easily, a warm room, and whether the food supply there isn't too bad. Concerning food, Mama really doesn't need much, she eats virtually nothing. Mama doesn't know anything about this right now; the idea comes from Dr. Tobler.

Consider the matter and write us about it soon. If the transfer should take place, it would have to happen soon, before the onset of cold weather. Miss Tobler and I would accompany Mama.

Fond greetings and kisses to all of you from your

Maja.

# Vol. 9, 145a. To Elsa Einstein

Stendal, [19 October 1919]

Dear Else,

We're already stuck here in Stendal[1] because the locomotive has taken sick. How will things continue? As I opened the tin box, I thought gratefully of the two industrious ladies, who have provided for me so excellently.[2]

---

All went smoothly.[3] About half past 10 o'clock I will arrive in Utrecht, where I (unfortunately!) have to spend the night. Tomorrow morning it's onwards to Leyden.[4]

Warm greetings, yours,

Albert.

# Vol. 9, 145b. To Elsa Einstein

[Leyden,] Monday. [20 October 1919]

Dear Else,

This morning attended Lorentz's lecture.[1] Then he was here and ate with us; a great, fine fellow! I've become very friendly with Ehrenfest's children.[2] Sweet little company. In the evening, went on a walk with Ehrenfest in the dark over open fields along a large canal, starry skies. Wonderful companionship! Otherwise few people. Your letter is being taken to heart.

Fond greetings, yours,

Albert

Greetings to the children, parents, and Mrs. Hellberg.[3]

# Vol. 9, 145c. To Elsa Einstein

[Leyden,] Tuesday. [21 October 1919]

Dear Else,

Today was a quiet day. I chatted and played music with Ehrenfest.[1] Otherwise nothing important. The weather is splendid and the area here is still in magnificent autumnal dress. My condition is excellent, considering that both Ehrenfests keep a strict watch over me.[2] This evening, with some trouble, I got some forbidden—and thus that much more wonderful—tea in a glass.[3] What are all of you up to now?[4] Has an answer arrived from Switzerland?[5] I received no sign of life from you yet. Soon I'll be given the nickname "the prolific"!

Heartfelt greetings to all of you, yours,

Albert.

# Vol. 9, 148a. From Mileva Einstein-Marić

Zurich, 22 October 1919

Dear Albert,

I'd like to respond to your letter[1] with the following: I can't very well decide to pick up and leave suddenly for Germany now, because my health is still far too unstable and I fear that the move and everything that goes with it would be too much for me; and then, in a strange town where I don't know a soul or my way around, I couldn't help myself at all if I were bedridden again for a longer time.[2] This approach could lead to a grim calamity for us. Also, I am stocked up with coal here for the whole winter, which nowadays is no trifling matter, likewise with potatoes, etc.;[3] leaving all this behind and possibly having to freeze there would also be a bitter thing indeed, for you always have to keep in mind that I am simply *not healthy* yet by a long shot, and that I really can't walk much and going shopping is a very difficult thing for me.

On the other hand, I do completely understand the reasons impelling you to make this suggestion to me, and after much consideration I thought of the following procedure:

I'd very much like to spend this winter still in Zurich; I will restrict myself to the utmost and cover these bare essentials with the remaining money I have left here, which isn't very much, but I hope to be able to make ends meet. I'd like to request that you deposit the money you have designated for us[4] in my name at a German bank so that I can withdraw it in an emergency, or if I don't need it, it could stay there for later. But please do let me know roughly what sum you have determined, in order for me to know approximately what I can count on so that we don't run into any difficulties later. By the spring I hope to have recovered better, and then it will be easier for me to make decisions.

I discussed these matters with Dr. and Mrs. Zürcher,[5] whom you also have written, and they also think that this solution of the problem is probably the most reasonable. Dr. Zürcher will be writing you himself, of course. I think you will agree to this as well.

The children are as well as ever. Tete[6] started playing the piano 2 months ago and is developing much understanding and love for this enterprise; he's making very rapid progress and seems to be much more musical than one would have thought earlier. Albert[7] is very skillful and is always building something. Now he even earned a few francs through the sale of a little model of his own making.

With amicable greetings,

Miza.

## Vol. 9, 148b. To Elsa Einstein

[Leyden,] ⟨Tuesday or⟩ Wednesday ⟨?⟩ [23 October 1919]

Dear Else,

This stay is extraordinarily charming. Tomorrow we're driving out to see De Haas, who is ill.[1] Ehrenfest[2] is a very conscientious impresario; he's keeping everyone off my back this time. I play music, talk shop, and converse with him about all sorts of things. He is an incomparably inspiring person. Today at the colloquium he delivered a talk that was simply masterful. The plates of the expeditions have now been measured (definitively). My theory has been verified exactly with the greatest precision conceivable. Eddington reported it here.[3] Now no rational person can doubt the validity of my theory anymore. This evening two letters of yours arrived at once. I didn't send you a telegram upon arrival because it was already terribly late and I was lucky enough to get a room through the help of a Dutch fellow passenger. Planck is also going to Rostock; he wrote me here.[4] Many thanks to Ilse for the message; I will send condolences.[5]

Warm greetings & kisses from your

Albert.

## Vol. 9, 149a. To Elsa Einstein

[Leyden,] Friday [24 October 1919]

Dear Else,

One week is almost over already, a really wonderful and harmonious time.[1] And the glorious, rare weather of brilliant sunshine here! Yesterday I was with Ehrenfest in Delft, where we visited de Haas, who unfortunately fell quite disturbingly ill (tenacious bronchitis, had already been tubercular once).[2] There I also saw Lorentz's magnificent daughter again[3]—these Lorentzians are superhumans in the noblest sense. Then we wandered about in Delft—a very picturesque city.

Ehrenfest's lively mind and versatility are admirable. We play music together daily. Tomorrow we're going to the Academy in Amsterdam.[4] I already told you that Eddington found the theory exactly confirmed.[5]

Hearty greetings to all of you[6] from your

Albert.

My health is splendid. I don't even feel a trifle.[7] I intend to be with you again on Sunday (2nd Nov.).

# Vol. 9, 151a. To Elsa Einstein

[Leyden,] Sunday, 26 October 1919

Dear Else,

Yesterday Lorentz spoke at the Academy in my presence about the foundations of grav. theory and Eddington's results.[1] I wrote the latter from here. Ehrenfest[2] is taking touchingly good care of me; I don't even have to deliver a talk. On Sunday I'm returning home. Yesterday I walked around much with Ehrenfest in Amsterdam, also the Jewish quarter. In the evening we visited a female bachelor, prof. of botany, where it was very relaxed and hoydenish.[3] She had also studied in Zurich (& Berne). We spent today comfortably here.

Now, after your postcard, the matter with Mama is decided.[4] So we'll take it upon ourselves. You are a brave woman. But I am sorry that it will now have to be so hard for you; Maja is behaving quite strangely. She acts as if her conduct were self-explanatory. For all her erudition she doesn't do her share as I had expected of her. Planck wrote me very nicely about Rostock that we should spend the time together.[6] He's also going there. I hope my pretty card to Frau Hellberg[7] will arrive.

Heartfelt greetings to you and to all of you from your

Albert.

# Vol. 9, 152a. To Elsa Einstein

[Leyden,] Tuesday [28 October 1919]

Dear Else,

I still have nothing from you but definitely believe that it got stuck somewhere. Today we're going to Utrecht to tour an institute.[1] The day after tomorrow I'll be with Lorentz.[2] It continues to be unusually nice at our place; yesterday evening we visited Kamerlingh Onnes[3] in his study and talked shop. Everyone is unusually hospitable and friendly.

Wednesday

It was very nice in Utrecht at Prof. Julius's. He showed me his instruments and his theory.[4] After that we played music with his very musical and attractive daughters.[5] It was an unusually fine day, perhaps the finest since I came here. Today is colloquium. Tomorrow I'll be alone at Lorentz's for the whole day. But now I'm looking forward to home again. It's nice that Ilse experienced such a liberation from penury[6] and picked up the stitch again. We'll be needing it. But now, don't be too thrifty. It's going to be hard enough for you as it is.[7] It's good that Uncle J[acob][8] has promised—I hope he keeps his word in the long run.

Over here I'm being nicely spoiled[9] (so I don't get out of the habit). I didn't hold a single lecture. Freundlich published something a bit foolish again, as I gather from information given by his colleagues here.[10] One can only judge once the entire observational data have been mastered. He simply isn't reliable enough, after all. I'll be relieved once he has his position in Potsdam![11]

Greetings & a kiss from your

Albert.

# Vol. 9, 166a. To Ejnar Hertzsprung

16 November 1919

Dear Colleague,

I thank you sincerely for the effort you took to inform us of the findings of the expeditions.[1] The result is truly gratifying. But it is peculiar that the observed deviations seem to be somewhat larger than the theoretical ones. Although it only involves a tenth of a second, it does look systematic. What do you think?

Cordial regards, yours,

Einstein.

# Vol. 9, 183a. From Eduard Einstein

Zurich, 30 November 1919

Dear Papa,

How are you? I am fine, except I'm not happy that Mama's going away.[1] I'm probably going to Aegeri.[2] I'm not keen to leave, either, because we would have had a little performance at school in which I would have participated.[3] I enjoy playing the piano very much.[4] If it's possible, send me the score: "Notebook for little Magdalena Bach" by J. S. Bach. Mama said it would be nice for me. Many greetings to you from your

Teddi

# Vol. 9, 183b. From Hans Albert Einstein

[Zurich, 30 November 1919][1]

Dear Papa,

How are you? Well, I hope. How's your stomach?[2] You've asked me so often now about the little plane.[3] I never actually let it fly outside yet and I still don't have any proper wheels underneath the frame and so therefore it can't pick up speed. Furthermore, the propeller became faulty from various trials on the flat roof,

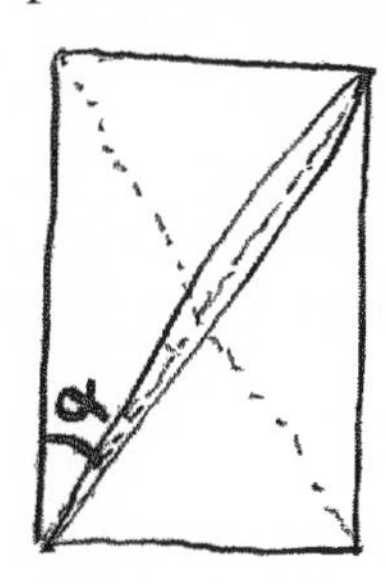


so I decided to make a new one, particularly because it seems to me that it has too little thrust because the sides don't slope enough. If the rectangle shown here on the side is the side view of the propeller with the propeller blade: could you write me how large $\angle \alpha$ has to be to have the best success for about 1,000 rotations per minute? As landing wheels I have been using with best success until now small train wheels on a knitting needle, for which I specially made extra wooden rails to run it along.

Also, I recently had the mumps. I was very heavily swollen on both sides and surely must have looked very funny; it just hurt very much and I couldn't eat anything, but now it's over. Teddi didn't get it, even though it's very contagious. Maybe he too will come to grief.

At school[4] various teachers changed, so e.g., in mathematics I now have our headmaster Amberg;[5] it's completely different with him than with the other teachers. I can't say why, but I like it a lot. In physics we have, instead of Stierlin, Mr. Seiler, "the rope" as we call him.[6] It's very funny in his class and you have to be very careful that you don't laugh out loud, because he pulls such funny faces while he's speaking and takes on such funny poses that often one of us vanishes under the bench to laugh.

Many greetings from

Adu

# Vol. 9, 183c. From Mileva Einstein-Marić

Zurich, 30 November 1919

Dear Albert,

I haven't been able to answer your letter[1] until today because I had too much to do. I'm muddling along without a maid in the house and Albert was sick. 10 days in bed, so it just became a little too much for me.[2]

I can't decide to go to Germany now; if you knew our circumstances better, you would understand. I'm still too unhealthy and weak to live in an unfamiliar place.[3]

It would have required, even in former, better times, much effort before one became accustomed to it and knew where to go for everything etc.; now it would simply be impossible for me and certainly so in Germany, where it isn't easy even for the healthy.[4]

After much thought and consultation I decided to lease out the apartment,[5] accommodate the children here, and travel to see my parents in Novi Sad, who are ill and expressed the wish that I come; so I would prefer to grant them this wish as I am the only child who can fulfill this wish for them.[6]

The apartment will cover part of the costs for the children. I would also be infinitely sorry if Albert had to interrupt school, because important though it is to save money, it is infinitely more important that at least one of my children become independent as soon as possible.[7] And changing schools would definitely mean a loss of time for Albert. I am now and later going to be very much in favor of having at least Albert stay here. I think that you will agree to this.

If there is no other way out, I would come to Germany in the spring with Tete,[8] but that won't be so simple, because I would already have to have a permit from there from the authorities before I received a passport here.

I'd like to request of you urgently again that the money meant for us be sent to me personally. For this quarter you sent us 1,000 francs and told me that you have money designated for us at Karr's.[9] But I have no business at all with Karr.

Please do send me the money earmarked for us; I won't cash it now under any circumstances, but it is proper that you give it to me. If I'm not here next quarter, then please send it to Zangger or Zürcher;[10] they will keep it for me until I'm back. I hope that you will grant this wish of mine.—You write me about an annual sum of 12,000 marks. I can't estimate what this means, of course, because I'm not familiar with the situation over there. But I ask you please, under all conditions, to have the money paid out to me and not to Karr; you have promised me, per your wish, such independence.

Please agree with all this. At the moment I can't do otherwise. If you were more familiar with the situation you would understand [this], and it is anyway all the same to you where we are, so long as the costs are reduced, and I'm willing to try. With friendly regards,

Miza

## Vol. 9, 206a. From Maja Winteler-Einstein

Lucerne, 10 December 1919

My Dears,

Are you waiting for news from us? We, in any case, are yearning to hear word from you. I imagine you are having major difficulties about vacating the room.[1]

Meanwhile, nurse Frida, because she doesn't like it here anymore, has offered her services for the same conditions as the other nurse you hired to come along to Berlin.[2] She did this completely spontaneously and we aren't *obligated* in any way to retain her. It would, of course, be convenient for Mama if she didn't have to make a change. So if you could retract the hiring of the other nurse without too many problems or without doing any kind of injustice, she would continue to care for Mama. She's coming along to Berlin anyway, also at her specific wish, because she would like to visit an old lady of whom she's very fond. And for Dr. Tobler[3] and me it is naturally very pleasant to have her assistance during the journey, not to mention Mama, who would find it a great reassurance to have her familiar nurse during the transfer. Additional expenses would not be incurred by it because we have to buy 12 tickets anyway. But there is also the question of whether the German authorities will grant the residence permit. I have already obtained an entry visa for her without difficulty.

Dear Albert, now you are downright popular. Recently a local teacher at the Cantonal School was asked by a 13-year old boy what the Einsteinian theory was. And in a Lucerne paper (!) an article was published about you. I imagine this causes you much unpleasantness, now that so much is being written about you. I was heartily pleased at the time about the confirmation of your theory by the English expeditions[4] and picked up your "children's book" again.[5] But the "general" one still remains very obscure to me and I would be much happier if I could understand the importance of your ideas and conclusions properly. Mama is having a difficult time; she is very nice and patient nevertheless, much more than when she was feeling better.

Fond regards to all of you from your

Maja.

# Vol. 9, 206b. From Paul Winteler

Lucerne, 10 December 1919

Dear Albert,

It turns out that Dr. Curti[1] made available to me at the beginning of November inst. the earnings including [credit?] coupons of 27,500 francs[2] instead of just half of it, because the fiscal year lasts from 1 July–30 June and per 1918/19 just half should have been paid out corresponding to the 6 months January 1919–30 June 1919, hence 13,750 francs. Dr. Curti writes me that he had been mistaken and so I'm going to send the 13,750 francs back to him. The occurrence of this error had initially been unclear to me, but then I thought it comes to the same thing in the end; the remainder, after deduction of our credits, would also arrive in the right hands through you as well.

So after letting the 13,750 francs find their way back to Dr. Curti, the following balance results:

Receipts: Fr 13,750.–
To consortium member Einstein:
5,500 [Fr 500 to me per written agreement, whereof I was to pay out to Elsa 750.– *marks*, which has happened.][3]
to Koch[4] 937.50
to Winteler[5] 937.50
7,375.–
remainder (to be delivered to K[oppel][6]) 6,375.–

Now I should soon have the definite instruction of where and how K. would like to receive the remainder, whether through you directly, but am still without any reply. Would you therefore, since I have winter holidays at the end of the month, like to report to me beforehand, whenever possible, pertaining to the amounts so that I can fulfill my duties as representative of the consortium?

– What Guste Hochberger wrote you about us was surely well meant but does not describe matters properly.[7] I don't understand why she writes about things that Maja can best tell you directly soon and, as I believe, also objectively. We have been living very harmoniously, as before; it was, in my opinion, always just our dear friends who, with the best of intentions, put sour mushrooms in the wine, not least but rather first and foremost, Dr. Häfliger, who was, of course, there to see you.[8] Now Guste Hochberger is starting with it as well. The only truly reasonable woman was Azzolini.[9]

Cordial regards, yours,

Pauli

# Vol. 9, 217a. To Erwin Freundlich

Thursday [after 15 December 1919][1]

Dear Mr. Freundlich,

Upon reading your article a few pressing questions occurred to me. We have probably already touched upon all of them, but I would like to have the answers in one place:[2]

1) How did the radii of the B-stars result from occultations?[3]

2) Since the parallax of the Orion nebulus seems to be properly known, one could determine from the data[4]

Apparent brightness
surf. temperature
and parallax
the radius of each of the individual stars.[5] How do they come out?

3) After determining the radii by methods (1) and (2), how large are the resulting real masses from these and from the measured gravitational effects?

4) Do the thus obtained masses agree, *within the margins of error*, with those obtained from Newton's law and Doppler's principle?[6]

---

Regarding your paper, also the following. The determination of the parallax does seem to me to be based on shadow-boxing, because the gravitational effect does not seem to enter into the magnitude $\left(\frac{M}{d}\right)^{2/3}$ at all, but only the *radii* determined by the occultations. $\frac{M}{d}$ is simply a volume, and therefore has virtually nothing to do with the gravitational effect.[7] Furthermore, I suspect that your assumed value $\frac{1}{10}$ solar density is an inconsistent assumption; can this "conventional" assumption be upheld independently of the "conventional" opinions about the masses of B-stars? Am I not sitting on the branch I myself have sawn off?[8] You must definitely enlighten me about this.[9]

Best regards from your,

A. Einstein.

# Vol. 9, 239a. From Paul Winteler

31 December 1919

Dear Albert,

I received your letter, according to which I was supposed to settle accounts with SAG,[1] and paid the amount of about 4,937 francs, which was due according to the statement of current account submitted to me by the SAG. I transferred the remainder of about 63 francs to the Zürcher Kantonalbank under your name.

I had hesitated with these transactions, and when no further report arrived I never carried them out. Now I still have the amount outstanding for Ko[ppel][2] here under my name, namely 6,400 francs *at K's disposal*. Since the repayment to SAG, as mentioned above, has already taken place, I would think you could get in touch with K. to have him cede a part of his credit as advance payment for your future earnings from the SAG.

I have already had to use a very small amount of it anyway, Fr 260 *in advance payment* for Mama's trip,[3] because as a consequence of Uncle Jakob's stupid behavior, Mama's trip would have been completely impossible for lack of money right to the last moment.[4] It is anticipated that this amount of Fr 260 would be accessible again, because part of it had to be advanced merely as a deposit fee (200 francs). So much for these money matters; so I await your instructions. If I hadn't advanced in total Fr 1,060 from foreign and private funds, Mama's trip would have become completely impossible. I then balanced accounts with Jakob on Dec. 27th and he then credited back to me Fr 800, so the Fr 260 still originate from the advances.

My opinion now is that you should be in demand as little as possible during Mama's stay. Uncle Jakob should simply keep his promise and Ogden, Alice, Robert, and Alfred theirs.[5] The [stupid thing] is that they have assigned Jakob as their representative. For Jakob always acts as if everything came out of his own pocket; also, the accuracy of what he says cannot be relied upon at all; he doesn't care if there's one lie more or less. On the other hand, I did manage to arrange relatively easily that he at least return the Fr 800 to me. Best would be if you didn't give Uncle Jakob any slack out of exaggerated generosity; given his entire disposition, he doesn't deserve it in the least. It would, however, only be right and good if the relatives who have some money contribute. Didn't Jakob likewise depend on the relatives in the past when the business in Genova went downhill?[6]

What you observe of Mama is the effect of the drugs and the long illness. I can imagine that you are suffering severely under the circumstances, and I would have liked to see you be able to pursue your work undisturbed, and that was why I was actually against Mama's trip to Berlin. But now that it has taken place, I hope you make do with the state of affairs as best you can.

Cordial greetings to all, yours,

Pauli

## Vol. 9, 240a. From Hans Albert Einstein

[after 1 January 1920]

Dear Papa,

I'm terribly sorry that I didn't write earlier but maybe you'll understand later why not. How are you, by the way? Are you coming to Basle or not?[1] Until about a week before Christmas I didn't write you simply due to laziness. I thought during the 3 weeks vacation it would surely have come about sometime. But this vacation was much less vacation than I had thought. The packing started right at the beginning, even though no one had been found for the apartment yet.[2] We didn't actually pack very much, but we toiled away; that's how it went until late in the day on the 29th of Dec. '19. Then 2 American ladies suddenly came by and had a look at the apartment. It appealed to them, they took it: for the 1st of Jan. '20. I tell you; that was a hurry and scurry, back and forth. Albert upstairs, Albert downstairs; in short, it was the first time that I was tired to the bone. On the 1st of Jan. '20, 4 o'clock in the afternoon, the ladies appeared in full train—grandmother, mother, and child, as the verse goes,[3] and, very satisfied and happy, we shoved Mama into the hotel, Teddi to Unteraegeri to Dr. Bosshard, and me to the Zanggers', and now everything's still the same except that Mama is in Novi Sad.[4] Just today I conveyed my greetings to Prof. Amberg.[5] I hope you forgive me now for the long delay in writing and will soon send news of yourself.

Many greetings from your

*Adu*

P.S. My pen is "spluttering" [*sprutzt*][6] horribly. I think you'll understand the expression.

# Vol. 9, 288a. From Hans Albert Einstein

8 Zürichberg St., Zurich, 28 January 1920

Dear Papa,

I was very happy to hear from you again. So that you don't have to agonize over my scribbles again, I thought I'd write with the typewriter. Because you ask what's happening with the piano, I'd like to ask you please to send me the Mozart sonatas for violin & piano; you have them in duplicate, you know.[1] I'd like to play them with my chum, Häusler, you see. Otherwise I play the Beethoven sonatas, also Mozart for piano, a little bit of Brahms & Schubert.[2] Häusler & I want to try Handel sonatas. I'd be very grateful if you would have the Mozart sonatas sent to me.

In mathematics, with Prof. Amberg,[3] we last solved problems with quadratic equations. In geometry, i.e., stereometry, we had very little at first, but that little bit was very interesting & nice, e.g., to prove that the sum of the corners plus the sum of the surfaces was equal to the sum of the edges plus 2. In trigonometry we're deriving formulas for dear life. I still have to relay greetings from Prof. Amberg, by the way; so I already relayed yours.

Besides school I've been reading plays by Shaw. I like it very much because he acquaints you with the individual characters very soon and through & through. E.g., you get to know Bluntschli in *Arms and the Man*, or Caesar or Napoleon much better than in history, for example.[4]

In hope of hearing from you again soon, I send you many greetings, yours,

*ADU*

# Vol. 9, 328b. From Eduard Einstein

Unteraegeri, 25 February 1920

Dear Papa,

We'd like to be together again in Zurich in spring.[1] Mama is coming back again soon.[2]

Many greetings and kisses from your

Tete

See you soon!

# Vol. 9, 351a. From Hans Albert Einstein

Aegeri, 14 March 1920

Dear Papa,

I'm here with Teddy right now in Aegeri, and now we're writing you together.[1] Meanwhile I've had a little bit of the flu but am completely well again. Mama is coming back soon.[2]

From the 1st of April we can go back into the apartment.[3] At the moment I'm staying with Besso, because Prof. Zangger is on the Riviera.[4]

I have various questions to ask you again about flying machines, etc.[5] Also please send me your plans for the summer holidays.

Many greetings from

Adu

# Vol. 9, 351b. From Eduard Einstein

[Aegeri, 14 March 1920][1]

Dear Papa,

I'm feeling well, but I really want to go home.[2] How are you? The weather is dreadful.

Many greetings from your

Tete

# 1. To Ludwik Silberstein[1]

[Berlin,] 1 May 1920

Esteemed Colleague,

I live permanently in Berlin.[2] The Stokes-Planck ether leads to a hopeless heap of independent hypotheses. Such an unfinished theory is, of course, intrinsically irrefutable.[3]

The problem of connected masses can be treated to first approximation according to Newton, of course;[4] Newton's theory is a first approximation, you know. A more precise calculation is only permitted on the basis of continuum mechanics (relativistic elasticity theory). For the stress-energy tensor of the connection has an influence on the motion and does not allow one to regard the connection as merely kinematical. I find, though, that a rigorous treatment of this problem would not be worthwhile in any way.

With kind regards,

A. Einstein.

# 2. From Paul Ehrenfest

[Leyden,] 1 May 1920

Dear Einstein!

As I already wrote you two–3 weeks ago, Onnes sought to accelerate your travel permit *directly*.[1]—He requested that it be arranged such that you obtain the visa *directly* upon inquiry at the Dutch Consulate in Berlin.—[2] *Did you ask at the Consulate* whether such an authorization had been issued from The Hague? At the *Embassy* perhaps?![3] I immediately telephoned Onnes and he, in turn, will immediately urge them on at The Hague and again ask that they give the Consulate authorization by telegram.—If upon receipt of this postcard the matter is still not settled, then please telegraph me again.—Everything is so terribly slow and sluggish.—One has to wait endlessly for everything.[4] We are all expecting you with great impatience.—The flowers will be blooming just when you come.—Julius too is very anxious to see you.—[5] If the train connection is so unfavorable again—then spend the night at Julius's in Utrecht.[6]— Do let me know sometime at what time you are arriving in Utrecht!!! Greetings to all, yours,

P.E.

# 3. From Gottlieb Haberlandt[1]

Dahlem, Berlin, 1 Königin Luise St., 1 May 1920

Esteemed Colleague,

Upon returning home from a many-day excursion, I discovered your friendly letter, in which you state your willingness to sign if other Socialists also give their signatures.[2] This condition is already satisfied, in that Prof. Cunow of Berlin is among the signers of the draft appeal, which I am sending out to you again herewith.[3] Our Colleague Meinecke, with whom I discussed this,[4] informed me that other Socialists will certainly receive the appeal for signatures as well. That is the business of the men responsible for the individual universities. In any event, it is far from the minds of the appeal organizers to wish to exclude any Socialists who firmly back our Constitution.

Thus I believe that you can safely sign the appeal without the least sacrifice to your political stance and convictions. Your name would be the more welcome to us since the number of colleagues in complete agreement with the content of the appeal but who, for opportunistic reasons, are unwilling to sign is unfortunately larger than we initially thought.

Amicable greetings from your devoted colleague,

G. Haberlandt.[5]

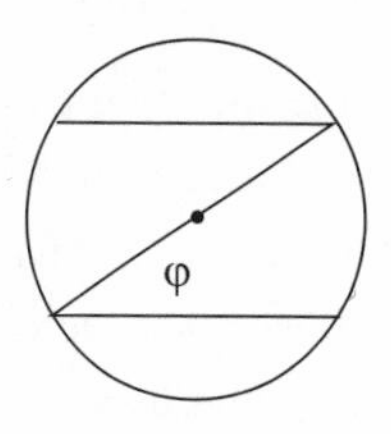


$$mv \cdot 2\pi r = n_1 h$$

$$mv \cdot 2\pi r \cos\varphi = n_2 h$$

$\varphi$ between 0 and $\pi$ $\quad |n_2| \leq |n_1|$

$n_1 = 0 \quad n_2 = 0 \quad \frac{n_2}{n_1} = \cos\varphi$

$n_1 = 1$ — $n_2 = 1$, $n_2 = 0$

$n_2 = -1$

$n_1 = 2$

$n_1 = 2 \quad$ 2, 1, 0, −1, −2 $\quad 2n + 1$

# 4. To Niels Bohr[1]

[Berlin,] 2 May 1920

Dear Mr. Bohr,

The splendid gift from the Land of Neutrality, where milk and honey still trickle, gives me a welcome opportunity to write you.[2] My cordial thanks to you. Not often in life has the mere presence of a person given me such pleasure as you.[3] I understand now why Ehrenfest loves you so much.[4] I am studying your great papers now and—when I happen to get stuck somewhere—have the pleasure of seeing your kindly, boyish face before me, smiling and explaining. I have learned much from you, especially how you feel about scientific matters.[5]

The way in which you derive individual quantum states successively with the aid of quantum states ("by means of the Riemann surface") remains somewhat opaque to me. For it seems to me that the *reversal* of the process that yields state $I_2 = h$ from state $I_1 = 2h$, generates from state $I_1 = h$ state $I = \frac{h}{2}$, which clearly is not supposed to be a quantum state. Somewhere one passes through a discontinuous change in integration time, if I have understood you correctly.[6]

I am very much looking forward to our discussions in Copenhagen.[7] In the meantime, wishing you a happy time and blessed efforts, yours,

A. Einstein.

# 5. To Hans Wittig

Berlin, 3 May 1920

Dear Sir,

I am very willing to check your work. Provided I find it clear and to the point, I shall be glad to accept your dedication,[1] although I would like to note from the start that I do not consider myself competent in philosophical matters. The pompous vocabulary and revolutionary gestures in your letter make me suspicious; Truth tends to appear modestly and in simple dress. As concerns your intention to attempt a doctorate, I don't really know, since I personally am not a member of any faculty[2] and am just better acquainted with the Swiss situation. But this much I do know, that in these affairs officialism blossoms and thrives just as it did under the old regime.[3] But that is tomorrow's worry [*cura posterior*]; the main thing is, of course, the work itself. If it is interesting, with God's help a full professor in philosophy will be found who will help you to get your doctorate. In the meantime,, just aword of advice: use a sober style, also as concerns the theory of relativity; there is already too much quite repulsive sinning in this regard.

In great respect,

Deleted fragment in draft: "That you, too, are not quite free of this is proved, incidentally, by the circum. that you [make use of] the silly doctor's title . . ."

# 6. To Paul Ehrenfest

[Berlin,] 4 May 1920

Dear, dear Ehrenfest,

I am still sitting here, while my ether sermon is surely already in your possession.[1] Every day, though, the Dutch Pythia at the consulate shakes her little head and says with sweet regret, "Still nothing."[2] She can't be blamed, of course, for the sluggish red tape. It's silly that I promised to be in Halle on May 29th, where a relativistic/philosophical synod is taking place, for which the holy limbs are required.[3] And for the beginning of June, I promised to preach in Christiania.[4] Healthwise I am feeling so well that this time you can really pile it on me. I would like to do something to prove myself worthy of the Dutch professorship, but am clearly conscious of the superfluousness of what I could lecture on.[5] Should I perhaps come only in the second half of June instead of now? If so, then send a telegram. I am going to bring one violin along. It's magnificent.[6] Herglotz is not coming; he declined the professorship here.[7] Soon the local chairs will be peddled like old clothes at the secondhand dealer's—times change [*tempora mutantur*].[8]

Bohr was here and I am in love with him as much as you are.[9] He is like a highly sensitive child and wanders about in this world in a kind of trance.

# 7. To Elsa Einstein

Leyden, Friday, 7 May 1920

Dear Else,

Arrived here well in splendid weather. Both Ehrenfests are in town; I came completely unexpectedly.[1] This morning I talked shop with Julius at the institute in Utrecht.[2] He's a real "monomaniac," but still a fine fellow. So the violin is stuck in Bentheim.[3] Provided you haven't done anything with this matter yet, you should best do as follows:

1) Go to the Reich Commissariat for Export Permits and apply for the permit to export the violin lying in storage at Bentheim.

2) Send the export permit to Gerlach & Co., Bentheim,[4] with the instructions to send the violin here to me and to charge all expenses cash on delivery.

The journey proceeded excellently. I made the acquaintance of a few young men who are emigrating to South America, former officers. The trip third class was excellent. I did not need the cushion. Little Paul Ehrenfest is playing with me, such a cute fellow, but he won't let me write.[5]

So, for today, just heartfelt greetings from your

Albert

Also to the children.[6] Liberate the poor violin soon!

# 8. From Willem H. Julius

[Utrecht,] 8 May 1920

Dear Colleague Einstein,

Permit me, in connection with yesterday's conversation,[1] to summarize briefly once again the main points of our thoughts on line displacements;[2] this could perhaps be useful in some way toward solving the problem.

The conclusion that Grebe and Bachem reach did not appear to us sufficiently supported;[3] since, first, the mean 0.56 km/sec is noticeably smaller than the theoretical value 0.63 and even drops down to 0.475 if the *one* line 3866.960 is omitted (which in the earlier paper had been described as unusable);[4] and second, it was not shown beyond doubt that the observed shifts did not furnish the result of a superposition of many causes. Possible photographic effects and evaluation errors have been taken into consideration by G. and B.; however, pressure and Doppler effects or anomalous dispersion still could distort the gravitational shift even of a line standing by itself.

There can be no doubt that various causes of the same order of magnitude play a part. Surely, the considerable limb-center shifts (0.0068 Å as the mean for 476 lines)[5] are interpretable as a gravitational shift only with difficulty;[6] or as a pressure effect (as Evershed has adequately shown);[7] but even less likely as a consequence of a repulsive force that the Earth specifically, or the British Empire, were exerting upon the solar gases.[8] And the fact that the shifts are so *very* different from line to line (to the degree that well-established violet shifts occur) without showing any relation to the experimental pressure shifts, whereas, if one takes a *mean,* a simple relation to *line intensity* results as predicted by dispersion theory—this fact seems inexplicable unless one conceives Fraunhofer[9] lines essentially as dispersion lines.

The following reasons, in brief, speak additionally for this interpretation: (1) the generally small width of the Fr. lines and the great similarity between spectra from the center and limb of the solar disk; (2) the type of relation between Fraunhofer lines in the limb spectrum and bright lines of the chromospheric spectrum; (3) regularities in the line distortions in solar-spot spectra, generated by a radially positioned slit, which phenomena are only very unsatisfactorily explained by means of the Doppler principle as a consequence of radial emissions, but on the contrary, quite easily by the effect of refraction; and (4) the influence that total limb-center shifts experience by the presence of neighboring lines. A neighbor on the red side reduces the redshift, a neighbor on the violet side magnifies the redshift, namely, ceteris paribus, the first effect is larger than the second, exactly as required by dispersion theory.[10] (Among 656 lines measured by Adams, Evershed, and Royds,[11] we found 44 lines with red companions, 49 with violet ones. The former gave as the mean shift 0.0044 Å, the latter 0.0077 Å, while the mean value for all 656 lines was 0.0063.) For solar-center-arc shifts, the same seems to apply, as emerges from the detailed paper by Albrecht (*Astroph. Journ. 41,* 333 [1915] and *44,* 1 [1916]).[12] Royds, who contests Albrecht's results (*Kodaik. Bull. 48*),[13] nonetheless himself finds that 17 lines with red companions furnish a redshift ($\odot$-arc) of only 0.0032 Å; by contrast, 30 lines with violet companions yield such a shift of 0.0079 Å. He does not provide the total mean of all lines he had taken into account.

[St. John, Evershed, and Larmor took pains to show that no mutual influence of adjacent lines exists and, theoretically (Larmor), is not expected to any noticeable degree;[14] however, I believe I can attribute the reason for their statements to an incorrect conception of the nature of dispersion lines.– We at the laboratory are now occupied with the problem of thoroughly analyzing, visually as well as microphotometrically, closely lying line pairs, using artificial lines.][15]

Thus, if it can be shown that spectrum lines actually do exhibit noticeable displacements by anomalous dispersion in the Sun, both toward the violet as well as toward the red, it ought to be possible to explain the lack of a gravitational shift of the relevant magnitude in many lines by a violet shift due to dispersion, which partially conceals the gravitational effect in precisely these lines.

That is why I hoped it would be feasible to interpret all the material on the measured redshifts quite completely as a superpositioning of two main effects: a gravitational displacement—increasing proportionately with wavelength—of the eigenfrequencies and a—strongly fluctuating—displacement of the enveloping dispersion lines.

The following data regarding ⊙-arc shifts have been available to us to date: The tables by Grebe and Bachem (June '19), Evershed (*Kod. Bull.* 36), Royds (*Kod. Bull.* 38), Evershed and Royds (*Kod. Bull.* 39),[16] in total 446 mean values for individual lines. We divided the spectrum into three parts, each covering 800 Å, and took the mean $\delta$ for the shift in each part. Against that we set the theoretical gravitational shift $\delta' = \Delta\lambda = \lambda \times 2 \times 10^{-6}$ for the mean wavelength of each part. The differences $\delta - \delta'$ would then have to represent the average dispersion shifts (possibly mixed with pressure effects, etc.).

| | $\lambda$ | Number of Lines | Observed $\delta$ | Gravit. shift $\delta'$ | $\delta - \delta'$ s |
|---|---|---|---|---|---|
| I | 3650–4450 mean 4050 | 287 | 0.0050 | 0.0081 | –0.0031 |
| II | 4450–5250 mean 4850 | 118 | 0.0042 | 0.0097 | –0.0055 |
| III | 5250–6050 mean 5650 | 41 | 0.0065 | 0.0113 | –0.0048 |

All three differences are undoubtedly negative, i.e., as a result of anomalous dispersion more violet shifts must occur than red throughout the whole spectrum. (Such large *negative* mean shifts could not be attributable to the pressure effect.)

Now, that may not be unthinkable on its own; it would mean that $n_0 < 1$ would apply in many places in the solar spectrum—although it would be a somewhat ticklish affair to have to set the mean refractive index of the solar gases lower than 1.

But even if we disregard this reservation, it still looks bad. For dispersion effects must, in general, increase from the center outward to the limb, the violet shifts just as the red. Therefore, the majority of the limb-center shifts should be violet shifts as well. (This consequence could be avoided with the aid of the Doppler principle only if one accepted the nice hypothesis of a selective repulsion by the Earth of solar gases.)[17]

Observation instructs us, however, that the limb-center shifts are all quite decidedly toward the red (with two or three exceptions among 476 cases), and to be precise, according to Adams,[18] exhibit the mean value 0.0068 Å, which is even larger than the mean ⊙-arc shift.[19] (That the center-arc shifts show much greater fluctuation than the limb-center shifts probably stems from the uncertainty regarding many of the arc-line observations.)[20]

(Hence, if we assume the existence of a gravitational shift, we arrive at conclusions that are in contradiction to experience so far.)[21] On the other hand, it seems to be possible to explain consistently the whole available observational data concerning the displacements of Fraunhofer lines, if one assumes that there is no gravitational shift.

But I still do hope that you will find a solution to the problem that retains the magnificent General Theory of Relativity.[22]

Most cordial regards to you, as well as to our colleague Ehrenfest and his esteemed wife, also in the name of my family.[23] You will, I hope, do us the great favor of coming again, as last year on 28 October,[24] to visit us and make music? For the sake of convenience, I note down here the days of this month of which we have not yet otherwise availed ourselves and on which you would consequently be most welcome: 11, 15, 17, 18, 21, 22, 23, 25, 27, ⟨28⟩, 30, 31.

If there is any desire for discussion, I shall be glad to come to Leyden sometime as well.

Yours sincerely,

W. H. Julius.

Recipient's note: "Solar sp[ots] perceived [with] indiff. light."

## 9. To Elsa Einstein

[Leyden,] Sunday. [9 May 1920]

Dear Else,

This time even Ehrenfest says I'm somewhat shabby, à la Berlin; for the festive dress coat is a bit moth-eaten.[1] We're very merry here, even without the violin. I'm curious what the further developments with it will be. You'll carry it through, all right.[2] My appointment is encountering difficulties, thanks to my ticklish polit. renown. But it's sure to come.[3] The peace here is wonderful, likewise the conversations with Ehrenfest and the games with his children.[4] Ehrenfest plays magnificent things by Bach for me. Yesterday I was with Kamerlingh-Onnes at the institute and heard a nice talk by him; saw interesting experiments.[5] And the cigars!! In this respect it's good that you're not here. It's raining hard and is cold. There too? The children [6] aren't letting me write anymore. It's rollicking fun.

Warm regards to all of you, yours,

Albert.

How was the wedding? I was lucky.

# 10. From Elsa Einstein

[after 9 May 1920][1]

My dear, darling Albert,

I'm at the export office daily; that's a pretty tale to tell. Applying now for the permit for *one* violin and petitioning for an export permit for *another* violin a short time afterwards again is not allowed.[2] One can't get a permit so often. It looks as if we are in the musical instruments trade; then, misfortune would have it that Margot[3] is held up over there still, too, with the exportation of her lute. Now the violins have to be registered and transported separately, because one of them is being sent from Bentheim,[4] the other from Berlin. A lovely law came into effect three days ago. 10% of the price of the object is to be paid as export tax.[5] So I'll value the violin at 1,000 marks; I'll put everything into having the one in Bentheim go through as is, since it's been in storage there for 10 days already, before the law came into force. For the one here I'm paying 100 marks, postage and crate cost around 30 marks. Insurance for the violin around 15 marks, therefore we have another 145 marks in addition to the total I indicated recently in my postcard. I'll cover it for the time being.– What does Planck want of you? And Haber? It makes me uneasy; why are they writing?[6] So your socialist disposition is being held against you everywhere![7] Even in Holland![8] Do me a favor and don't act like such a furious Socialist; you're not one any more so than Ehrenfest[9] and many others! Please finally put an end to this stupid talk; at last, you're regarded everywhere as a raging revolutionary, even in England.[10] You'll harm your public image more than you can appreciate. If this inquiry were true and well-founded, then I wouldn't have anything against it! But this way it's nonsense, enough to get fed up with. I know perfectly well that in England, at least, they use your name somewhat nervously. It's bad enough that you aren't getting the Nobel prize because of it; it shouldn't go any further.[11] A critical mind like you is not a communist!

Since your departure the weather has been ideal, never rainy, just sunshine and clear skies. Now it's beautiful even in Berlin in the Bavarian quarter, the puny little front gardens between the rental blocks are [bursting] into beautiful blooms and so it's possible to see some finery even on our street.[12] Everywhere gold rain [Laburnum], lilacs, and flowering almond.– I'm so glad that you're having a good time. And I'm looking forward to your coming back. But I'm also relaxing during your absence. I'm so often outside and frequently sit out on your balcony. What a pity for the fine asparagus that can't be eaten by you. I feel so sorry every time I'm peeling it, wanting to give it to you, good, dear fellow. You should use the suit that everyone thinks is so shabby *just for traveling, never wear it in Holland*, do you hear? I beg you earnestly. Don't make yourself ridiculous in your dress coat, which is good for train travel but not for anything else.[13] Change your socks regularly, otherwise they get too large holes. And give a shirt and a nightshirt to the laundry now; you took hardly enough along with you.

Take care and all my love, your

Wife.

## 11. From Ernst Cassirer[1]

Hamburg, 26 Blumen Street, 10 May 1920

Highly esteemed Colleague,

Please accept my cordial thanks for your kind willingness to glance briefly through my manuscript now, while you are still traveling.[2] I am having the manuscript sent to you today and hope that it reaches you safely: no confirmation of receipt is expected, but I would request a brief note only if, contrary to expectation, it does not arrive.[3] As far as the content of my text is concerned, it evidently does not propose to list all philosophical problems contained in the theory of relativity, let alone to solve them. I just wanted to try to stimulate general philosophical discussion and to open the flow of arguments and, if possible, to define a specific methodological direction. Above all, I would wish, as it were, to confront physicists and philosophers with the problems of relativity theory and bring about agreement between them. That in doing so I took pains to make wide-ranging use of the physics literature and to learn from the writings of great physicists of the past and present—this you will gather from my exposition. But with the various conceptual mind sets and different languages that physicist and philosopher speak, even the best of intentions does not always suffice to avoid misunderstandings. So here your verdict would be of exceptional value to me: I remain open to your criticism and instruction, all the more so since while writing my work I was not thinking of publication at all, initially, but rather only undertook it because I felt a growing inner need to arrive at deeper clarification of these questions. Wherever your judgment may fall, this clarification will be substantially advanced by it, either way.

I am, in expressing my thanks and utmost respect, yours very truly,

Ernst Cassirer.

# 12. From Moritz Schlick

Rostock, 23 Orléans St., 10 May 1920

Dear, esteemed Professor,

This morning I received a few copies of the English edition of *Space and Time.*[1] The translator presumably already sent you a proof or is about to do so;[2] should this not be the case, however, please do be so kind as to let me know—I would then dearly like to send you one of my copies. Owing to the peacetime supplies, the translation presents itself much more nicely than the original.[3]

Through a slight indiscretion I learned that an inquiry about my character has reached here from Giessen. It evidently involves the chair that Medicus was supposed to take.[4] I don't believe there is much chance of my being favorably placed among the list of applicants. But in any event we can draw from this that Medicus must have finally decided to stay in Zurich.[5] It must be a fine situation there now, after Debye accepted the call to Zurich.[6]

Here the summer term has started with promising student attendance, but the faculty is being shaken by convulsions, a consequence of the putsch in March during which some apparently "Kapp"-itated themselves.[7]

In sending my compliments to your wife, I remain with most cordial greetings and best wishes for your health, also from my wife and children, gratefully yours,

M. Schlick

# 13. To Elsa Einstein

[Kathijk?] Tuesday [11 May 1920]

Dear Else,

It's splendid weather here now. We're outside a lot, talking shop. It is very interesting with Ehrenfest.[1] Yesterday was Lorentz's lecture.[2] Today it's little Paul Ehrenfest's birthday. We festively went shopping. Last night at Kamerlingh-Onnes's.[3] The gentlemen are bending over backwards to accelerate my nomination in every way.[4] I'm in no hurry. I'd wish all of you such a nice little stay as well! This scribbling is so illegible because I'm lying in the sand[5] as I write.

Warm greetings to all of you, yours,

Albert.

# 14. From Mileva Einstein-Marić

[14 May 1920][1]

Dear Albert,

The bank informed me that a few more securities on deposit here are soon being paid out, and they would like to know what they should do with the money.[2] They suggest buying mortgage bonds on real estate because everything else is insecure (if one isn't precisely familiar with the business). Please write me as soon as possible what you think of this or make another suggestion. I ask you please to write me about this and not simply sign the sum over to your name like with the 2,000 marks last year;[3] you won't eventually want to have this present to your children back again as well, I hope?–

The boys are doing excellently well; Albert is 16 years old today. He's turned into a dear and jolly fellow. Tete wrote his first essay today with much diligence and enthusiasm.[4] He feels a little left out that you always write to Albert and not to him.–

Kind regards,

Miza.

# 15. From Hans Albert Einstein

[14 May 1920][1]

Dear Papa,

I haven't written for so long because first it was vacation and then I was in Ticino with the school.[2] We lived a very soldierlike existence and needed very little money. It was very nice and we saw quite a lot. It was very funny how we got by with German and French; it worked out quite well. Then school started, and so there wasn't much time because you first have to get used to it again. But that's over with and so I'm writing you now. Today happens to be my birthday.[3] Mama's[4] home-baked cake was excellent; what a pity you couldn't try it, and potato salad upon special request. Yum!–

Thank you very much for the photograph and the nice scores.[5] I've been accompanying violin more, recently, and playing 4 hands with Mama, so I had less opportunity to play 2 hands. But I did have a look at the things.

Then I have another request of you, or better said, of your wastepaper basket! By this I mean no more and no less than some stamps, especially those that didn't exist before, like those from Weimar or the new Bavarians or any war or peace stamps you have. I'm writing this less for myself than for Teddy,[6] who pounced on my stamp album.

You probably still remember the "double-rubberband plan" for my little plane with the ratchet wheels aft.[7] I tried it out but the ratchet wheels I obtained didn't do their job; there's too much friction. Would you happen to know of any other solution, for I haven't given up on the whole plan with the double-rubberband yet.

I'm going to have the little plane photographed soon and then take it on a test flight. That'll be an expedition, all right!

Despite all the improvements the catch is still in the "legs." On the one hand, the wheels don't turn freely enough, on the other hand it's not elastic enough.

I'm supposed to tell you also that we are anxiously awaiting the money that's due now.[8]

How are you doing otherwise? Write me sometime soon.

Many greetings from your

*Adu.*

P.S. You have had the "English Course" by Langenscheith. Could I have it for at least some time? I could use it very well in addition to class.

*Adu.*

## 16. From Max Wertheimer[1]

[Berlin,] 15 May 1920

Dear Mr. Einstein,

Hearty greetings! (–How much I would prefer to be able to say good day to you once in person again–!) I wish you very enjoyable days indeed in Holland![2]

Today I have to write a few words about the—peculiar—Halle matter. Dear, esteemed Mr. Einstein—well, how strange people are–! And what have you, in your boundless good nature, let yourself in for with these people?![3]

In April, in Prague, I heard about the "upcoming *important congress about Einstein, where Prof. Kraus*[4] *(!) has been assigned the leading role,* which is now (finally) going to uncover in public before a philosophical tribunal the elementary absurdities of Einst[einian] theories, so that thenceforth it will be clear, how –."

Here I find waiting an invitation by the Kant Soc[iety]: Mr. Einstein, too, will be in Halle! Then, a publication, a supplement of the *Kantstudien*, from someone who, on the other hand, seems to me to share the tendency characteristic of Kant circles.[5] *They* surely want to present themselves roughly as follows: ingratiatingly, interesting ideas—which, however, cannot collide in the least with the genuine philosophical problems (tendency of the somewhat fearful limitation: "everything in Kant legitimately stands" and—where possible—exploit what you can–)

Now, i.a., in the paper *Vossische Zeitung:* "In connection with the plenary session of the Kant Society . . . a number of prominent German scholars, i.a., *Einstein, Abderhalden, Kraus!* (*Prague*), *Vaihinger,* extend an invitation "*to a scientif. discussion about Positivistic Idealism along the lines of the 'Philosophy of As If'(!)*" in Halle—Prof. Einstein will be attending the proceedings."—[6] And I receive a printed invitation: "A number of members of the K.S., who are interested in Positivistic Idealism along the lines of the Philosophy of As If . . . The *signers,* in part members, in part nonmem. of the Kant Soc., herewith invite you to this discussion . . . Members of the Kant S. also have the right to participate *in the As If Conference–*" *Signed:* Abderhalden, Becker, Bergmann, Einstein!, Feldkeller, Fliess, [Gocht], Knopf, [Koffka], Kowalewski, Kraus (Prague),—[6] Müller-Freienfels, R. Schmidt, [J.] Schultz, Vaihinger, Wichmann, H. Wolff.–!![8]

Good grief! Into what sort of publicity campaign are you being recruited?! Would physicists of commensurate caliber dare to do such a thing?! For the greatest part, feeble-minded, languidly regurgitating, squabbling mediocrity and in part like Kraus: insolent; and with such an obsession for publicity—Well, for God's sake, if one could at least imagine there were *some sort of purpose:* that anything serious could be advanced by the "conference" or even just seriously treated—but you, good fellow, do not know what these people are like and what they are aiming at?! And even if there were only an unmentionable dullness of these persons – – With persons of earnest intention—for example, Cassirer[9]—it would perhaps be possible; but even with these, better not at the center of *such a corona*–! And how can such a "conference" turn out? The people will present their thing in their characteristic psychological habit and perform feats of disputation—and you will say a few kind words, and then, smiling a little, hold your peace—and the people—ugh. This is no pretty affair and no good can come of it.

That's how things are with these philosophers—Then someone like you comes along—and look what the people do–! Good grief!

(*If* you really are considering going there, then I'd be inclined to go too.)[10]

–So; this I had to get off my chest; and I still positively cannot quite imagine that you are really going there (a sunny day in the woods or a workers' lecture[11] instead would be a much finer and better thing)—and now I send you, you good, overly good fellow, my best wishes and am annoyed that I had to write to you over there about such rot as this!

Good wishes to all! Yours,

Wertheimer.

# 17. To Elsa Einstein

[En route to Leyden,] Monday. [17 May 1920]

Dear Else,

I motored around magnificently with Ehrenfest.[1] Saturday & Sunday on the country estate of van Aardenne's friend de Ridder,[2] today at Prof. Julius's with solar physics, nice daughters and much music.[3] Now trip homewards with chocolate. It's late already. The day after tomorrow I'm giving a children's lecture on relativity = appearance of the newspaper lions.[4] I'm considering traveling directly from here to Christiania, dropping Halle and only giving a talk on my return trip in Hamburg.[5] The boat connections are supposedly good and I feel like making the sea voyage.

Best regards to all of you from your

Albert.

The bad handwriting comes from the wobbly ride.

Today I was given a finely painted picture by the artist [from] there in Ede.

# 18. From Lucien Fabre

Paris, 55 Amsterdam St., 17 May 1920

Professor,

By way of the kind intermediary Mr. Oppenheim[1] I took the liberty to send you the paper by Varcollier on displacements in vector fields and their relation to the theory of relativity.–[2]

I consider this paper, like the one by Guillaume,[3] with which you are already familiar, as one of the most serious and original attempts to afford extremely solid new foundations to your ingenious theories and capable of harmonizing with them, augmenting their intellectual attraction with a new element of certitude.–

I likewise permit myself to send you today, also by the obliging solicitude of Mr. Oppenheim, a study I conducted of your theories entitled "*Une nouvelle figure du monde*" ["A New Personality of the World"].[4]

This study is destined for one of the major magazines of fine intellectual culture in Paris and consequently had to give as philosophical, general, and synthetic a survey of your work as possible, while avoiding the use of mathematical formulas.–

It was, of course, impossible to employ in this study the usual comparisons one finds in the popular brochures published on your theories. I believe, Professor, that the subject was extremely difficult and if it does not ring true to you, I hope you will excuse the failure by reason of the difficulty that it posed.–

In order to lend the broadest cultural value possible to this study, I have given a very brief and general historical survey of mathematical physics since Newton. If you would like to glance through this paper, I point out that I start discussing Lorentz's work on page 19; the experiment by Michelson and Morley on page 23, whereas the Einstein theories are not really treated before page 30.–

I naturally proceeded with the greatest care, so as not to clash head-on against any of the scientific opinions held by the readers; and I had to hide, to a large part, the immense intellectual sympathy which your discoveries inspired in me.–

There remains drawing the article's general conclusions as they pertain to its proximity to truth, considered from the triple philosophical perspectives of its relation with space, time, and the categories of reason.–

I would be very glad, so as not to commit any errors on this subject, to know your view if possible; for it is difficult for a man without genius to penetrate the intimate mind of Einstein.–

I dare to hope, without attaching any undue value of discovery to this essay, that you would find its endeavor of interest (which had hitherto not been tried) to give intelligent and educated readers, albeit lacking mathematical knowledge, an impression of what the progress of science could hold, when it is guided in a particular direction by a man of genius.–

If in this paper certain assertions may seem erroneous, doubtful, or inconsistent with your point of view, I would be extremely pleased if you would point them out to me, so that I and my readers may profit from them.–

I extend to you, Professor, my profound admiration.–

L. Fabre.

## 19. To Elsa Einstein

[Leyden,] Wednesday. [19 May 1920]

Dear Else,

This evening I have to give a popular sermon.[1] Yesterday I attended a talk by a scientific opponent; it interested me even though it was not particularly profound.[2] I've become very good friends with the Ehrenfest children and play with them very much.[3] I also have to study Cassirer's manuscript, which is less amusing.[4] These philosophers are peculiar birds. I think I'll call Halle off because the blather would make me sick.[5] In the coming days I'll be going to the seaside with Mrs. Ehrenfest.[6] Both E[hrenfests] cannot be away from the children. We are living here so very pleasantly that the time seemed to fly by. It would probably be best if I went directly to Christiania and go to Hamburg only on my return trip.[7]

I have received nothing from any of you for almost a week, but it probably is due to the bad conveyances. I haven't written as diligently as usual, because I had little time. But from now on it will be different. I want to look for a pretty card for little Margot; I haven't found one that was pretty enough yet.

Greetings & kisses to young & old[8] from your

Albert.

# 20. From Elsa Einstein

[before 20 May 1920][1]

My dear Beloved!

Your languishing violin will be liberated in the speediest way possible.[2] You silly, overly truthful and helpless child! Such a thing shouldn't [ever] happen. It was *your* violin that you have been playing for half a year now already and that you take along with you on every trip. Didn't you say so with any natural force of persuasion? Really! (I got the export permit for the second violin at the same time.) It will be dispatched from Bentheim. All's well with us. I'm putting the house in order and had much sewing to do. In addition I go out into the countryside more now, I make time for it. Yesterday I was perched on the shores of Grunewald Lake, gazing out over it; now there are a whole series of wooden benches directly by it, [those] natural benches, a novelty that impressed me. The view from there is wonderfully beautiful. Meanwhile you have received a laundry-basket full of correspondence. Half the world is now buying the Struck picture and is sending it to you so that you can immortalize yourself on it.[3] That will keep you nicely busy! Besides that, you are being featured again in various humorous periodicals. I assembled a splendid quartet for you: the two Weissgerbers, fine professional musicians,[4] and a talented female pianist. They are worthy of you!

I won't write you any rules of conduct; they aren't of any use, anyway. I gave up. I'm glad that you are enjoying so many nice things among dear people,[5] but now I'm already looking forward to your coming home.

Kisses! not greetings, like you send. Yours,

Elsa.

# 21. From Paul Winteler

[Lucerne, before 20 May 1920][1]

Dear Albert,

According to Elsa's postcard you supposedly said that ignoble haggling was going on over Mama's estate.[2] Maja's character is exactly yours and yours exactly hers, and you may therefore really be assured that within your *closer* family any trace of greed, any attachment to objects per se is lacking; you really ought to know that. Since you didn't want to take care of the distribution of the estate (*you* really ought to have taken the matter in hand; it really was your duty), you have no license to judge Maja's conduct. *You simply weren't present* and Maja is not near you, aside from the fact that by her whole nature she isn't capable of protecting herself against wrong judgment. That's why I consider it my duty to remind you what a nasty report you allowed yourself to be told by Miza that time;[3] a second rendition is all we need. If you can't believe husband Paul, who hasn't perceived the slightest little blemish in Maja throughout 20 years, then at least draw your conclusions from your closer family (father,[4] mother, and you yourself) about Maja's temperament and, for all I care, from your other relations to the others, but not vice versa.

More I won't say; I personally don't care a fig about the estate issue; it's just important to me that you don't soil your own nest, as once before.

Regards to you as a brother-in-law and friend, yours,

Pauli

## 22. To Elsa Einstein

[Leyden,] Thursday [20 May 1920]

Dear Else,

Yesterday evening I delivered a talk before a large audience at the university, which took place very ceremoniously.[1] Today I was even invited by the German envoy in The Hague.[2] Just now I received two postcards from you.[3] *I am not going to Halle.* It would be senseless.[4] Besides, on the 29th I still have to be at the Academy in Amsterdam, where I was appointed foreign member.[5] Yesterday I stepped in for Ehrenfest, who had to travel to Delft, in his lecture course.[6] On the 31st I'm coming home again and look forward to it very much. I'm anxiously thinking about how I should obtain my passport visa for Norway so quickly. Do I have anything at all in writing? Without that the visa is unattainable.[7] The violin still hasn't arrived.[8] Your friend in Amsterdam is sending me a snack package for you which I'll be taking along.[9] I intend to write Maja. Pauli wrote me a somewhat crazy and nasty letter.[10] I'm sending him a postcard.

Kisses to all of you from your

Albert.

# 23. To Max Wertheimer

[Leyden, 21 May 1920][1]

Dear Wertheimer,

You are entirely right with your warning, and I think it is very nice of you not to let me fall into the trap.[2] I am having a fine time here. On the 31st, I am returning to Berlin. I hope we shall see each other soon then.

Cordial regards, yours,

A. Einstein.

# 24. From Erich Regener[1]

Stuttgart, 13 Wiederhold St., 21 May 1920

Dear Mr. Einstein,

May I ask for your help in two small matters, which I hope will not take up too much of your precious time.

The first point regards Dr. Reichenbach, whom I would like to help with his habilitation degree as quickly as possible.[2] He has submitted a thesis, "Die Bedeutung der Relativitätstheorie für den physikalischen Erkenntnisbegriff" [The Significance of Relativity Theory for the Physical Concept of Knowledge]. He told me that you have already read the work, so I do not need to send it to you.[3] Well, I must say that my own experience does not suffice to appreciate the work enough and I believe, furthermore, that your evaluation of the work would have very much weight here at the university, so I have favorable hopes for a smooth passage of the habilitation. It would be particularly valuable, because the work really is to a large part of a philosophical nature, but Mr. Reichenbach is supposed to habilitate in physics. Now, I do believe that you will agree with me that particularly with the current development of physics such works in particular are very useful for physicists, so his thesis could also serve Mr. Reichenbach toward his habilitation.

I would therefore be extremely grateful for a few lines in this reg[ard].

The second point concerns the filling of an Extraordinary Professorship in theoretical physics at the local university. The position is pending at the State Parliament and should be available by autumn. I already discussed this with you in Berlin and you mentioned Mr. Reiche.[4] I am still considering him primarily but must have three other names on the list of candidates. The requirements imposed on theoreticians here are approximately the following: In the first place, he must be well versed in mathematical calculation. Because the matters in which his advice will be sought from the other departments will be less from the conceptual side of physics than, usually, the subject of some tricky calculation. Personally, however, I would not like to have a theoretician of the old school by my side, obviously, but one at home in modern topics. For I would very much like him to nurture contacts with the institute's experimental research, so that he can supplement me, for whom mathematical calculation was never an end in itself. He should also be able to lecture well, of course.

I would be very grateful to you, dear Mr. Einstein, if in a few brief lines you would give me your kind advice on both the above-mentioned matters. I hope you are enjoying your sojourn in Holland and that the news that I heard shortly before my departure from Berlin, that you were not in the best of health, is not confirmed.[5] We have here, incidentally, at the institute what we refer to as the overnight room, which you are welcome to make use of anytime, if you happen to be passing through Stuttgart. Mr. Reichenbach and I, and certainly very many others, would be immensely pleased if you would make use of it very often and for very long.

I remain with most cordial regards, yours very sincerely,

Regener

## 25. To Elsa Einstein

[Leyden,] Saturday. [22 May 1920]

Dear Else,

On the 31st I'm coming home again and am eager to see you all. My dear Ilse shouldn't be cross with me for not writing her a whisper.[1] I haven't been able to get to it yet. Today I talked shop with Kamerlingh-Onnes for 6 hours.[2] Yesterday I was at his brother's and his son's, both painters, who showed me their magnificent things.[3] Yesterday evening I was in Katwijk with Mrs. Ehrenfest, while *he* watched the bunch of kids.[4] I was nominated a member of the Amsterdam Academy.[5] On the 29th I'll still be at the session.[6] Your friend L. Deng sends you a snack parcel and has invited me. But I have no time and wouldn't drive there even if I did have time. I'm terribly sorry that you are having so much trouble with the violins. Ehrenfest is whining a lot about it. Where did you get the idea of sending the second one out as well?[7] I won't write any more about it, in order not to confuse things even further. Kisses from your

Albert.

# 26. To Hendrik A. Lorentz

[Leyden,] 22 May 1920

Highly esteemed Colleague,

First I would like to thank you and your wife for the warm reception you both gave me recently.[1] The walk among the dunes will remain an unforgettable memory for me. Yesterday the announcement arrived of my nomination as external member of the Academy;[2] I do not even need to tell you how it thrills me to be accepted among this circle of outstanding men, in which I feel half at home now already. My talk on Wednesday was intended only for more casual onlookers who want to throw a quick glance on the relativity issues, and not for experts; it is good that you were not present. It thoroughly embarrassed me that you should write me specially about that.[3]

Occasion for this letter is provided by a letter from Planck in which he requests that I encourage neutral academies to support Germany's efforts to stay current with foreign scientific publications.[4] He writes:

"According to reports, a Swiss scientific society made the proposal of holding a conference in Berne that is supposed to be attended by delegates of the neutral academies and aim at discussing international scientific problems." I should point out that there is an established neutral office for scientific reporting[5] in Berlin (address, Prussian Academy of Sciences, care of Mr. Kerkhof) and that it places great value on receiving scientific journals *possibly by exchange.*[6] Any suggestions in this regard would be welcomed with gratitude.

Our colleague Planck did not instruct me to present these matters to *you* specifically, rather he just asked that I mention it at a suitable opportunity. (I tell you this so that you do not wonder why Planck did not write you himself.) Additionally, I wanted to communicate this to you right away, because directly after the Academy meeting on the 29th, at which I hope to see you, I must return homewards again, as my passport visa is expiring.[7]

In heartily thanking you and your wife[8] again for the unforgettable hours that I was allowed to spend with you, I am with cordial greetings, yours very sincerely,

A. Einstein.

# 27. From Max von Laue

Zehlendorf, 17 Albertinen St., 22 May 1920

Dear Einstein,

As I am not going to be seeing you here before my Stockholm trip, and because after my return the matter is easily forgotten, I would now like to write you about it. A short while ago you said the way in which I presented the consistency of the definition of ⟨time⟩ synchronization, in my book (p. 51 of the third ed.), was not elegant.[1] Under the condition that light followed each path there and back in the same time interval, it would be trivial. But this condition was supposedly not obvious.

I now had another look at the matter, particularly since I have to prepare the 4th edition soon. My response is: On p. 51 I am speaking exclusively of frames of reference in which the propagation of light occurs at the same velocity in *all* directions, consequently also in the direction opposite to the same path. This appears explicitly directly before the consideration in question, namely, on l. 3 from the top on page 51. Therefore the consideration is acceptable. Now, you call it trivial.[2] That is a matter of intellectual ability; and you must excuse me if I do not consider you as quite the norm in that. So I am thinking of leaving the relevant section as it is.[3]

I hope to see you on 9 June at the colloquium.[4] I have a couple of very minor and perhaps very stupid questions about general relativity.

With warm regards, yours,

M. Laue.

# 28. From Carl H. Unthan[1]

Charlottenburg 9, 3 Linden Avenue, on Whitsunday, 23 May 1920

Esteemed Professor,

You will kindly pardon me if I must again take up your valuable time, with some delay, owing to pressing and unpostponable obligations. For me one letter is no cause for great effort; considering that in the last period I have been spending many a day sitting at the typewriter for 12 to 14 hours and working through up to 36 folios of translations in 5 languages.

I too have been a pacifist for a long while now.[2] No platonic relationship toward pacifism satisfied me; I involved myself energetically in the polemics between Popert, Otto Ernst, Siemering, etc., versus Fried, von Gerlach, von Ossietzki and the rest of the scum, who cannot outdo themselves in fawning submissiveness and disingenuousness in laying the blame of the war on us alone.[3] The truth can only be made known to the people with a fight; and both Viereck[4] and I are striving toward this; and only knowledge can lead to pacifism. Our ideal, a World Parliament,[5] can be attained only through hard struggle; passive idealism is no help there; the only way to peace leads through battle and war. And once the goal is reached, once the League of Nations and a World Parliament are created, there too negotiations will be carried out on a nationalistic basis, the interests of the nations weighed against one another, because there simply is no other basis. Yet my pacifism does not extend to eternal peace and fraternization between wolf and lamb; my goal for the time being is the elimination of anarchy among states and nations.

In utmost respect,

C. H. Unthan

## 29. To the Royal Academy of Sciences in Amsterdam

[Leyden,] 24 May 1920

Highly esteemed Secretary,[1]

In your letter of May 21 you informed me[2] that your Academy has resolved to accept me as a member.[3] I gratefully accept this nomination and consider myself lucky to number among this circle of researchers, among which some members in my discipline share common ties born of years of intense collaboration, long-standing friendship, and feelings of high esteem.

In great respect, yours very sincerely,

A. Einstein

## 30. From Elsa Einstein

[Berlin, 24 May 1920]

Dear Treasure,

Your agenda is revealed to me always first by the newspapers, before your messages arrive. It always reads: "As a private telegram from The Hague informs us," pompous![1] How glad I am that you aren't traveling to Halle! All that fuming, for what purpose? You won't be able to convince that sort anyway.[2] But Schlick and Freundlich are going.[3] An official letter arrived from the Norwegian students.[4] The Foreign Office would purportedly send everything connected with your trip to you; you don't have to worry about anything. The entire trip 1st class from Berlin. Where is the appointment as Dutch professor now?[5] You are [in the middle of] a ridiculous situation. At Springer's the midnight oil had to be burned in order for the speech to be ready by the 5th.[6] I had to harass Berliner.[7] Now a whole edition is finished off the press; it costs a few thousand marks. Are you supposed to pay for it? That's an [audacity], of course. It will never sell, because it was never [delivered]. A very crazy affair. You have to admit. It really was a rash matter. I was on a long visit with Haber today, said my mind about all those historical things.[8] This has to be, once a year. You know that Haber is now a good friend of mine; we understand each other well and [put our heads together]: we both just want what is good and nice for you. We discussed many things. If my girlhood friend sends you a package, then do send her a couple of nice words of thanks.[9] *Do it nicely*, then she'll be pleased. Please do so, I beg you. Do you think the violin will arrive so quickly in Leyden?[10] The document is traveling from place to place, it changes nothing [about] my eloquence. But I surely would have brought it over the border as *my violin*.[11] I'm as eager as a child for your return soon; I can hardly wait. Fond kisses, your

Wife.

## 31. From Robert Fricke[1]

Bad Harzburg, Lorenhöhe, {17 Kaiser Wilh[elm] St., Braunschweig}, 26 May 1920

Dear Colleague,

As current Pres. of the German Math. Association, I recently contacted Schönflies, the Opening Chairman of the Ist Sec[tion] of the Nauheim meeting, about the framework of the meeting.[2] I then made the proposal that, for the joint session of the Ist Sec. with the Sec. for Mathem. Physics, the Thursday morning be reserved exclusively for talks on the theory of relativity. I take it as certain that you will speak on Monday or Tuesday at the general meeting of the main section of sciences. This would surely not exclude that you would again take the stand before the more limited circle of specialists in your discipline on Thursday. Apart from you, I also wanted to approach Messrs. von Laue, Hilbert, Sommerfeld, Weyl, and Born with the same request.[3] If I should succeed in organizing such a morning session, I would think it could become one of the greatest successes of the Association and could show the world what Germany can accomplish in the field of science, even in such deeply troubled times as these.

In great respect,

Dr. R. Fricke
Privy Councillor.

# 32. To Elsa Einstein

[Leyden,] Thursday. [27 May 1920]

Dear Else,

Now this sojourn here is coming to an end.[1] It was nice, but I am looking forward to being home again as well. I am coming on Monday. I let Halle go without me.[2] I won't be traveling directly to Norway because the passport difficulties would probably be even more inconvenient to settle than from Berlin. I have nothing in writing in hand, you know. Nor do I know whom I am being invited by.[3] My nomination here is still lying in the lap of the gods, to the great discomfort of my colleagues. So the speech remained undelivered.[4] Next time! The violin unfortunately hasn't arrived, either. I will continue to look after this business from Berlin.[5] I'm gratified to see that I'm not the only slovenly bungler in the world; that's a consolation.[6] Tomorrow and the day after tomorrow I'll be in Amsterdam (at Zeeman's and at the Academy, whose member I have become (don't say anything so that nothing gets in the papers).)[7] On Sunday we'll still be visiting Julius in Utrecht to play music.[8] Then it's homeward bound. Ilse kept me very nicely up to date. But I was slovenly.[9]

It was nice, but now I'm looking forward to seeing you all again very much. Kisses from your

Albert.

Greetings to Ilse, Margot, and the parents.[10]

# 33. To Ilse Einstein

[Leyden, 27 May 1920]

Dear Ilse,

I'm a heel for not having answered your letters and for writing Mama[1] so little; but you are a dear, irreproachable monkey, as you yourself acknowledged.[2] I'm coming back on Monday, so help me God; then we can discuss everything I had neglected from here. This evening I had a war of words at the colloquium with an antirelativist (not an anti-Semite, because he's called Polak).[3] Apologize for me to Mama for having written her so little this time—she also was quieter than usual; but I hope and confidently expect she will make up for it abundantly in person.

Kisses to all three of you,[4] yours,

Albert.

# 34. To Heinrich Zangger

Leyden, [27 May 1920]

Dear Zangger,

I received both your letters, also the genuinely Italian effusion by Enriques, forwarded here,[1] but not Habicht's letter, so I still don't know what he is up to.[2] I am eager to know whether Weyl will be coming to Berlin now; it would not be smart of him, since Berlin would take away his peace and contentment. Göttingen, though, would be good for him in every respect; Zurich too, of course, if he stayed there.[3]

I'm supposed to receive a guest professorship here, and be here 3–4 weeks each year. So I counted on sending my wife 2,000 francs from here, but can now only do so from Berlin because my appointment is happening too late.[4] I telegraphed Albert that he should simply tap you. The money will be returned to you at the latest in a fortnight. Many thanks in advance for the help.[5]

Much in your letters I cannot read, possibly often precisely what is important. Typewritten would be much preferable.

Here I spent wonderful weeks with Ehrenfest, also gave a few lectures,[6] had very many scientific discussions with him, Lorentz, and the other local physicists.

I very much long to see my boys again,[7] but still don't know when I can come to Zurich again at last.

Warm regards from your

Einstein.

# 35. From Hendrik A. Lorentz

Haarlem, 27 May 1920

Dear Colleague,

I really ought to tell you—and should have done so earlier—how heartily pleased I am that you now belong to our Academy of Sciences as a corresponding member.[1] You can be sure that all members share this joy and regarded your election a foregone conclusion.[2] I hope that we are going to see you very often at the meetings. What a pity that you have to travel again so soon and that your inaugural speech must still lie waiting.[3] I am very sorry that Onnes's applications to the government could not speed things up.[4] If it had been in our power, all would have been ready on February 10.[5]

With cordial greetings and in hope of seeing you the day after tomorrow,[6] yours truly,

H. A. Lorentz

# 36. From Konrad Haenisch[1]

Berlin, [28] May 1920[2]

To Prof. Einstein,

With reference to the report of 5 May this yr. submitted to me by Dr. Freundlich at your instruction in re. the release of state funds for the advancement of your researches,[3] I humbly inform you that at the second and third consultations on the planned state budget for 1919 in December 1919 the Constituent Prussian Assembly resolved "to apply to the state government to make available the funds in agreement with the Reich government, in order to make possible Germany's continued successful collaboration with other nations toward developing ⟨your⟩ Albert Einstein's fundamental discoveries ⟨and you personally⟩, and to further his own research (printed matter no. 1612 section 233)." I have opened initial contacts with the Prussian minister of finance,[4] in order to promote this matter jointly with him and to approach the Reich government together with him. ⟨With regard to the general state of the finances, it will scarcely be possible to place the eventual funds earlier than into the draft of the state budget for 1921.⟩ I shall inform you of further developments in this matter.

# 37. From Greti Moser[1]

Bad Wiesee, [Bavaria,] 28 May [1920][2]

Esteemed Professor Einstein,

I was very surprised when your photograph came from Berlin; I often think of the nice time. The two villas are still standing as they were on Adolzreiter St. Just a short while ago I looked at them again; the beautiful garden is gone, of course. How is your sister, Mrs. Maja Winteler?[3] I used to correspond with Mrs. Einstein, you know. I lost the address during the war; then I visited Mr. Elinger and heard, unfortunately, that the good Mrs. has died.[4] Does Mrs. Maya have no children? Is she still in Lugano?[5] She often invited me, of course, but I never managed to visit her & now all has been lost. I shall treasure the picture; I have the pictures with both children on them. Mrs. Einstein wrote me your son also used to be like that. Permit me to convey my wishes to Mrs. Maja as well. Also best wishes to your esteemed wife, in all due respect,

Greti Moser.

# 38. From Paul Epstein

Zurich, 6 Physik St., 30 May 1920

Highly esteemed Professor,

Permit me to direct a few questions to you regarding the business of Miss Edith's dissertation.[1] Miss Edith did not exactly expedite the matter; it is also correct that due to Meissner's illness she was completely prevented from working and studying for months on end. No doubt she has a share in the success of nursing Meissner back to health,[2] but obviously this cannot be combined with progress in the project; and that is how it came about that only now did she complete the calculations that you fully described. It is not much, of course, but at least something.

With reference to some notes she made following your instructions,[3] I would, nevertheless, still like to request some explanation. It involves what can be regarded as constant in each cross section. Let us look at the integral $m\iiint_{-\infty}^{+\infty} f(\xi, \eta, \zeta) \cdot \xi^2 d\xi d\eta d\zeta$; the same thus gives the momentum that is carried through the unit of area perpendicular to the $x$ axis.[4] When we take a parallel plane, no momentum can form between the two if the collisions satisfy the laws of mechanics and, in the stationary state, the same momentum must be transported through the unit area of the second plane. If I understand Miss Edith's notes correctly, however, she regarded a different quantity as constant, namely (if we denote it as $\sigma_{xx}$), $\sigma_{xx} - \frac{1}{3}(\sigma_{xx} + \sigma_{yy} + \sigma_{zz})$. This, as far as I can see, would involve that the quantity enclosed in parentheses be constant, which, however, scarcely allows agreement with the condition of constancy of $\sigma_{xx}$. In hydrodynamics and elasticity theory, constancy of this quantity is required, but there the terms stemming from the temperature gradients are simply not taken into account.

I am completely aware that my mind has been a little dull recently: it seems that the nervous tension from which I can't escape here in Zurich has gone to my head and has impeded my ability to concentrate. It is therefore easily possible that I have been writing nonsense and, in the latter case, I ask you please not to read my further questions, since they are all based on the former train of thought. For if it is correct that $\sigma_{xx}$ = const., then the reader could ask why we introduced just three secondary conditions: number constant, energy = $\frac{kT}{2}$ (or, to be precise, $\frac{dU}{T} = ds$), and energy flow = $\varphi_x$ and not also the fourth $\sigma_{xx}$ = const., which would alter the form of the function $f(\xi, \eta, \zeta)$ even further.[5] The reason obviously lies in that the state of the system is already completely determined by those attributes, but perhaps Miss Edith ought to work this out a bit more. For as it stands, although one can always arrive at a constancy of $\sigma_{xx}$ by suitable choice of the density distribution, it is not convincing to everyone that this alternative is the right one. The complete theory would have to be able to indicate from a constant $\varphi_x$ the full distribution of temperatures, densities, etc. I fear, though, that Miss Edith will not be equal to this task, and think that an explanation that only one more constant was involved here can make the matter plausible.

How would it be, though, if one regarded instead of a single layer the whole breadth between the boundaries? The entropies of the various layers cannot be summated, of course, as no equilibrium, in the normal sense, subsists. But one can certainly sum together the heat amounts that are introduced into the various layers with an infinitesimal rise in all the temperatures and obtain

$$dQ = \sum dQ_n \; ; \; TdS = \sum T_n dS_n$$

It thus seems to me that a correct minimum principle is obtained when one multiplies Boltzmann's $H$ for each layer with the relevant temperature and integrates over all the layers[6]

$$\int dx \int T(f \log f - f) d\xi d\eta d\zeta = \text{minimum}$$

The advantage would be that the density distribution would be obtained directly. Yet, even if you should express approval of this consideration, I would not recommend it for Miss Edith, because it would take forever and she must finally finish up.[7]

This semester I am lecturing on "boundary value problems," or, in other words, partial differential equations, and am very pleased with my class because they are working with real diligence and interest.[8] I am sorry that I shall probably not reap the fruits of labors of this course, for if all indications do not deceive me, this will be my last semester in Zurich.[9]

Meyer and Weyl were in Germany and returned the day before yesterday. You probably spoke with Weyl in Berlin;[10] Meyer was in Tübingen. He was able to relax well but is appalled by the reactionary spirit at the university.[11]

Lately I delivered some popular lectures on the theory of relativity to psychiatric circles. It is strange why precisely these people are suddenly so extraordinarily relativistically concerned. Most interesting is Bleuler,[12] who is really trying to get to the bottom of the issues.

With respectful greetings, I remain, yours very truly,

Paul Epstein.

## 39. From Anton Lampa[1]

Hadersdorf-Weidlingen, 30 May 1920

[Not selected for translation.]

## 40. From Adriaan D. Fokker[1]

Arosa, 2 June 1920

Dear Professor,

I would like to congratulate you heartily on your nomination to the Amsterdam Academy.[2] I hope that a good many favorable things spring out of that for you and that you will want to frequent our country often, very often. I was happy to see that you and Rutherford were nominated at the same time:[3] you and he are persons who gave me a strong impetus to do physics, which helped me to keep my interest intact throughout military service.[4]

Sometime not too long from now, I am hoping to come to Holland with my wife for the summer and autumn, but do not dare to think that you will still be there then.[5]

Today I read the article by Majorana in the *Phil. Mag.* on his experiment, which purportedly showed that a lead sphere of 1 kg was less heavy when surrounded by 100 kg of mercury than when not.[6] I believe that experimental errors will surely be revealed: even only displacement of the mercury will have deformed the ground noticeably. Even so, the question arises whether such a thing is possible. Is the acceleration of the fall of an infinitely small test piece just as large as for a body of finite weight that itself could influence the field of gravity? Is it larger or smaller? I could not say right off.

I would certainly have much more to ask. Particularly the reasoning behind the relativity of inertia; the way you understand it is still a puzzle to me.[7] But it can only bore you to treat these topics by letter. You can make much better use of your time. And so I will do without it for today.

When are you going to invite an outstanding experimenter to demonstrate that moving clocks run more slowly than ones at rest, or resp., moving atoms radiate redder than resting ones? It really is almost insufferable that the theory of rel. still has to limp along on a single leg: Michelson for the Lorentz contraction. The inertial mass of electrons has stood the test,[8] but the true counterpart to Michelson and extension of the theory's foundations are still missing. I have already written to

Guye and Rutherford about this,[9] but you yourself really ought to work toward that too!! It *can't* be impossible, surely, to furnish experimental proof!!

I hope very much to meet you still when we come to Holland. So, hopefully *auf Wiedersehen*!!

With most cordial greetings, yours most sincerely,

A. D. Fokker.

# 41. To Hans Vaihinger

Berlin, 3 June 1920

Highly esteemed Colleague,

Unfortunately, it was absolutely necessary, due to my nomination as a member of the Amsterdam Academy,[1] that I be present in Amsterdam for the meeting of May 29. That is why it was unfortunately impossible for me to come to the As If Conference, which certainly was very interesting.[2]

The thought of your intending to publish an "Einstein Issue" gives me a creepy sensation of embarrassment. Perhaps the issue could be dedicated to the theory of relativity instead of to me personally. So much recognition as is being showered on me is a heavy burden indeed for the living to bear. I am not in possession of a manuscript for the speech held at Leyden, hence I cannot have it printed either.[3] Time is completely lacking for drafting a separate essay. But I would like to inform you that Prof. Cassirer in Hamburg has written a very interesting essay on the theory of relativity from the philosophical point of view that is as yet unpublished.[4]

In great respect, yours very sincerely,

# 42. To Paul Epstein

[Berlin,] 4 June 1920

Dear Colleague,

An epidemic urge to flee Zurich seems to have got hold of everyone.[1] It is a pity how little skill or good will the Zurichers are able to muster to make use of such a propitious time for developing their universities. I hope you finally get that long overdue professorship now; there are—as far as I know—many options.[2] But in any case, if all else fails, you will take in a very comfortable temporary subsistence in Leyden, in which you can wait worry-free. Ehrenfest told me about it.[3]

My comprehension of your remarks is hampered by my not rightly knowing anymore what I wrote down for Edith at the time.[4] It appears certain to me, too, that

$$m\iiint f\cdot\xi_\mu\xi_\nu\cdot d\xi_1 d\xi_2 d\xi_3 = T_{\mu\nu}$$

should be interpreted as the components of the pressure tensor (or momentum tensor), and that therefore, in the stationary state

$$\sum_{\substack{\nu\\1-3}}\frac{\sigma T_{\mu\nu}}{\sigma x_\nu} = 0$$

must apply. Otherwise the principle of momentum conservation is violated.[5]

In carrying out the variation, under no condition may $T_{11}$ be regarded as *given*.[6] The question that the variational principle is supposed to solve is the following:

What is the most likely velocity distribution within an unmoving gas of a given density and a given energy, when it is known that the same gas is transporting a given heat flow?[7] This then yields the pressure anisotropy. Once this distribution is obtained, it furnishes the $T_{\mu\nu}$'s, by which the problem is solved. One can arbitrarily define the temperature, you know, by the equation $3\kappa T = \overline{\xi_1^{\ 2}+\xi_2^{\ 2}+\xi_3^{\ 2}}$.

The ruse is, of course, that—instead of determining the anisotropy's dependence on the inhomogeneity of the gas's motion through subtle calculations in statistical mechanics, using Maxwell's method—one introduces an arbitrary hypothesis (most likely distribution for the given energy flow).[8] But I am convinced that by strict scrutiny of the outcome, this ruse can subsequently be vindicated.

I no longer remember now what I wrote down for Edith at the time, and it is certainly possible that I was somehow mistaken in the last part of the consideration, which I recall the least clearly. However, I would proceed in the following manner:

The variation argument delivers as the final result for a heat flow ($T$ = abs. temp.) parallel to the $x$ axis:

$$\left.\begin{aligned} T_{xx} &= \alpha+\beta\left(\frac{\partial T}{\partial x}\right)^2\\ T_{yy} &= \alpha+\gamma\left(\frac{\partial T}{\partial x}\right)^2 \end{aligned}\right\}\quad(1)$$

where $\alpha$, $\beta$, and $\gamma$ no longer contain $\dfrac{\partial T}{\partial x_\nu}$. Additionally, it generally holds that:

$$T_{\mu\nu} = A\delta_{\mu\nu} + B\frac{\partial T}{\partial x_\mu}\frac{\partial T}{\partial x_\nu} + C\,\mathrm{grad}^2 T\delta_{\mu\nu} \quad (2)$$

it follows from (1) and (2):

$$A = \alpha$$
$$B + C = \beta$$
$$C = \gamma.$$

Thus the pressure tensor is calculated.

It is clear, incidentally, that terms of the pressure tensor that contain second derivatives $\left(\frac{\partial^2 T}{\partial x_\mu \partial x_\nu}\right)$ of the temperature, according to location, cannot result out of this consideration.– *

⟨As I am writing all this down, I suddenly have second thoughts about the whole theory. The length of path does not enter into the expression, rather, e.g., $T_{xx} - T_{yy}$ must be expressed by a formula of the form $\kappa\,\mathrm{grad}^2 T$, where $\kappa$ can now only depend on molecular mass *m*, temperature *T*, and molecular density *n*. $\kappa$ has the dimensions $\frac{1}{\text{energy} \cdot \text{length}}$, which in the available quantities is only in the form⟩

Nonsense! Heat flow still has to be incorporated, of course! This thing is correct.[9]

Please put a little pressure on Edith to finish off the matter. Specifically, she ought to calculate an example to a quantitative and, if possible, an experimentally verifiable level. I do indeed believe that she was held back and think it very nice of you to defend her.[10] Give her my regards.

With best regards to you and the rest of my Zurich physicist friends, I am, yours,

A. Einstein.

*The paragraph in brackets is struck out.

# 43. To Klaus Hansen[1]

Berlin, W. 30, 5 Haberland St., 4 June 1920

Dear Sir,

I found your letter of May 15 upon my return from Holland.[2] I am now ready for the trip and only await more detailed information from you. Because I am somewhat overstrained, it would be a great relief for me if I could take my stepdaughter, who is at the same time my secretary—Ilse Einstein—along with me on the trip.[3] So as not to incur any additional expense for you, I request that you please procure

2nd class tickets and accommodations in a very simple hotel (they may be very small rooms), considering that I don't place any importance on amenities. Please also arrange for the travel permit on the part of the Norwegian state and take care that I be immediately notified when I may obtain the Norwegian visa for me and my daughter. I shall depart as soon as all the necessary formalities for the trip have been settled.[4]

In great respect,

A. Einstein.

## 44. To Ernst Cassirer

Berlin, 5 June 1920

Highly esteemed Colleague,

I studied your treatise thoroughly and with very much interest and admired, above all, how securely you master the essence of relativity theory.[1] I made brief comments in the margin where I was not completely in agreement. E.g., I could not accept your opinion about the Kant-Newton relationship with reference to space and time.[2] Newton's theory requires an absolute (objective) space in order to be able to attribute real meaning to acceleration, which Kant does not seem to have recognized.

I can understand your idealistic way of thinking about space and time and also believe that one can thereby arrive at a consistent point of view. Not being a philosopher, the philosophical antitheses seem to me more conflicts of emphasis than fundamental contradictions. What Mach calls *connections* [*Verknüpfung*] are for you the ideal names that make experience possible in the first place.[3] You, however, emphasize this aspect of knowledge, whereas Mach wants to have it appear as insignificant as possible. I acknowledge that one must approach experiences with some sort of conceptual tool in order for science to be possible; but I do not think that our choice of these tools is constrained *by virtue of the nature of our intellect.* Systems of concepts seem empty to me, if the way in which they are to be related to experience is not laid down. This seems to me highly essential, even though we often find advantage in theoretically isolating purely conceptual relations, in order to have the *logically* secured interdependencies come more cleanly to the fore. With the interpretation of *ds* as a result of measurement that can be obtained in a very specific way by means of measuring rods and clocks, the theory of relativity stands and falls as a *physical* theory.[4]

I think that your treatise is very well suited to clarify philosophers' ideas and knowledge about the physical problem of relativity.

Best regards, yours,

# 45. From Adolf Smekal[1]

Vienna, 5 June 1920

Highly esteemed Professor,

Perhaps you still remember that, during the discussion of my photophoresis report at the Berlin Colloquium,[2] I asserted at the time that the Stokes-Cunningham law of falling bodies was one of the best-secured laws of physics, owing to the splendid agreement between the particle radii drawn therefrom and those from Ehrenhaft's[3] "optical" size determinations.[4] You, Professor, rightfully countered then that the theory of Brownian motion was actually merely grounded on probability assumptions, and that the lack of agreement with the particle radii calculated from Brownian motion did not speak favorably for this law of falling bodies.

Well, not long ago, Dr. E. Norst demonstrated in a talk before the Viennese chapter of the G. Phys. Soc. that all the previous results from "optical" size determinations are useless.[5] Since I am thus proven wrong, in your favor, Professor (which with a bit more humility I might perhaps have been able to foresee!), I consider it my duty to communicate the details to you immediately. Mrs. Norst found the following errors:

1.) The irradiation curves are in p[art] wrongly calculated; the number of points, from which they are determined, is too small to exclude arbitrary factors.

2.) Instead of the *arc-lamp* spectrum (= glowing carbon + arc), G. Laski, J. Parankiewicz, and M. Schirmann used the carbon-*arc* spectrum (!);[6] Snow, whose measurements they used, had *carefully screened out* the carbon![7]

3.) In applying the physiological theory, König's basal perception curves, which relate to the solar spectrum, must be used.[8] Laski had neglected to perform this "Sun" conversion.

4.) Even in Laski's work (without her having noticed it, partly also as a consequence of computational errors), *many different* particle radii belong to a *single* wavelength (albeit, the colors are of various saturations), so the equivocal correspondence between color and size (or the rate of fall, which according to the observations exists qualitatively in any case) does not obtain.

Mrs. Norst avoided the above errors by using Parankiewicz's irradiation curves for sulfur and calculated the relation between particle radius and physiologically effective color identification, on the one hand, neglecting whether the intensity distribution in the arc-lamp spectrum was the same as in the solar spectrum, and on the other hand, taking into account the Sun : arc-lamp ratio, which had been defined for this purpose by Kohlrausch.[9] In both cases, in general, *larger* particle radii resulted; but here too the difficulty mentioned under (4) arose (lack of a definite assignment of color to size), so that "optical" size determination is presently completely unusable and hence all the conclusions hitherto based upon it as well.

Perhaps a considerably more careful calculation of the irradiation curves (even the first of Ehrenhaft's curves for Au are faulty, as Fürth has already pointed out) will lead to useful results.[10]

The fact that Mrs. Norst arrived at up to 100% larger particle radii than did G. Laski seems to me to be a remarkable approximation of the particle radii of Brownian motion. In addition, there are some neglected factors in the law of falling bodies, whose elimination seems to lead to a larger radius as well. All these things thus are not exactly favorable to the "subelectron."[11]

Quite recently, now, I overheard a conversation between Von Mises and Zerner,[12] from which I gathered that the elementary quantum has now finally revealed itself to Ehrenhaft himself. Prof. von Mises, using observations that had been carried out by Ehrenhaft, apparently carried out calculations according to Weyl's cyclical error theory which furnish the electron charge as the mean value. Naturally, Ehrenhaft does not quite believe it yet; to me, at least, he said that new measurements would still have to be taken. If one adds to this that Bär has purportedly brought about the fall of the "branching method,"[13] one can, I think, predict the "official" end of the "subelectron" in the not too distant future. For I do grant Ehrenhaft that much objectivity, of eventually turning away, under the weight of the evidence, from his own idea, which seems to have been misleading him for over 10 years now. At least I hope that he will allow the electron to live on as a "statistical" "mean value."

Soon Thirring will be reporting on the new paper by Majorana (absorption of gravitation) at the local colloquium.[14] The general theory of relativity does indeed foresee a "screening effect" of gravitating masses.[15] Could this Majorana effect perhaps involve the first effect detectable in the laboratory of the theory of gravitation?

With my best compliments to you, Professor, I remain as ever your sincerely grateful,

Adolf Smekal.

## 46. To Paul Ehrenfest

[Berlin,] Sunday. [6 June 1920][1]

Dear Ehrenfest,

I have been here already for almost a week and it's only today that I'm writing you, rascal that I am. The journey was quite comfortable and easy. I did not fuss with the violin. We already have the export permits for both of them. Soon you'll receive the first one. Then I can take the bow and case along myself.[2] I've calmed my wife and Springer down about the inaugural lecture.[3] Here I have already become acquainted with some interesting things of a physical nature, specifically, a thing by Hettner who showed that the infrared gaseous spectra HCl, $H_2O$, etc., can be explained completely by (quasi)elastic natural oscillations along with combination tones. If, e.g., $\nu_1$ and $\nu_2$ are absorption frequencies, then there are also $2\nu_1$, $2\nu_2$, $\nu_1+\nu_2$, etc., as the quantum theory leads us to expect for incompletely elastic oscillations.[4] Furthermore, work has been done at Regener's on the Ehrenhaft business with droplets, which shows that the apparent lowered level of elementary quanta are produced by too small degrees of mobility, that probably stem from gas layers that enlarge the particle's hydrodynamically effective radius; the existence of such gas layers has long been established by weighing and optical experiments.[5]

My thoughts are often still with all of you. Those were nice times that I spent with you. I already very much miss the children with their cheerful chattering,[6] also our conversations and the music making. I thank your wife very particularly, and Aunt, too, for the touchingly feminine care they bestowed on me under such difficult domestic circumstances.[7] In a few days, I have to steam off for Norway; I am taking Ilse along.[8] Meanwhile, the nomination might come, nice and slowly, from The Hague. It's good if it takes a little while longer, since it's getting almost too much with all this traveling in rapid succession. There's still time in July, you know.[9]

Dear Ehrenfest! All of you are so good and warmhearted toward me, without my being able to explain it, spoiled and overrated heel that I am. But I am grateful to all of you from the bottom of my heart and know how very precious this is. It's so remarkably good for both of us, too, that we are together more often, because it's as if nature made us for each other.

Warm greetings to all from your

Einstein

It was supposedly quite silly in Halle. Once again you were right.[10] Fokker sent me a very elegant paper on bound electrons; he is a fine analyst.[11] Greet particularly Van Aardenne for me, whom I did not see again,[12] also Lorentz and Kamerlingh Onnes, who took such pains for me, quite unnecessarily.[13]

Dear Ehrenfest! She [Elsa] doesn't have time to finish writing now, because she has to leave about the violin.[14] But I want you to get a sign of life from us, at last, if only an unfinished one.

# 47. To Moritz Schlick

[Berlin,] 7 June 1920

Dear Mr. Schlick,

This morning I received your friendly letter and your manuscript.[1] The situation surrounding the invitation to the philosophers' congress is considerably different from what the cunning Vaihinger led you to believe. He wanted to know whom among those knowledgeable in the theory he could also invite; so I naturally gave him your name. But there is no question of my having *requested* your or any one else's attendance in Halle. The whole business had little attraction for me and I was glad to have a valid excuse to avoid all that palaver there.[2]

Now some remarks about your wonderfully clearly written manuscript.[3] I agree almost, but not quite completely, with your interpretation of causality.[4]

Assume for a moment that we were familiar with gravitation only through the movements of comets that passed by in (single) hyperbolic orbits diverted by the Sun. Let it furthermore never happen that two comets have even approximately the same orbital elements, hence that repetitions of the same events do not occur. Couldn't we then conceive of the event as causal? Certainly! One would draw the laws, for inst., those corresponding to Kepler's laws, hypothetically from a few cases. Then they would subsequently be confirmed and every scientist would assign to these laws the character of a natural law, even though any repetition of the *same* occurrence was never observed. The whole empirical world could, in principle, be thus composed, without our having to give up our causality principle, although we might have been less prone to test it.

Furthermore, the problem of the law of inertia violating the causality postulate. You legitimately pointed out in your little book that I had gone too far in that exposition.[5] But I cannot concede to your current construction of the facts.

According to my view, it would be correct to say: Newtonian physics must acknowledge the objective reality of acceleration, irrespective of the system of coordinates. This is possible only if one regards absolute space (or the ether) as something real. Newton consistently does this as well.

You, however, simply say: *Form* is not an event. It is not "form" that is involved, but the "persistence of a form." I must reply: persistent equilibrium in a specific form certainly is an event in the physical sense. Rest is a dynamic event in which the velocities are constantly zero, one that for our consideration is, in principle, equivalent to any other event of motion. As a matter of fact, dynamic events do take place differently with respect to both rotating celestial bodies as well. (E.g., Foucault's pendulum, a moon's orbit, etc.)

Whether, in order to lend reality to acceleration, you call that which you need absolute space, the ether, or a preferred system of coordinates, is all the same (although one would probably not want to incorporate the last of these into the causal series as something real). The unsatisfactory situation remains that this something enters *only one-sidedly* into the causal series. Whether one can declare the law of causality satisfied or not depends on the subtleties of the definition of the law of causality. Newton's absolute space is independent, uninfluenceable by anything, the $g_{\mu\nu}$-field of the general theory of relativity is subject to the laws of nature, (not just determining, but) determined by the properties of matter. This you expressed masterfully, by the way, on page 27.[6]

Re page 28 top. It seems to me unjustified to state that gravitational fields should not be regarded as observable in the same sense as masses;[7] the "process characteristic" of the latter seems unessential in this connection. What is essential is that one absolutely cannot speak of "*all* characteristics" of a body (because there are $\infty$ many of them); if it belongs within a theoretical system, there are always properties that are a consequence of the others, regardless of whether this system operates with "processes" or whether it makes do with static considerations (this difference does not seem to me one of principle).

The restriction of causality to the ability to continue the givens within a spatial section is not my meaning; however, that viewpoint is in any case admissible.[8] It is not necessary to describe an extension of the natural laws beyond that—should it ever prove possible—as an extension of causal knowledge. But why not do so? Just in order to single out time? It may very well be possible that a free choice of initial conditions that leave room for more developed natural laws will be much more limited than seems to be the case from the current state of our knowledge. Then one would also explain the lawfulness within the time section as a "causal" one, in order not to make an unnecessary distinction between temporal and spatial extension.

Your invitation pleased me very much.[9] If I can arrange it, I shall pay you and your family a short visit. I do not think, though, that it will work, because I have to be "stingy" with my time like a real European.

Cordial greetings to you and yours, from

A. Einstein.

## 48. To Robert Fricke

Berlin, 9 June 1920

Highly esteemed Colleague,

Your letter of 26 Jun. arrived in my hands late because I was away on a trip.[1] Many thanks for your kind invitation.[2] In my opinion, though, our professional colleagues are sufficiently well informed on the basic contours of relativity theory, so there does not seem to be any need for the talk you are proposing. By contrast, it might be of some interest to organize a general discussion on the subject. To such an event I would gladly come and answer all questions posed to me. However, attention must somehow be given to have the questions screened in advance so that the discussion not be disturbed by low-quality questions. In case you think my suggestion is a good one, please do send me further information.

In utmost respect, yours very truly.

# 49. From Hendrik A. Lorentz

Haarlem, 9 June 1920

Dear Colleague,

After having already briefly mentioned it to you verbally, I now have the pleasure of inviting you, in the name of the science committee[1] of the "Solvay International Institute of Physics" [The committee members are currently [Edmond] van Aubel (Ghent), W. H. Bragg, [Léon] Brillouin, Mrs. [Marie] Curie, [Heike] Kamerlingh Onnes, [Martin] Knudsen, Lorentz, [Augusto] Righi, [Ernest] Rutherford], to the "Physics Conference" to take place next spring in Brussels[2] and to ask you to prepare a report on the effect predicted by you and observed together with de Haas[3] (Ampère's molecular currents), and its consequences.[4] Should you want to participate again in this little congress, you would be doing us a great favor.[5]

The meeting will begin on April 1st and will last one short week. As a general theme we have chosen the electron theory and its most important applications, atomic emissions and radiation phenomena (*not* black-body radiation); it is our intention to prompt an exchange of ideas specifically on the fundamental concepts, general ideas, and still-open questions. We thus would like to limit the discussion, e.g., on Bohr's theory, to the fundamentals and the simpler cases, without delving too deeply into the details of complex atomic structures.

Division of the material, indicating the gentlemen whom we are going to ask to draw up reports:

1. General topics on the theory of electrons. Consequences of the theory and difficulties. *Lorentz.*

$2^a$. Structure of atoms. Constitution of the nucleus. Isotopes. *Rutherford.*

$2^b$. A few things on quantum theory. Specifically, your photoelectric rule. *De Broglie.*[6]

3. Bohr's theory. Electron arrangement within the atom. *Bohr.*[7]

$4^a$. Electrons and magnetism. Gyroscopic effects. *Einstein.*

$4^b$. Attempts to explain para- and diamagnetism. Magnetism at low temperatures. *Kamerlingh Onnes* or *Langevin.*[8]

We are going to invite the following physicists to the conference:

Barkla, Bohr, de Broglie, Ehrenfest, Einstein, de Haas, Jeans, Langevin, Larmor, Millikan, Perrin, Richardson, J. J. Thomson, Weiss, Zeeman; additionally, either Siegbahn or Vegard; I am asking Bohr to decide that.[9]

Hence, with the members of the science committee, altogether 25 participants; we do not want to exceed this number so that the meeting stays of as intimate a nature as possible.

It would please me very much to hear from you that we may count on you. If that be the case, I shall take the liberty of returning, in the coming weeks, to the arrangements and scope of the reports.

With cordial regards from both of us, yours faithfully,

H. A. Lorentz.

## 50. From Arthur Schoenflies

F[rankfurt-am-]Main 59, 3 Grillparzer St. [between 9 June and 28 July 1920][1]

Esteemed Colleague,

Unfortunately it is probably quite certain that we will have to relinquish our Max Born to the Göttingers. Thus the question arises, who should become his successor?[2] We do have Stern in the first place, whom you already know from your Zurich time,[3] and about whom I have an excellent opinion. Other parties are pointing in particular to Kossel in Munich. Finally, Lenz should probably also be under discussion.[4] In any case, may I be so free as to turn to you with a request that you give your opinion on the three gentlemen named. It is, of course, destined for the faculty. But I would like to hear in what order the three should be nominated, according to your judgment, and what you think of the individuals in particular. It goes without saying that I would be very much obliged to you for any other recommendation of any distinguished person within our reach. No consultations have taken place yet; I am writing entirely at my own volition. As you may know, I am always for the younger gentlemen who allow room for promising development and after some time may be counted among the leading figures. I shall be very grateful for anything you could advise me in this regard.

You have learned from Prof. Fricke in Braunschweig that we are planning a meeting of phys. and math. on relativity for [Bad] Nauheim.[5] I personally am driven by the thought that the Entente people are planning for September—apparently almost simultaneously with us—an internat. math. congress in Strasbourg and that we are therefore honor bound to draw up the most prominent program possible for Nauheim.[6] You offered, as Fricke writes me—I am the introductory speaker—to answer questions that are posed to you in advance.[7]

Perhaps the formulation of such questions is not so simple. But would it not be possible for you, guided by the above-mentioned consideration, to communicate something of your own about a current aspect of modern relativity? A report on recent researches—limited to some direction or a specific focus—would surely be very warmly welcomed by everyone and would not cause you any substantial effort.

Weyl recently announced a talk "Electricity and Gravitation" and Laue also wants to present something on optics.[8] So I ask you also for this reason to please reconsider again whether you could not announce a talk. I would then request the topic as soon as possible.– In the hope of not extending my request in vain, in view of the good cause, I am with best regards, yours very truly,

A. Schoenflies.

# 51. From Moritz Schlick

Rostock, 23 Orléans St., 10 June 1920

Dear, highly esteemed Professor,

Yesterday your friendly letter arrived.[1] Thank you very much indeed for having gone to such lengths with my draft![2] I only wish I had sent you the manuscript a little earlier; then I could still have changed a few things. But it is too late now, because it has long since been typeset for the *Naturwissenschaften:* the issue concerned is due tomorrow already, on the 11th. Please do not be irritated if I take up your time again today with the old causality principle, returning once more to some of the points. I would so much like to penetrate to the utmost clarity possible.

As concerns the possibility of causality in a world without uniformity, I fear I left a gap in the explanation of my view and I hope that no difference of opinion will remain after it has been filled. Surely, we could arrive at the discovery of gravitation, e.g., by observing all comets that orbit the Sun along different hyperbolic paths.[3] But I would like to think that without a certain repetition of the same in nature we would not even be in a position to describe the cometary course correctly nor to determine it quantitatively. To establish the various cometary locations, we evidently need certain instruments that can be set up the same way at different times; we must be able to take measurements with them. And the practical application of any scale and any dial seems to me to be based on the principle of repeating physically equivalent processes. When we say that the various cometary motions are based on the same law of gravity, then, it seems to me, the verifiable sense[4] of this statement can only be that the execution of very specific operations on comet observations somehow leads to the *same* experiences. Such considerations would seem to me to have very fundamental validity, and thus I would believe that without a recurrence of the same, one should not speak of a regularity. Am I mistaken in this? It would be nice to get some clarification still—hopefully in person even, in the near future!

About the problem of the causality postulate being violated by the old law of inertia (the philosopher H. Driesch, now full professor at Cologne, oddly asserted about it: The law of inertia is "the causality principle applied, nothing more"),[5] I am afraid to say I still have not reached final lucidity. For I still do not quite see how far your view actually deviates from the representations in my essay.[6] One part of your objections seems to be directed against the first assumption toward the solution, which in the article itself is considered only preliminarily and is later improved—obviously not enough. The initial purpose of the consideration is merely that absolute space, which Newtonian mechanics must obviously assume, does not need to be taken here as a *cause,* in the sense of the principle of causality. In o[ther] w[ords]: inertial resistance in certain motions need not be looked upon as *effects* of an absolute acceleration, but rather can be interpreted as its defining property. This statement does not seem to me to contradict your view, though, and if I have understood correctly, I am mistaken only in my explanation of the reason why Newton's approach is so unsatisfactory. I thought it was to be found in that the former mechanics stopped applying the causality explanation earlier than was *necessary;* am I right to conceive that it stopped sooner than it was *permitted* to at all? The latter seemed only to follow from the preconditions indicated in the paper, which are irrevocable postulates for current science, of course. I naturally must admit, though, that it was an impermissible schematization to speak of an object's properties as if there were only a finite number of them; likewise, that some of the properties were always a consequence of the others as soon as the object belonged within a "theoretical system." It just seemed to me that in experience no other theoretical systems existed than those which operate with *processes* (four-dimensionality of all real things).

I was probably not right with the assertion that a gravitational field was not observable in the same sense as masses. This does apply, at very most, in the very rough sense that one may say: I do perceive two objects but not the gravitational field midway between them. It obviously does seem to me to be a debatable point whether in examining *Mach's* ideas the word "perceptible" may be taken in the broadest sense.[7]

It was somewhat unphilosophical and dogmatic of me, of course, to think that lawfulness within a time segment should not be denoted as causal.[8] My reasons for it were simply (1) the fact that in *conscious* reality, time plainly does seem to play a preferred role; and (2) that those regularities would have to be of a different character from the ones in the direction of time. But these are only subjective reasons, which may perhaps even be brushed aside upon closer consideration.

I do hope it is still possible to hear a few words verbally about these things! We heartily hope that you will do us the pleasure of stopping in Rostock, if your time in any way allows.[9] You will be traveling via Warnemünde, in any event, won't you? In the very worst case, I ask you please at least to inform us of the time of your transit so that there is a possibility of greeting you in the train. We are hoping for very favorable news and have the warmest wishes for your welfare.[10] Once again, sincere thanks for your letter! In extending best compliments to your esteemed wife, I am truly gratefully yours,

M. Schlick

P.S. I am going to send a copy of the English translation of *Space and Time* to you today.[11]

# 52. To Arthur S. Eddington

[Berlin,] 11 June 1920

Highly esteemed Colleague,[1]

It is an inescapable necessity for me to clear my conscience by writing you a brief letter. You know that I did not visit you this spring despite having informed you of such plans.[2] But first of all, I was prevented by far too many extraneous obligations, which restricted my time very much; on the other hand, I had the feeling that a trip to England at this time would be perceived by our English and German colleagues as an ugly attempt to fish for favor [*captatio benevolentiae*];[3] this I wanted to avoid. In such a case even an internationally minded person must curb himself if he does not want to do harm to the good cause.

I have now received from you the detailed report of your expeditions,[4] for which I thank you kindly. I greatly enjoyed being able to witness in this way, retrospectively, so to speak, the successful analyses by English astronomers. The problem of the line displacements is becoming more exciting by the day. Now a personal friend of mine, our colleague Julius in Utrecht, has also come to the result, upon careful review of all the data, that the Earth-Sun line displacement does not exist;[5] but I have no doubt in my mind that in the end this consequence of relativity theory will find verification as well. Perhaps a careful comparative study of the terrestrial light sources will bring clarity.–[6] I have the impression that Julius's theory on the emergence of asymmetric line broadenings through the influence of dispersion in relation to local density fluctuations in the solar atmosphere is too little appreciated by astronomers; in any case, I believe that Julius describes many things more naturally with his theory than is commonly achieved on the basis of the Doppler principle.[7]

With best regards, yours very truly,

A. Einstein.

# 53. From Moritz Schlick

Rostock, 23 Orléans St., 12 June 1920

Dear, most esteemed Professor,

When I wrote my last letter,[1] I unfortunately did not have the causality article itself at hand, only in my memory.[2] Just now I received the rel[evant] issue of the *Naturwissenschaften* from the publisher and I find upon rereading the article that my formulations were indeed more unfavorable than they had appeared to me in my imagination. The essentials of this article are supposed to be incorporated into a future natural philosophy; on that occasion I shall be able to add the necessary improvements.–

In hope of favorable news conc[erning] your visit[3] and with cordial regards also from my family, yours in sincere gratitude,

M. Schlick

# 54. From Willem H. Julius

[Utrecht,] 13 June 1920

Dear Colleague,

That was a nice surprise! We are all immensely pleased to be in possession of this excellent picture and thank you very cordially for it.[1] It will take turns adorning the music room and study in our new apartment.

I still cannot abandon the idea that, by a suitable introduction of selective radiation pressure on gaseous molecules, the apparent contradiction between the general theory of relativity and the results of observations of line displacements will somehow be removed.[2] But I do not want to molest you again with my further considerations before they are somewhat matured, and so will first speak about it sometimes with Ehrenfest or Lorentz; they are better accustomed to my blunders. Ornstein considers the idea appropriate and feasible, as far as he can see up to now.[3]

I do not know how far your nomination to Leyden has come;[4] I hope we shall see you again soon.

Best regards to you, also in the name of my wife and children.[5] Yours truly,

W. H. Julius.

## 55. From David Reichinstein[1]

Leipzig, Flossplatz $29^{II}$, 14 June 1920

[Not selected for translation.]

## 56. To Hendrik A. Lorentz

[Christiania, Grand Hotel,][1] 15 June 1920

Highly esteemed Colleague,

Already in Haarlem I told you that it will be a great pleasure for me to accept the invitation to the Solvay congress. I shall likewise be glad to take on the report.[2] It is a heartfelt need of mine to see our French friends in particular after such long and difficult times and to shake their hands. It also pleases me that it is still possible, even today, to be treated as an internationally minded person, without being compartmentalized into one of the two big drawers.[3]

The time spent in Holland is still vivid in my mind's eye. I only regret the great effort that I caused you and Mr. Kamerlingh-Onnes. I should have thought of such an option as having my trip to Holland postponed by a couple of months, which I would have had no difficulty doing.[4]

In Berlin, Mr. Hettner[5] recently showed that the observed infrared eigenfrequencies (absorption) of vapors (excluding Bjerrum's rotational influence)[6] can be described in the form

$$\nu = m_1 \nu_1 + m_2 \nu_2,$$

where $\nu_1$, $\nu_2$ are the actual eigenfrequencies; $m_1$, $m_2$ are low, positive whole numbers (or zero). This fits nicely with Bohr's theory if one assumes that in the potential energy the quadratic terms in the deviations from the state of rest are the primary, but not quite the only, critical factor.[7] The basic frequencies $\nu_1$, $\nu_2$, etc., furnish interesting touchstones for molecular models.

With cordial greetings to you and your esteemed wife, I am, your devoted servant,

A. Einstein.

# 57. From Hans Reichenbach

Stuttgart, 13 Wiederhold St., 15 June 1920

Dear Mr. Einstein,

I must approach you with a big request. In the near future my work on *Relativitätstheorie und Erkenntnis a priori* [*Relativity Theory and a priori Knowledge*] will be appearing as a booklet (similar to Freundlich's in breadth) with Springer.[1] I would now like to ask for your permission to dedicate the work to you. You know that with this work my intention was to frame the philosophical consequences of your theory and to expose what great discoveries your physical theory have brought to epistemology. By placing your name at the head of the text, I would like to express how greatly philosophy in particular is indebted to you. I know very well that very few among tenured philosophers have the faintest idea that your theory is a philosophical feat and that your physical conceptions contain more philosophy than all the multi-volume works by the epigones of the great Kant. Do, therefore, please allow me to express these thanks to you with this attempt to free the profound insights of Kantian philosophy from its contemporary trappings and to combine it with your discoveries within a single system.

With this dedication, however, I also want to extend to you my very personal thanks as well, which I owe you. For I was permitted to hear out of your mouth the deepest truths I ever encountered in physics; I shall never forget the great inspiration you gave my intellectual efforts and this text in particular.[2]

I am, in cordial devotion, yours,

Hans Reichenbach

# 58. From Ernst Cassirer

Hamburg, 26 Blumen St., 16 June 1920

Highly esteemed Colleague,

Accept my cordial thanks for your letter and for the great effort you expended on a thorough study of my manuscript.[1] It was of prime value to me to know from you personally that my understanding and rendition of at least the mathematical and physical *content* of your theory of relativity is essentially correct. As regards your critique of individual epistemological *consequences* that I had drawn, I do not need to say that it likewise was exceptionally helpful to me and caused me to look over my whole exposition again thoroughly and to revise many points.

In particular, I have now given stronger emphasis to the theory of relativity's purely empirical point of departure which, when set against the analysis of the theoretical assumptions, certainly did get the shorter end of the deal.[2] I now intend to publish the work with the alterations and additions prompted by your comments and objections: not in the opinion that it could offer a final solution to the difficult epistemological problems that relativity theory leads to—rather, simply to draw more interest among philosophers in general toward these issues than has hitherto happened. I myself hope to learn from the discussions about these questions—especially also from objections that may arise on the part of physicists toward my conclusions. I am, in expressing my gratitude and my admiration, yours,

Ernst Cassirer

## 59. To Hedwig Born

Christiania, 18 June [1920]

Dear Mrs. Born,

The news about the bitter experience that you had to live through affected me very much. I know what it means to see one's mother in death throes, without being able to help.[1] There is no consolation. We all must bear such tribulations, for they are inseparably bound with life. One thing does exist, though: loyal friendship and mutual support in carrying the burden. We do share so many beautiful things together that we do not need to succumb to numb brooding. Dead elders do live on in the young. Don't you sense it now when you, in mourning, look at your children?–

I am here with Ilse and am giving a few lectures for the students,[2] sprightly, likable young folk. Add to that the marvelous natural surroundings and downright formidable heat, which one shouldn't expect up here.

Warm greetings to you and Max, yours,

Einstein.

[. . .][3]

## 60. From Leonhard Grebe and Albert Bachem

Bonn, 18 June 1920

Highly esteemed Professor,

In the attached we forward to you the revision of our paper on the gravitational shift, wherein we added, as you wished, all the microphotogram data used. We request that you please be so kind as to arrange publication in the Academy's

Reports.[1] The reason for such a long delay in sending it out is that, as you know, we wanted to use the latest microphotometric recordings for a new *measurement* of the shift. But unfortunately this endeavor did not succeed because the measurements revealed that time-lag effects in the potassium photocell are included,[2] so the thousandths of a unit Ångström, upon which it rests here, cannot be guaranteed. We would not like to postpone publication of our earlier data until success of the new analyses, however.

We thank you heartily once again for your multifarious support.[3]

In great respect, yours very devotedly,

L. Grebe
Dr. Alb. Bachem.

# 61. From Heinrich Zangger

Zurich, 8 Zürichberg St., 19 June 1920

Dear friend Einstein,

I beg your pardon. I wrote the letters very quickly, partly while traveling, so I understand why you could not read some things easily.[1] Langevin sends his regards and suggests you come to Geneva sometime.[2] Then he too would travel from Paris to Geneva. He offered to stress Weyl's importance to the Swiss government. He seems to value Weyl extremely highly and to expect very much more from him. *I naturally immediately passed this on to Prof. Gnehm.*[3]

Your son came to see us today. He told me that he was going to French Switzerland for the holidays; he wants to learn French. I was not at home and thus could not ask him about your wish to spend time together with him in the summer;[4] but the most sensible thing surely would be for Father to give way and come to Switzerland when the chance arises.

Victor Henry is coming to Zurich in the next day or so. He is very interested in the physical chemistry position. I also think that would be a proper solution; [he is] a motivating, richly talented person with a quite different temperament. I am already considered quite presumptuous, but not easily dismissed, since Einstein and Debye have conferred such a good reputation to my activities in Zurich.[5]

We are now going to take up the spectrophotographic research again on a broader basis, with my large Hilger instrument, which I bought for myself 10 years ago.[6] In Brussels, I agreed to give a report about this field for next year.[7] It is extremely interesting how typical the results are in the ultraviolet; and most interesting of all is that the typical absorption bands in layers of 1/100 mm of serum solution are so clear, because absorption in a series of alkaloids at a slightly shorter wavelength than serum absorption yields typical spectra. You know that I already examined spectral displacements by absorption with Shepperd in Paris.[8] It now involves determining the displacements by the colloids of blood; then we would have a very far-reaching physical method for detecting alkaloids and characterizing physic. solutions and poisons in part to a considerable degree. You remember still how I managed to identify a blue dye and its origin. I imagine that, in a short while, we shall have come so far in the area of poisonings, as with our ultraviolet instrument for traces, that we shall have proof in hand [. . .] for the main groundwork of evidence of the presence of a foreign substance, the distribution, and at least the groupings, with a serial recording of 20 individual spectra, which we can easily fit on one plate. This is exceedingly important because poisonings from new substances are extraordinarily widespread. The introduction of mass poisons is perhaps the greatest misfortune that the war caused, along with reckless money-making with all the so-c[alled] substitute products.[9] As soon as any old filth can be presented in some form or other so that the public can identify a particular property in that object, which it must otherwise pay dearly for, and as soon as it is just a little cheaper than the valued one, it is bought. It started with celluloid and honey; and there will be no end to it.[10] The workings of knowledge about isolated causal relations are such that it easily becomes a means of power.

It will be of further interest to you that the very latest is that we must investigate suspended sentencing and its psychological repercussions. I find the suspended sentence an essential factor in the responsibility of justice, as recourse against the stifling framework that looms behind every law.

With the wish that you are doing well,

Zangger

[. . .][11][12]

## 62. From Vladimir K. Arkad'ev[1]

[Moscow,] 22 June 1920

Highly esteemed Professor,

With great interest I read your new papers, which fell into my hands through National Commissar of Culture and Education Lunacharsky.[2]

At the University of Moscow a first convention of the Russian Physical Association will take place at the beginning of September 1920. The program will address, among oth. things the question of relations with foreign scholars.[3]

With best regards, yours very sincerely,

W. Arkad'ev.

My address: Moscow, Pretschistenka 27, apt. 7 (via Berlin, Ministry of War U. 5), for Prof. W. Arkad'ev, Secretary of the 1st conven. of Russ. Phys. Association.[4]

# 63. From Hendrik A. Lorentz

Haarlem, 23 June 1920

Dear Colleague,

I think I must tell you once again how things stand with your nomination, or, more accurately put, with the obligatory royal approval. In the final days of your sojourn in Holland, I could still inform you that no more obstacles remained and that there was a real hope of seeing this affair dealt with roughly on the 12th of this mo.[1] Unfortunately, the matter has dragged on again, so it was good that you were not waiting for the results. Meanwhile, we have now come close to the goal. The State Council has handed down its decision (obviously, in the favorable sense) and thus the last formality is met.[2] Prof. Oppenheim, a member of this council (as well as of the curators of the University of Leyden), told me he heard from the Minister's general secretary that the royal decree could be expected very definitely at the beginning of July at the latest.[3] Since vacation starts on July 10, you may *perhaps* still be able to hold your inaugural speech during the last week of the semester. There is absolutely no doubt that, in any event, you will be able to do so directly after the vacation, end of September.

As concerns the question of whether you would like to deliver your speech before or after the vacation (given that the former is also possible), you should do entirely as you see fit. I can only say that the sooner you come, the more we, all your friends in this country, will like it.

Howsoever, as soon as the royal decree is published, Ehrenfest or I will telegraph you "properly"; we shall include on which day you can give your speech, if you so desire. The decision is then still left entirely to you.

---

I was very pleased to hear from you that we are going to see (and listen to) you at the physics conference in Brussels.[4] I gathered from your letter that you accepted the invitation to Christiania.[5] I hope very much that you will find the time to see something of that beautiful countryside; you have worked so intensely for years that you can allow yourself some relaxation.

Yesterday, I received the fine and stimulating book by Eddington: *Space, Time, and Gravitation: An Outline of the General Relativity Theory.*[6] There is much in it I would like to discuss with you; I would be delighted if that could happen soon.– With cordial regards, yours faithfully,

H. A. Lorentz

# 64. From Niels Bohr

[Copenhagen,] 37 Stockholmsgade, 24 June 1920

Dear Professor Einstein,

I cannot tell you how great a joy it was for me to hear that you are coming to Copenhagen and with what expectations we are all looking forward to your lecture.[1] For me it was one of the greatest experiences ever to meet you and speak with you, and I cannot tell you how thankful I am for all the friendliness you showed toward me during my visit to Berlin, and for your kind letter, which I am ashamed not to have answered yet.[2] You do not know how great a stimulus it was for me to have the long-wished-for chance to hear from you personally your views on the problems with which I have been occupying myself. Never shall I forget our discussions on the way from Dahlem to your house,[3] and I very much hope that during your visit here, occasion will present itself to take up that discussion again. Unfortunately, my wife has been at the clinic for the last few days to give birth to a son,[4] which luckily went well, and we must, therefore, deny ourselves the pleasure of seeing you at our home until your return to Copenhagen. If, however, during your current sojourn here you should have the time and inclination to take a walk alone with me in the beautiful environs of Copenhagen and along the lake, or perhaps have a meal together with a few close friends, it would be a specially great pleasure for me.

With most cordial greetings also from my wife, yours very truly,

Niels Bohr.

# 65. To Hans Thirring, Adolf Smekal, and Ludwig Flamm

Copenhagen, 25 June 1920

My dear Friends & Colleagues,

First I would like to say that by no means did I put in my word about the ticklish Ehrenhaft problem at my own initiative but felt compelled, after a number of inquiries, to offer my opinion.[1] It revolved around the issue of whether Ehrenhaft should receive or retain the possibility of doing independent scientific research. To me there could be no doubt about that, for E. is diligent and enterprising and has undoubtedly helped advance progress.

Now, it seems to me from your letter that the faculty has found a quite fortunate solution, in that it has given E. an autonomous sphere of activity without the danger of his personality or approach having too much of an influence. I never did doubt the inaccuracy of E's interpretations of his own experiments, and I also believe that his obduracy—if it is not compensated by other influences—can become a disadvantage for the institute; in this I agree with you. You, Mr. Smekal, do know that I have always thought of Mr. Ehrenhaft's theoretical attitude in the way I indicated here.[2]

It would be particularly bad if Viennese theoretical physicists found their options narrowed.[3] Among the informed, there can be no doubt that in Vienna today theoretical physics outweighs experimental physics. Under no circumstances should Vienna neglect to nurture its exceptional traditions in theoretical physics.[4] It seems to me to be your right and, indeed, your responsibility to insist resolutely upon this. What I had feared, on the other hand, in the initial phase of the Ehrenhaft affair was, on the contrary, that fostering experiments could fall far too much into the background if no genuine, spirited experimentalist received the vacant chair. It seems to me, though, that the way out[5] devised by the faculty already reduces this danger substantially.

With best regards, I am yours,

Einstein.

## 66. To Hans Reichenbach

[Berlin,] 30 June 1920

Dear Mr. Reichenbach,

I am really very pleased that you want to dedicate your excellent brochure to me,[1] but even more so that you give me such high marks as a lecturer and thinker.[2] The value of the th. of rel. for philosophy seems to me to be that it exposed the dubiousness of certain concepts that even in philosophy were recognized as small change. Concepts are simply empty when they stop being firmly linked to experiences. They resemble upstarts who are ashamed of their origins and want to disown them.

Pardon my brevity; my correspondence debts are immense. With best wishes and cordial greetings to you and Regener,[3] yours,

A. Einstein.

## 67. To Moritz Schlick

[Berlin,] 30 June 1920

Dear Mr. Schlick,

I did travel past, after all, unfaithful me, albeit with a heavy heart. But I had my daughter and much hand luggage with me, so it probably could not have been otherwise.[1] Your letter was again a masterpiece of clarity,[2] and I let myself be quite completely convinced by you, especially as far as the fundamental role of the repetition of similar events is concerned.[3] I really could not quite see the forest for the trees there. Only on the question of Newton's law of motion-causality, which treatment you, on your part, are not quite satisfied with either, do we still not quite see eye to eye.[4] How nice a private conversation about it would have been. It is a complicated affair. If I take the equation

$$\text{mass} \cdot \text{acceleration} = \text{force},$$

then "force" is something "absolute" ([independent] of the frame of reference), likewise with mass, if only the units (also of length) are fixed. Thus an absolute meaning must be assigned to acceleration too. This itself is defined by length and time, by the expression $\frac{d^2x}{dt^2}$; so one is not permitted, on the other hand, to define acceleration by the law of inertia either. One rather has to decide to define $x$ and $t$ themselves as absolute, or physically meaningful quantities. For $t$ this succeeds with a clock if one ignores the problem of simultaneity, $c$ = practically $\infty$; but for $x$ it won't work. One has to resort to ascribing a mysterious, i.e., empirically inaccessible reality to space. But the special principle of relativity in mechanics again speaks against this.

Besides, even according to the gen. theo[ry] of r., physical space has reality, but not an independent one, in that its properties are fully determined by matter.[5] It is incorporated into the causal nexus without playing a one-sided role in the causal series. It is to the credit of his logical conscience that Newton decided to create absolute space (and absolute time, which, however, was less necessary). He could just as well have called the absolute space the "rigid ether."[6] He needed such a reality in order to give objective meaning to acceleration. Later attempts to do without this absolute space in mechanics were (with the exception of Mach's) only "playing hide-and-seek."

With cordial regards, also to your wife, and in wishing (nonetheless) to see you soon again, I am yours,

A. Einstein.

## 68. From Edouard Guillaume[1]

Berne, 30 June 1920

Dear Einstein,

I gather that you are in possession of my latest paper[2] and that you have read it—which is probably not the case . . . ![3]

As various persons have explained to me, however, it would be very fruitful if you could substantiate your position sometime; and for me personally it has become absolutely necessary that you speak up. For in a controversy in the press (in the *N[eue] S[chweizer] Z[eitung]*), your friend Grossmann wrote that my "views about relativity had been completely repudiated (?) by you."[4] When I thereupon asked Grossmann when and where you had expressed yourself in this way,[5] he replied to me that in a letter, you had written that my "views on relativity were nonsense."[6] This groundless judgment, which is apt to hurt me to the highest degree, forces me to insist that either it be given a basis or retracted. That is why I must ask you, dear Einstein, for a reply that I can publicize.[7] This you really can't refuse your old Office colleague![8] It will be even easier now that the *experimentum crucis*—the spectrum-line shift—which is decisive for my interpretation, has been established in my favor by Prof. Julius on the basis of measurements.[9]

Briefly, my interpretation can be summarized as follows:

1. As I have proven in my last paper, time can be *singly* or *multiparametrically* described. 2. This descriptive equivalence requires that in the quadratic forms

$$ds^2 = dU^2 - dX^2 - dY^2 - dZ^2 \; ; \; ds^2 = g_{44}du^2 - \ldots - g_{11}dx^2 ,$$

the quantities $\frac{dU}{c_0}$ and $\frac{du}{c_0}$ are not periods but different masses of the same timespan; $\underline{dU}$ and $\underline{du}$ have only *one* physical meaning: they represent "light paths." You can find the proof in the mentioned paper and on the enclosed printer's proof.[10]

This interpretation can be illustrated very nicely with the help of the Lorentz transformation. For a light signal, you have for an infinitesimal time interval

$$du_2 = \beta(du_1 - \alpha dx_1) , \; dx_1 = du_1 \cos\varphi_1 ;$$

hence, if we set

$$du_2 = \varepsilon , \quad (1)$$

where $\varepsilon$ is an infinitesimal quantity independent of $\varphi_2$,

$$du_1 = \frac{\varepsilon}{\beta\sqrt{1 - \alpha\cos\varphi_1}} . \quad (2)$$

(1) represents an elementary light sphere and (2) shows that it appears, "judged" from $K_1$, as a *rotational ellipsoid* with one focus at the origin.

This thoroughly simple consideration shows how the principle of *relative* constancy of the velocity of light, which I have introduced, must be understood. For if one puts

$$du_2 = c_2 dt ; \; \varepsilon = c_0 dt$$

you thus obtain:

In $K_2$ a sphere: $c_2 = c_0$

"Judged" from $K_1$, it appears as an ellipsoid:

$$c_1 = \frac{c_0}{\beta(1-\alpha\cos\varphi_1)}.$$

If you wish further explanations, I am naturally always gladly prepared to provide them for you.

In anticipation of a prompt sign of life, I remain, yours truly,

E Guillaume

## 69. From Gaston Moch[1]

3 July 1920

[Not selected for translation.]

## 70. To Hans Albert and Eduard Einstein

4 July 1920

Dear Albert and dear Tete,

I long so much to see you again.[1] Since you, d[ear Hans] Albert, are away from the middle of July until the middle of August, nothing can be changed now.[2] After that, you have school again there.[3] From September 10th to 25th, I have obligations in Germany.[4] But then I am free. So I suggest that we meet at the end of September for one or two weeks at my friend's house, the priest Camillo Brandhuber in Benzingen near Sigmaringen.[5] It would be too expensive and complicated in Switzerland.[6] For you, it's just a short trip and physically refreshing at the same time. There we are in excellent hands and it's very casual. If you don't have holidays at that time, then we must simply apply for leave from school for this short time. The rector[7] will understand that a father also has to see his boys from time to time. It's rare enough as it is. Then I'll pick you up in Sigmaringen once I know exactly when you are arriving.

I hope you have received the 1,000 francs.[8] In one month I hope to be able to send remittances again. Under the current conditions, it's not so easy; but I'm hoping to be able to pull you through in Zurich, after all.[9]

Make inquiries at the German consulate, d. Albert, about what is required in order for you to obtain the travel permit. But also say that you are the son of the Berlin resident Prof. Einstein; otherwise nothing can be achieved.

Warm greetings to you both and to Mama,[10] and do write back soon, yours,

Papa.

# 71. To Edouard Guillaume

[Berlin, 4 July 1920]

Dear Guillaume,

My behavior must appear unkind and unloyal to you; it may seem that I wanted to condemn you by my silence.[1] In reality, though, it is entirely otherwise. I often tried to understand your statements but always utterly failed. So if I wrote Grossmann that it was nonsense, this must be understood with reference to me, or better yet, to the present state of my brain; nonsense[2] is what one calls whatever one cannot grasp; there is no other criterion. Now you must excuse me all the more, since papers and manuscripts are raining down on poor me like whip lashes on a cab horse.[3] But since you attack me so energetically, I have no choice but to try to find elucidation through repeated correspondence on the basis of the enclosed little notice[4] and your letter.

You say, $\theta$ and $\theta'$ were periods of clocks, hence things, not numbers, and write[5]

$$\theta\tau = \theta'\tau'.$$

Are these somehow symbolic equations? I cannot comprehend it as an equation between numbers. That is why I cannot understand all that follows and think that you have temporarily fallen for a forbidden mysticism, in that you forgot that equations deal with numbers. When I speak of "length $l$" or "period $\tau$" of a clock, then I always mean "magnitude $l$ of the length upon measurement with a given unit of length" or "*magnitude* $\tau$ *of the period* upon measurement with a given standard clock."

I can think further only once you have removed this digestive complaint for me. Warm greetings to you, our old colleagues, and your wife, from your old

Einstein.

# 72. From Joseph Petzoldt[1]

Spandau, 6 Wröhmänner St., 6 July 1920

Highly esteemed and dear Professor,

For a long time already, I have been cherishing the wish to be able to discuss the epistemological aspect of relativity theory with you more often, which is especially close to my heart. I just did not dare to express it because I was afraid of bothering you, whose time is probably subject to high demands from very many quarters. Now, however, the epistemological problems are becoming increasingly pressing and need clarification more than ever. I could see this at the conference of the Kant

Society in Halle, which I went to at Vaihinger's and Kraus's invitation.[2] The philosophers there were not even clear about the theory's experimental bases.[3] Furthermore, Mr. Riese mentioned to me that you would surely be inclined to debate with relativity theorists interested in philosophy. In addition, the young Dr. Winternitz[4] from Prague has come over here. He has a vibrant interest in the matter, given the opportunity. Finally, I think that we should seek to have epistemological articles accepted on an equal footing in the *Zeitschrift für Physik* (with particular caution and strict censorship). I have already spoken with Privy Councillor Scheel[5] concerning this point; he would also like to get in touch with you about it after your return.

How essential all of this is emerges, for inst., from Helge Holst's[6] paper, with which you are probably already familiar. It wishes to reduce the th. of rel. to the level of a calculation aid, and similar endeavors came to light in Halle. I took up the opposition. Aside from the article about Holst, which I am enclosing herewith, another more comprehensive one on Holst's work is going to appear in the Kant Society's *Annalen*.[7]

Even though I cannot agree with you on the issue of the world's finitude,[8] you will see, nonetheless, that I still conform in all essentials with relativity theory. I just expressed this again in no uncertain terms in the new edition of Mach's *Mechanics*.[9] Full clarity on all these problems will only be possible, however, if physicists and philosophers discuss them together. Mr. von Laue, and oth., will certainly be willing as well.[10] Thus if you would be inclined to suggest such a meeting in a few weeks (I expect to be away until the beginning of August), you would be doing a very good deed.

With cordial greetings, yours very truly,

J. Petzoldt.

# 73. From German League for the League of Nations

Berlin [3/8?], 7 Unter den Linden, 8 July 1920

Esteemed Professor,

After long hesitation, we decided to ⟨bother⟩ impose on your precious time with the following request: In close consultation with Dr. Elisabeth Rotten,[1] we arrived at the conviction that there is a pressing need to lay before the German public the plight of German science, what help has been given until now, and what can still be done.[2] This could best happen with an article in our newsletter, *News from the*

*German League for the League of Nations*, which is sent out to hundreds of newspapers for reprinting. In agreement with Dr. Rotten, we invite you, esteemed Professor, to write this article. We think it appropriate that the drive in aid of German libraries by English universities be placed at the focus of interest,[3] for the general public is only sparingly informed about this drive and its full import. If you would take on this task, we can be sure that neutral and enemy countries abroad would also take note of your arguments with the necessary attention.

Considering current space limitations in German newspapers, the article ought to be no longer than 100 lines.

As honorarium, we take the liberty of offering you 100 marks.

We would be very grateful for your kindly promptly notifying us whether you would like to undertake the task of writing this article.

In utmost respect, most humbly,

Deutsche Liga für Völkerbund
Press Department:
Müller-Ja[busch][4]

# 74. From German Central Committee for Foreign Relief

Berlin W. 8, 56 Mohren St., 9 July 1920

Re: Quaker Relief

To Professor Einstein,

In foreign countries, the new fund-raising campaign is now starting for the German Children's Fund [*Kinderhilfswerk*]. We can provide foreign press agents currently staying here with copious reports and statistics about the need in Germany, and descriptions about the distribution and receipt of donations.

People abroad do not want to see only governmental and official opinions, however; they also want to hear statements by generally popular and known personalities that fully acknowledge the relief organization's stature, and this not in the form of long speeches, but rather as incidental statements, scattered informative remarks in dialogues or letters.

At present, 632,000 children are being fed by the American Quakers alone. English Quakers are providing particularly for the youth and the poor middle class,[1] and generous charitable gifts have come from other sources to needy regions, segments of the population, and institutions. Continuation of this aid is urgently needed in the interest of adolescents, expectant and nursing mothers, the poor and the sick. I hope I do not turn to you in vain, in the name of the German Central Committee, with the petition to contribute toward executing the assistance so kindly offered and so willingly provided.

This request is being made completely unofficially, at the wish of the donors; thus I ask you please to direct your reply to this letter to me personally, possibly as a letter excerpt or in any similar form you wish, with an original signature.

I shall be happy to provide any additional information.

For the German Central Committee for Foreign Relief,

Elsa Herrmann, Jur[is] Dr.[2]

# 75. From Max Born

Frankfurt-am-M[ain], 16 July 1920

Dear Einstein,

It's highly probable that we'll be going to Göttingen, specifically if Franck receives the call and accepts; the faculty has nominated him.[1] Now the question of my successor becomes acute. Schoenflies wanted to write you and ask for your reference.[2] I want to have *Stern*, of course.[3] But Wachsmuth is not willing; he said to me: "I esteem Stern very much, but he has such a demoralizing, Jewish intellect!"[4] At least it's open anti-Semitism. But Schoenflies and Lorenz want to help me.[5] Wachsmuth proposes Kossel, a very cleverly designed proposal, since obviously nothing can be said against him— at most, that he has no mathematical proficiency; but that's not a fault.[6] Stern built up our little institute and thoroughly deserves the recognition. I don't have to point out his merits to you, obviously. Then Lenz and Reiche still come into consideration,[7] perhaps other outsiders too. An embarrassment of riches!– I asked Laue for an expert opinion;[8] perhaps it would be good if you discussed it with him so that your judgments do not conflict with each other.– I am very lazy now and hardly work at all; the only experiments I pursue with enthusiasm are those on the free length of path of atomic beams of silver. My assistant is doing the thing very well.[9] Our apparatus system is completely finished, but measurements are unlikely to start before the holidays, unfortunately. We are setting out for Sulden in South Tyrol (Italy) on Aug. 6th; I am tremendously looking forward to getting away from it all again and seeing something pretty. My wife has recovered again a bit from the trying time after her mother's death.[10] We go on many excursions; that does her some good. Tomorrow we are traveling to the Rhine, with which she is as yet unacquainted. The children are well.[11]

Unfortunately, the decision about Göttingen is drawing out endlessly; we still don't have an apartment there. Next week my wife wants to drive there and look for accommodations.

Will you be coming to southern Germany sometime? We would so much like to see you and talk.

With warm greetings to your dear wife and the young ladies,[12] yours,

Max Born.

## 76. To Paul Ehrenfest

[Berlin,] 19 July 1920

Dear Ehrenfest,

The violin has finally arrived![1] How can it possibly have become so damaged? They must have handled it barbarically. But it was insured by me. (Something like 1,000 marks.) So go to a good instrument maker immediately and have the damage assessed, so that you can be reimbursed. If you have the violin repaired, consult with an expert beforehand so it doesn't get ruined. Send back the case and the bow as soon as possible, because it is my only case and my only bow. How eager I am to have you try the instrument out; it really is fine.

The nomination is now happily settled.[2] So I am coming in October, in order to deliver the stale old, but duly withheld, sermon. September is impossible for me because of the science convention.[3] Afterward, I intend to meet my boys in southern Germany.[4] What's the latest on the Hall effect?[5]

I received the photograms from Grebe and Bachem and do indeed find their method of line selection well founded.[6] Comparison is possible only on the basis of photograms. I am confident now that the line displacement problem will soon be resolved satisfactorily. Koch has improved his photometer substantially again.[7] I was there yesterday and the day before, delivering the promised talk.[8] Some propaganda for theoretical physics had to be undertaken so that Epstein can be appointed there.[9] There really is a chance. The mathematicians Hecke and Blaschke are there as well,[10] so it would be very nice indeed for Epstein.

Excuse my brevity; I'm working myself to pieces. Greetings to you, your wife, the learned daughter, and the dear little rompers,[11] from your

Einstein

Kind regards to Aunt and our big boys.[12]

# 77. To Edouard Guillaume

[Berlin,] 19 July 1920

Dear Guillaume,

In the formula for the light vectors,

$$A\sin\frac{2\pi}{\theta}\left(t-\frac{lx+my+nz}{c}\right)$$

$t$ means the standard clock reading at the specified location relative to the resting coordin. system, $\theta$ the progression of readings of such a clock as a complete wave passes by it (period of oscillation). Hence, both quantities are numbers that are obtained as measurement results with standard clocks at rest. Both are obtained by counting the periods by means of a standard clock.[1] I do not understand what you conceive as a "period" in your letter; with me, it is not a number but the designation for a cyclic process that repeats itself.

Your deduction of the relation $\theta t = \theta' t'$ seems to me, quite frankly, completely crazy. In the consideration that leads to the relations[2]

$$\theta = \frac{\theta'}{\beta(1+\alpha l')}, \text{ etc.,}$$

$x, y, z, t$ are variables, for whose total values the equations must be satisfied identically. In the second part of the consideration, $x, y, z, t$ are the coordinates of a plane moving with the velocity of light ($x = ctl$, $y = ctm$, $z = ctn$, $x' = ct'l'$, etc.), thus something quite different from the first consideration. It makes absolutely no reasonable sense to compare the period length $\theta$ with the arrival time $t$ of that plane moving with the velocity of light, which passes through the origin of the coordinates at time $t = 0$.– Your relation $\theta t$ = invariant does not relate to all world points but only to a three-dimensional manifold![3]

My above remark about the meaning of $\theta$ and $t$ as results of measurements taken from standard clocks is just valid for the special theory of relativity, whereas in the general theory of relativity only $ds$ is defined as a measurement result. In the gen. th. of r., $dt$ initially has a purely conventional meaning. In the consideration about line displacements, however, $t$ again receives an absolute meaning in that the 4 coordinates are chosen so as to have the field of an isolated mass-point become *static*; thus the number of wavelengths that are traveling between the sun and the observer cannot depend on $t$.–

In any case, I must emphasize that it seems absolutely senseless to establish a relation between the quantities $\theta$ and $t$ appearing in your consideration. Upon closer reflection you will also find it so. If Hadamard and Levi-Cività are capable of attaching any reasonable sense to this relation $(t\theta = t'\theta'\ldots)$, then I cannot understand them; if anything, only envy them.[4]

So, once again. In the special theory of relativity both $\theta$ and $t$ are defined as numbers that indicate how many periods of a standard clock have elapsed, that is:

$t$ between the space-time epoch nil and the observed point in (space) time;

$\theta$ between the passage of two wave crests by the clock (at rest).

I consistently adhered to this interpretation in the special th. of r. and did not muddle anything up.

$dt$ is a fraction of one period of a standard clock. It is possible to introduce this concept because, in principle, there exist (short-period) clocks of arbitrary running speed, with whose aid one can execute that division.

It is different for the general th. of r.; but it is better if we do not go into this much subtler matter yet before the other one is cleared up.

You are going to think: "Einstein has become a loutish Boche." All the more amicable are my greetings to you, yours,

A. Einstein.

Repent, you hardened sinner![5]

## 78. To Gaston Moch

[Berlin,] 19 July 1920

Highly est. S[ir],

Your letter of 3 July was inordinately interesting and appealing to me.[1] I am convinced that you would have translated the booklet masterfully; besides, I must inform you that I have not heard from Miss Rouvière for a long time and that it has still not come to a contract with her. I find it very amusing that you know Miss R. personally so well.[2] If she does not insist on doing the translation, I would ⟨naturally⟩ very much like to leave it to you, under the same [conditions].

The man whom I quoted [and] who jokingly asserted "one should leave elegance to tailors and shoemakers," certainly did sport a big, bushy beard, but was no Vandal, rather an extremely subtle genius, namely, the Viennese Ludw[ig] Boltzmann, who discovered the relation between thermodynamic entropy and probability. Incidentally, it should be noted that h[is] lectures and other prose are also very amusing to read.[3]

I do not know what sort of a man Pfl[üger] is. (You probably mean the author of a not badly written popul[ar] work on relativity.)[4] I know him neither personally nor as a scientist.

I do find your judgment on Germans a bit severe. It seems that no nation is safe from falling victim to an imperialistic tendency, especially when its inner equilibrium is threatened by external success.

In sincere fellowship, I am yours tr[uly],

## 79. To Gösta Mittag-Leffler[1]

Berlin, W. 30, 5 Haberland St., 21 July 1920

Highly esteemed Colleague,

Multifarious obligations and particularly many business trips abroad unfortunately made it impossible for me to write the planned essay on Poincaré's position on the problem of geometry and experience.[2] I feared that in the little time available I could not have done full justice to the great master, since I would not have been in a position to study closely all his analyses related to the topic. With the kind request not to attribute this omission of mine otherwise than to the high regard I have for the task, I am, with great respect, yours very sincerely,

## 80. To Joseph Petzoldt

Berlin, 21 July 1920

Dear Colleague,

I also would find a gathering of the type indicated by you profitable, provided only those people join whom we ourselves invite.[1] If you would like to pave the way for something of the sort here, then I certainly shall appear. I am in Berlin during all of August and until about Sept. 10.[2] It would be my pleasure to have you and Mr. Winternitz visit me sometime in the evening so that we can talk about relativity.[3] I find Helge Holst's work weak;[4] it overlooks the fact that as soon as one relinquishes the relativity postulate, a hopeless multiplicity of possibilities remains before one, and that comprehension of the essential identity of inertia and gravity is lost at the same time. Your critique of Holst's work is legitimate.[5] Then we could also discuss the cosmological problem. I am firmly convinced of the finiteness of the world but do admit to not being able to furnish compelling evidence.[6]

Sincerely yours,

A. Einstein.

# 81. To Mileva Einstein-Marić

[Berlin,] 23 July 1920

Dear Mileva,

I've sent you 700 francs, the remainder of my holdings at the Züricher Kantonalbank.[1] From now on, for the time being, I'll be able to send you money regularly again. The matter seems definite at least for one year. We shall see what happens then. Nowadays everything is unstable and uncertain.[2] I fully agree to meeting the children on October th in Benzingen.[3] But in the future, you really should stop forbidding the trip to Berlin, at least for [*Hans*] *Albert*. It's simply ridiculous to treat a virtual grown-up like such a child.[4] My wife[5] will keep her distance from Albert; I could even take meals alone with him, if he specifically wants it. But these are silly trivialities. One shouldn't have to make such a fuss for you old women ⟨it's simply ridiculous⟩.

I'm sorry that Tete was ill again.[6] Do have him examined again for bacteria. If he isn't completely cured *before* puberty, then there's something wrong; that possibility does exist, though, in any case. You don't need to cross the border with the children, since they can travel home in *one* day, you know.[7] Our accommodations are private and very affordable. So unless you write again, I'll assume that the children are coming alone. I'll take care of the entry permit through my friend, the local priest.[8] I don't want to be his guest, three strong. But there is a woman in the village who supposedly would take us in. Albert is probably at the Chavans', as Michele wrote me.[9] Best regards also to the two children, yours,

Albert.

I have collected many postage stamps for Tete![10]

# 82. To German League for the League of Nations

Berlin, 23 July 1920

Esteemed Sir,

After considerable reflection and consultation with scholar friends of mine, I decided not to write the article in question.[1] I am sorry that in this case I am not able to be of service to your cause, which is so very near to my heart. Political points of view from inside and outside the country, which would have to be taken into account in the article's text, are of so complex a nature that I do not feel capable of doing justice to the task, especially considering that my understanding of the state of affairs is not sufficient. The circumstance that I am Swiss is another argument against it.[2]

In great respect,

## 83. From Paul Ehrenfest

[Leyden,] 24 July 1920

Dear Einstein,

I hope Lorentz has already informed you that now, at last, the pen pushers[1] have gotten so far as to approve the *professorship*;[2] the *automatic* consequence of this is that you can give your inaugural two months later, whether or not they have finished with their verification of your *character*.[3]– *So*— Your letter just came.[4]—Thank God that, finally, we can hope to have you in October—You can't possibly imagine how much we all, every one of us, like the thought of having you here periodically in Leyden. And you'll see—even for you it won't remain merely an unproductive bother!– Onnes is now trying to arrange for Langevin to come here at the same time as you.[5] Then let the four of us debate about magnetism. Now I believe I am *very* well prepared for this discussion from a study of the literature. I can clear away an enormous amount of rubbish. I am burning with impatience for this discussion.[6]

Tomorrow the violin case and bow are being sent to you, with maximum care taken in the packing and as quickly as possible.[7] It embarrasses me very much that you are stuck there bowless, for our sakes—forgive me!! Letter soon and greetings to you all—and never be annoyed with me!

P. E.

The violin seems to be attracting Tanya back to violin playing![8]

## 84. From Eduard Einstein

Zurich, 25 July 1920

Dear Papa,

Today I want to write you a little letter. We have vacation now. I always entertain myself well. Almost every day Richard comes and then we play together.[1] My first grade card in the 4th class goes like this:[2] (6 is the best) Arithmetic 5; language oral 5, language written 4–5, local history and geography 5, writing 3–4. In conduct, etc., I have "good." It would be nice if you were here, because then we could do many more things together. Other children, e.g., the Zürchers, can do many fine things with their Papa![3] On the 28th it's my birthday.[4] I think there's something nice coming, because Mama[5] is always doing something secretly and one room is always locked. I have a collection of cacti, which I tend every evening. They are very fun because they all have different shapes. Richard and I sometimes go to Prof. Heim's and play there in the garden.[6] It's fun there and there's no lack of fruit and berries. I'm reading "Götz von Berlichingen" by Goethe just now.[7] This spring I was allowed to go to the theater; they were playing "The Robbers" by Schiller.[8] Write me, too, sometime! Albert is in Geneva with Mrs. Chavan.[9] From what he writes, he's enjoying himself a lot.

Many greetings from your

Teddy

## 85. To Michele Besso

[Berlin, before 26 July 1920][1]

Dear Michele,

He [Guillaume][2] doggedly keeps writing the same rubbish, according to Napoleon's maxim that repetition is the most effective argument. Despite all the effort I take, I cannot find any sense behind his words and am (for my part) certain that there is none, either.[3] He seems not to be able to distinguish between *things* and *numbers* in his considerations. Julius's investigations certainly are interesting and speak against the existence of the redshift; but they *prove* nothing at all.[4] Until terrestrial light sources and both spectra have been analyzed properly with the spectrophotometer, accumulation of data is useless. One compares terrestrial lines with those of the solar spectrum under the assumption that those lines that are the least shifted against each other correspond, that is, under the precondition that no gravitational effects exist. With the immensely line-rich solar spectrum, this can lead to lines being inappropriately identified, as Grebe and Bachem have lately shown.[5] It is also important to prefer lines that do not exhibit the solar center–solar limb effect. Grebe [and Bachem] find the effect confirmed well by the cyanogen band, when they exclude asymmetric or distorted lines (photogrammatically established).[6] You will see that, in the end, a splendid confirmation of the theory results; I never doubted it for a single second. Weyl's theory cannot help here.[7] Either it yields the independence of measuring rods and clocks from their prehistories, then it's useless; or it yields nothing about this independence, then it is certainly false owing to the definiteness of atomic radii and frequencies. I was convinced of the inaccuracy (that is, the inapplicability) of Weyl's theory from the very start. Actually, there are almost exclusively factual reasons *against it*, but not for it.[8] But as long as you believe in [Guillaume],[2] it would be unfair to complain about Weyl, because these are so completely different orders of magnitude of sins; Weyl is always a profound, clear mind, sheer bliss to read, the other, however, is slovenly.–

That you take so much pleasure in [Hans] Albert warms my fatherly heart very much. I am going to see him and Tete in October over the fall vacation. I am having them come into the little land of Swabia, to Benzingen near Sigmaringen.[9] A joint stay in Switzerland would be too expensive. You really are right with your remark about the tractability of people; there actually isn't any superior guidance, though. All are puppets, moved by God knows what, but not by judicious people free from selfish motives. Only rarely is such a stroke of luck materialized in an enlightened despot (Marc Aurel or some Wilson,[10] if good intentions are also prudence and action). Your self-characterization is priceless, exaggerated, but not completely unfitting. If you had as much will and persuasive power as intelligence, you would have become a great leader of people.

My work is currently not up to much either. I split up my energy, have to deal with immense amounts of correspondence, evaluate, advise, act as protector, but make no progress on the larger questions. Planck has now accepted my derivation of his formula—are you familiar with it? (Quantum-like emission and absorption according to a statistical law, Boltzmann's distribution law.)[11]

The comment about Weyl's book relates to his theory of electricity. I do understand your view. You think: constancy in the relative extension of bodies does not need to be found in the theory's fundamentals; it would be nicer still if it resulted as a consequence, or were acceptable by finding a place in the theory as a *special* hypothesis. But don't forget, the theory is based on the geometry of measuring rods. Then it is assumed that the relative lengths of the measuring rods were a function of their prehistories. Then *real* measuring rods *should* come out as relatively invariant. That is why the measuring rods used as grounds for the theory would have to be just *theoretical* measuring rods, which behave differently from real ones. This is horrible. Added to that, the theory's previous accomplishments are lost. One has to take tensors of 4th rank instead of only of second rank, which carries with it a far-reaching vagueness in the theory: first, because considerably more equations come into consideration; second, because the solutions contain more arbitrary constants.

You yourself indicate another argument, the planetary system inside the pea. Ignore molecules for a while and assume that water has the same density throughout. Then one can introduce density, instead of mass, as the fundamental unit:

$$M \text{ (mass)} = D \text{ (density)} \cdot l^3 \text{ (length}^3\text{)}$$

Take Newton's law

$$k\frac{MM'}{r^2} = \text{accel.} \cdot \text{mass} = \frac{Ml}{T^2}.$$

Hence equation of dimensions:

$$k = M^{-1}l^3T^{-2} = D^{-1}T^{-2}.$$

Newton's law is therefore not invariant to measuring-rod changes if one also considers the law of the propagation of light. Then one must transform $T$ like $l$ (if in every system $c = \frac{l}{T} = 1$ is supposed to hold). So one cannot carry out any similarity transformation without altering the unit of density or the gravitational constant. The law of gravitation in connection with a constant velocity of light thus does not allow any measuring-rod transformations if the density of a substance can be regarded as something rigid (independent of its prehistory).

Now to Dällenbach's example.[12] The conductor rotating upon itself is naturally chargeless. In the case of the double conductor,

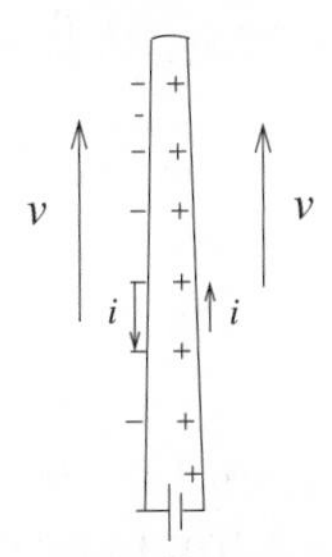


the *total charge* is obviously also nil, as the enclosed sketch shows. Nowhere do I see anything particularly paradoxical here, since no Galilean system rotating along with the rotating conductor exists. The latter case can only be dealt with by a transformation according to general relativity. *That a rotating conductor generates no charge is connected to the incompatibility between the system's conditions and the Lorentz contraction of the electron body relative to the matter.* The existence of an acceleration, in and of itself, has nothing to do with it, in my opinion.

Acceleration is, in a certain sense, absolute, because at every location there is an acceleration-free and rotation-free state (local system without a gravitational field), and relative to said local system one can define accelerations in this case. There is no question that charges must form on a rotating magnetic rod, completely disregarding every relativistic theory. These charges must lead to a field outside of the magnet, which must, in principle, be detectable. But there is nothing problematic about that.

You seem to be right about the measuring rod–clock concept; but to it we must add that the system's course of development is essentially stationary or periodic.

Governments are not based on legal conceptual systems but rather on power; that is, on the subordination of a multitude of essentially incoherent persons whose necessary organization it embodies and maintains. So they certainly do not depend on the stability of fundamental legal concepts.

I am sorry that Vero feels that way, because I believe that the source of this discontent is lodged in his emotions and not in his intellect.[13] I would wish for him some practical occupation as soon as possible, something like what we had at the Patent Office, so that he is always put before small, well-defined tasks.

Cordial greetings to you from your much harassed

Albert

Much harassed by excessive adulation and an oppressive hail of correspondence and other unofficial duties. Otherwise, I am doing very well, personally and healthwise.

# 86. To German News Agency for Foreign University and Student Affairs

Berlin, 27 July 1920

Esteemed Sir,

Despite an overwhelming workload, I cannot forgo applauding your endeavor most warmly.[1] On the occasion of a trip to Scandinavia, I saw the importance of the relations that German academia acquired through the hospitality of its universities before the war.[2] It is due to this that students of otherwise hostile Norway have an affinity toward Germans and continue, as before, mainly to use German books. In my opinion, one of your principal aims should be to convince young students that it would be of inestimable advantage to Germany if many foreign students received their academic training in Germany. I know very well that space and money shortages at German universities prohibit much, since one must, of course, first take care of the German youth. Yet many expulsions of foreigners happen out of political narrow-mindedness, which unfortunately developed as a result of unfavorable circumstances among local youth. Specifically, foreigners ought to be enrolled liberally in the more learned sciences, that is, studies not immediately directed toward practical applications, where there can be no question of overcrowding. As a Jew, I, e.g., often have occasion to hear about complaints by eastern Jewish students who too often are turned away with empty excuses out of mere anti-Semitism.[3] This embitters people who otherwise could form a bridge to the East which is so important for Germany's future economic development. But far from wanting to place economic considerations at the center, I see the moral and cleansing effect of your endeavor as its chief value. May your cause meet with ample success.

In great respect,

## 87. To Richard Fleischer[1]

Berlin, 29 July [1920][2]

Esteemed Mr. Fleischer,

We in Berlin are being plagued by flies. What to do? Very simple, you just catch all the flies in your room and then have some peace until new ones arrive.

Would a sensible person do it that way? He would rather let the flies buzz about, which are short-lived anyway, especially since he knows that in September they will disappear of their own accord. This is less troublesome than fly-catching and allows one to make use of one's time better, or at least more pleasantly. It is another matter when a wasp appears; it must be granted more importance. But one of those has not yet made its appearance.

In like manner do I respond to my critics and sycophants, who are excused by the fact that newsprint is paid by the line, regardless of whether it contains praise or censure.[3] Thus I think they all should be allowed to live on undisturbed. The judgment of those who cannot think or judge for themselves is of no consequence.

In great respect, yours truly,

A. Einstein.

## 88. To Friedrich Kottler[1]

Berlin, 29 July 1920

Esteemed Colleague,

Yesterday I talked extensively with Mises,[2] to whom I had handed your letter some time ago.[3] He told me that there is a considerable demand for teachers of applied mathematics, but he expressed doubts about whether you possessed the necessary experience in descriptive geometry, engineering mechanics, graphical statics, and graphical methods in general for such a teaching occupation. For such positions, he reportedly could only recommend fellow colleagues in the field who he knows have acquainted themselves seriously with these specialties. If you do so and can convince him of this, he will surely be happy to recommend you.

The results that you arrived at in your diffraction considerations have no relation to the quantum problem, in my opinion. I could not grasp where you find the formal kinship.[4] This nut seems so hard to crack that all your colleagues are wearing down their teeth on it; I too can sing a sorry tune about that.[5]

Do not be discontented if you are obliged to devote a large portion of your time to a practical occupation.[6] I also had to do so for a long time and still do.[7] I find that practical tasks protect one from becoming rusty and, contrary to research work, also give one a certain dose of self-esteem, which is so very necessary in life.

Amicable regards,

## 89. To Arthur Schoenflies

Berlin, 29 July 1920

Esteemed Colleague,

Pardon me for answering only now; it was impossible for me to get to it earlier.[1]

The most deserving theoretical physicist to be considered is, without a doubt, Paul Epstein, presently private lecturer in Zurich.[2] His achievements in quantum theory rank among the best that have been produced in our field in the last ten years.

Among the colleagues in the specialty whom you have named, I consider Stern the most suitable. I see his theoretical determination of the vapor pressure of solids as his most important accomplishment.[3] His knowledge is exceedingly multifaceted, his critical and pedagogical abilities extraordinary. In addition, he has both the inclination and a knack for experimentation. I do not think that my excellent opinion of Mr. Stern is essentially influenced by the circumstance that I know him well personally,[4] since just a short while ago, Mr. Bohr[5] spontaneously spoke in high praise of him.

Kossel is evidently a very original mind,[6] but his publications do not demonstrate that he has a command of the intellectual tools of theoretical physics. That is why I do not know whether it is advisable to confer on him the university's only chair for theoretical physics. As a teacher he cannot communicate scientific intuition to his auditors but can merely pass on known facts.

Lenz is undoubtedly a capable theoretician[7] who has a command of the tools of his trade. However, I would not consider it fair to place him ahead of Stern.

I surely cannot announce a talk in the field of general relativity,[8] because everything I found in this area is sufficiently known to my professional colleagues. But I still do think that public discussion would be of great interest and use.[9] The questions do not have to be presented to me beforehand.[10] Care would just have to be taken that no lay questions be posed, responses to which would inevitably bore the majority of those present.

With best regards, yours,

## 90. From Michele Besso

Berne, 29 July 1920

Dear Albert,

A *minimum* of nuisance in thanks for your letter:[1]

1. About G[uillaume]'s and W[eyl]'s[2] relative and absolute merits, I am in full agreement. I did not mean that from his letter you would see that it had any sensible meaning, but rather how he arrived at his tragicomical whim[3]—e.g., by a printing error or slip of the pen (as with Weyl, 3rd ed.,[4] p. 211, frequency appears instead of period, so also with you, some error may have been left standing somewhere).

2. As regards W.'s theory, you energetically brandish before my eyes what speaks against it. But what the transportability of measuring rods, etc., in this *uniquely* existing world is supposed to mean—above and beyond "plain" physical fact, which, e.g., would be explained by the enormous value of a constant—I cannot understand.[5]

3. About the Däll[enbach][6] example, I still do not understand the following: In your sketch (al[igned] pair of conductors), the Lorentz contraction determines the charges; for the rotating conductor, should they be "incompatible with the system's conditions"?—Däll.'s special problem deals with the *total charge* of the rotating magnet. *That* one stays 0, though, doesn't it?

Cord. greetings,

Michele

## 91. From Max von Laue

Zehlendorf, 29 July [1920]

Dear Einstein,

Regarding yesterday's discussion about Stern's paper,[1] I ask you please to consider:

$$\frac{\frac{1}{2}m\int_0^\infty u^3 e^{-\frac{mu^2}{2kT}}du}{\int_0^\infty u e^{-\frac{mu^2}{2kT}}du} = \frac{1}{2}kT \qquad \begin{aligned} &\int x^3 e^{-x^2}dx \\ &= -\frac{1}{2}\int x^2 de^{-x^2} \\ &= \int x e^{-x^2} \end{aligned}$$

holds.

So Stern is completely right.

With cord. regards, yours,

M. Laue.

# 92. To Paul Ehrenfest

[Berlin,] 30 July [1920]

Dear Ehrenfest,

Keep the case and bow there.[1] I want to buy myself new ones, so that the bother of shipping is avoided. The bow that you are getting there is hardly an excellent one. I believe I can buy both together for 700 marks. Do you actually want the second violin? Decide completely according to your needs![2] Make immediate inquiries about the damage so that the transportation insurance taken out by me can be made liable for the repairs.[3] I am very enthusiastic about seeing you all again in October, also about the prospect of seeing Langevin, who is *very dear* to me.[4]

I look forward to the discussion about magnetism at low temperatures. This topic is now fully ripe for theoretical analysis in connection with the spec. heat of $H_2$ and the Bjerrum spectrum of HCl.[5] Is pyrrhotite or any other, purer ferromagnetic crystal known at low temperatures?

Affectionate greetings to you and everyone, yours,

Einstein.

# 93. To Konrad Haenisch

Berlin, 30 July 1920

Highly esteemed Minister,

From friends among my fellow colleagues I am aware that there has been [no] budgeted chair for astronomy at the University of Halle since 1891, even though a specialist, acknowledged by all astronomers as a capable man—Prof. Hugo Buchholz—is working there as both teacher and researcher.[1] I consider it my duty to draw Your Excellency's attention to this truly indefensible state of affairs.[2] I find it very important that astronomical science, which is currently experiencing a dramatic rise, be granted more attention in Germany as well, especially since it cannot be denied that in this field we have been not insignificantly overshadowed by England and America throughout the last decades. Thus it is of great importance to support such a qualified astronomer as Buchholz and retain him in the field. I am very willing, at Your Excellency's wish, to write a factual report about Buchholz's scientific accomplishments.[3] I would deem it of exceptional urgency, in the interest of cultivating astronomy in Germany, that the provisional situation in Halle be put to an end and that the astronomer Buchholz, who has selflessly toiled for so many years already under difficult circumstances in loyal service to his science and the State, be given a permanent position.[4]

In expressing my utmost respect, I am Your Excellency's humble servant,

A. Einstein.

# 94. To Edouard Guillaume

31 July 1920

Dear Guillaume,

What appears on the first page of your letter can be expressed—if I have understood you correctly—in this way.[1] Let $t_s$ or $t_m$ be a specified time interval measured by a clock with a second hand or by a clock with a minute hand. Furthermore, let $\theta_{sec}$ or $\theta_m$ be the periods of the second or of the minute measured *by a different specified standard clock*, whose readings I want to denote as $\tau$. Then, for the observed time interval,

$$\tau = \theta_s t_s = \theta_m t_m \text{ applies.}$$

If you had defined $\theta$ this way earlier, I would have understood it immediately. It is strange and suspicious that you introduce many clocks at once into the consideration that are all supposed to serve for the measurement of time. But no one can forbid you from doing so.

I cannot understand the rest of the letter, because I absolutely do not know what you intend by it.[2] All my efforts were futile. I pondered for about two hours. Then I took up your letter of July 14 again, but I could do no better. Nevertheless, I would like to give you the argument in detail again about the influence of the gravitational field on the clocks.

*ds* is the time measured by a standard clock at rest relative to the coordinate system. Standard clocks are those that run at the same speed when positioned at the same place in time in the same state of motion. Time $t$ is *conventional*; but in the static case, time $t$ is universally given (with the exception of an *additive* correction), if it is given at *one* location. For in the static case, time $t$ has physical meaning to the extent that it is chosen so as to make static or stationary processes appear respectively as static or stationary. If I allow, e.g., monochromatic light to travel from the Sun to the Earth, then the time interval $\Delta t$ for generating 100 oscillations is equal to the time interval $\Delta t$ for receiving the 100 oscillations on Earth. With another time choice, the process of light propagation between Sun and Earth would not appear stationary.[3]

For resting standard clocks that are set up at two locations, one has the equations

$$(\Delta s)_1 = (\sqrt{g_{44}})_1(\Delta t)_1$$

$$(\Delta s)_2 = (\sqrt{g_{44}})_2(\Delta t)_2 \cdot$$

We now want to assume further that location (1) is the Sun's surface; location (2), however, is only so far away from the Sun that the special theory of relativity still applies precisely enough. Then $(g_{44})_2 = 1$ or $\Delta s_2 = \Delta t_2$ is true. In the conventional unit of time $(\Delta t)_2 = 1$, the standard clock hence strikes exactly once ($\Delta s_2 = \Delta t_2 = 1$). On the Sun, however, for the same conventional time unit $(\Delta t)_1 = 1$, which according to the foregoing case of the static field has direct physical meaning,

$$\Delta s_2 = (\sqrt{g_{44}})_1 \,.$$

Hence, in the conventional unit of time, the clock makes less than one stroke (since $g_{44} < 1$).

Alternatively, one can also reason like this. If the clock strikes once on the Sun, $\Delta s = 1$, then the elapsed conventional time (which preserves the system's static character)

$$\sqrt{g_{44}}\Delta t_1 = 1$$

$$\Delta t_1 = \frac{1}{\sqrt{g_{44}}} > 1 \,.$$

The equation you cited[4]

$$(dt)_{\text{sun}} = \frac{dt}{\sqrt{g_{44}}}$$

makes no sense to me either.

When you write back to me, please remain consistent in having the θ's be lengths of a clock's period measured by a special standard clock.

Best regards, yours,

A. Einstein.

I also tried to read your short exposition from the Société Suisse [1920].[5] First, $\theta$, $\theta'$ and $\tau$, $\tau'$ mean quantities that refer to different clocks and the same frame of reference. Then, quantities that refer to the readings of the *same* clocks in *different* frames of reference. Then you move on, in a manner completely puzzling to me, to the clock in the gravitational field, in which you utilize the same letters again. It all remains puzzling to me (in the chain of reasoning).

## 95. From Max and Hedwig Born

Frankfurt, 31 July 1920

Dear Mr. Einstein,

Max asked me to thank you warmly for your letter; your judgment is particularly important to him, as Wachsmuth is agitating against Stern for anti-Semitic reasons. So Epstein, as a Jew *and* a Pole, will be even less acceptable.[1]– Max is being very industrious, his experiment (atomic diameter of?) is finally working and he sits in the institute taking measurements until 8 o'clock in the evening.[2] It is our greatest heartfelt joy that you are coming to [Bad] Nauheim[3] and I hope that you will be staying with us then for a few days. Now—after my mother's death[4]—I am so very much in need of the genuinely close relationships left to me. The more distant the stroke of death becomes, the stronger is my yearning for what has been lost, the more incomprehensible and obscure is death's mystery. The cessation of such a strong personality and the sudden extinction of every one of her rights in life is such an agonizing problem that you wonder how you can simply live on without being constantly unsettled by it. But from it you learn to live more consciously and to feel more deeply and truly and to hold on to what you do have. If this were not so, one would have to sink hopelessly into the bitterly pessimistic philosophy of life of the ladybird comedy by Widmann[5] (are you familiar with it?), scenes of which constantly flashed by me in the first bitter moments of anguish: you talk yourself into thinking it is forever May and the whole world is always full of juicy, young, tasty vegetation, there just for you; and suddenly, in a split instant, you are sitting there with a leg torn off, weary of life, floundering in the mud of a rain-drenched roadway.–

My first thought was: ah, now I am stuck in the mud; but I see now that it still is May and that you shouldn't let it get you down.–[6]

So Göttingen is now decided, but there still isn't any prospect of residing there and we still may be staying here over the winter, because the Ministry is still dawdling.–

One more thing: Max wants to stay 2 days in Nauheim so that he can join his colleagues in the evening. Do you want to do the same? Or would you rather always drive over there (in 1 hour) from here? Should we alternatively reserve a room for you, and for how many days? But *in any event*, you will still be staying with us before and afterwards! No god can help you out of that! We are traveling on August 6 via Munich, Merano, Bolzano, to Sulden, Sulden Hotel, Tyrol (Italy), properly equipped with passports and liras. Your wife still wanted to write me[7] when you'll be traveling to southern Germany; how is she doing, and her daughters?[8]

Very cordial greetings to all of you, from your faithful

Max and Hedi Born.

*Very strictly confidential*: [Laue] seems to want to leave Berlin; what's going on? He asked [Wachsmuth] to appoint him back here because he's not happy with his position in B. He was purportedly just extraord. Prof. and with extraord. salary. That simply isn't true![9]

## 96. To Eduard Einstein

[Berlin,] 1 August 1920

Dear Tete,

Your long letter with the sweet pictures delighted me.[1] It also often hurts me that I have so little of both of you. But I am a very busy man and can leave here only infrequently.[2] Additionally, under the current difficult conditions it is too expensive for me in Switzerland. I certainly am glad, though, if I can support you there.[3] But now I am enthusiastic about seeing you both on October 5 in Benzingen near Sigmaringen.[4] We'll be staying either with the priest there,[5] who is a good friend of mine, or with a local lady, whom he has recommended to us. I am hoping that we can be together there for at least 10 days. Then I have to travel to Holland in order to give a lecture;[6] the exact time is not fixed for me, though; it all just has to be in October.[7] The two of us were so rarely together that I hardly know you at all, even though I am your father.[8] I'm sure you have only a quite vague idea of me too. But I'll make an effort to change that. In part it comes from your having been ill so much.[9] So we'll be meeting in Sigmaringen. If you both could come a couple of days earlier, that would be very nice. Find out sometime exactly when your fall vacation starts and then write me about it. It's nice that you are a good pupil; writing neatly was a weak spot for me too.[10] Don't be too ambitious in school. It does no harm when other pupils are better than you.[11] I'm very much looking forward to our conversations together. I know Albert's interests quite precisely, because I spent almost a month together with him last year, you know.[12]

I didn't send you anything for your birthday, because it's extremely complicated now.[13] If you have a special wish, though, do write me about it so that I can see about whether I can fulfill it.

Kisses from your

Papa.

## 97. From Théophile de Donder[1]

Brussels, 11 Forestière St., 3 August 1920

Sir and highly esteemed Colleague,

It was impossible for me here to procure the report that you published in 1919 (pp. 349–357) in the Berlin Proceedings;[2] I nevertheless cited this paper in the enclosed note: "Théorie nouvelle de la gravifique" [New theory of gravity] (p. 14),[3] because one of the results that I obtain there offers a strong analogy with a theorem that appears in the summary of your paper provided in the *Beiblätter*.[4] According to this summary, I concluded that you have abandoned Hamilton's general principle; in my analyses I retained this principle but I modified the generalized force which you adopt in the study of the atom [which] does not differ essentially from mine.[5]

I would be very pleased to have your opinion on the subject, and also to be able to peruse the above-mentioned report. Would it be possible for you to indicate to me the name and address of a bookseller who would be in a position to supply me with your publications of 1919, 1920 . . . ? If not, would you be obliging enough to send me these papers on loan?

In a recent article destined for the *Flambeau* (a magazine with whose directors Messrs. Grojean and Grégoire you are acquainted, having received them during their visit to Berlin), I attempted to present the handsome philosophy that emerges from your theories.[6]

Please, Mr. Einstein, allow me to assure you of my utmost respect.

T de Donder

## 98. To Hendrik A. Lorentz

[Berlin,] 4 August 1920

Highly esteemed Colleague,

If the delay in the appointment matter was painful, it was so only because I was aware that I caused you, Ehrenfest, and Kamerlingh Onnes an immense amount of effort.[1] Now, though, I am doubly pleased, first, because everything is finished, in principle,[2] so you won't have to exert yourselves anymore; and second, and particularly so, because now I have the welcome excuse for coming to Leyden more often. If all the formalities are settled by the end of October, I shall come at that time to deliver the inaugural lecture.[3] If it should take longer, however, it is all the same; I shall then simply come correspondingly later. The trip to Christiania was really nice; the hours I spent with Bohr in Copenhagen were the finest.[4] He is a highly talented and exceptional person. It is a good sign for physics that prominent physicists are mostly superb people as well. It is also encouraging that, despite the bad economic situation, interest in science has not diminished here in the least. Colloquium and physical society are always eagerly attended. The papers by Aston and Rutherford were reviewed by Rubens and received with enthusiasm.[5] It is good to see that there are still a few things that have not fallen victim to the political insanity.

To you and your wife cordial regards from your

Einstein.

P.S. I have now seen the photograms of the cyanogen spectrum by Grebe and Bachem; they will soon appear in the *Zeitschrift für Physik*.[6] One sees from them very clearly that nonphotometric spectra must not be used for analysis of line spectra. Although this paper does not conclusively prove the existence of the redshift, it does make it probable.

I am very curious what comments you have about Eddington's book.[7] I actually have objections only to his position on the boundary conditions (cosmolog. problem) but know that you will not concur with me in that.

## 99. From Paul Ehrenfest

[Leyden,] 6 August 1920

Dear Einstein,

Many thanks for your postcard.—It relieves me of a great worry, since I could not see how I could send the case and bow back to you.[1] But please do buy yourself a really *good* case and a *good* bow—just allow me to wait with the payment until you are here, because at the moment I am hopelessly stranded on the rocks, without any means of obtaining money for myself directly.

Now to your question about the second violin.[2] Provided you can pass it on *without loss*, it would probably be the right thing to do, because I really cannot see how I can get by in the near future without enlarging my debts, and do not want to owe you more money either.—But no matter what—*you* must not take any losses.

You ask for information about the damage to the violin[3]—well, the repairs are not at all major but are not yet completed at the moment because the man wanted to deal with the drying of the varnish first.—The repair costs won't be high enough to warrant demanding compensation.—Ugh—how boring this is!

---

Van Vollenhoven[4] and I are arranging everything so that (1) you get a visa valid for the whole year;[5] (2) that your talk can take place on October 6th or 13th[6] (you'll get a report from me about that in a few days). [[7] Your chair received 3 eminently capable men as Trustees, (1) Coebergh—a reputable Leyden notary; (2) Pateijn—a *very* big fish from the Foreign Ministry; (3) Zeeman, the physicist.][8]— In other words, Mr. Einstein can be counted on to be a man upon whom even "respectable"[9] people can depend not to mess around. I hope you won't prove yourself unworthy of "this bourgeois trust placed in you"! (Otherwise many a person dear to you will be directly hit!)—It's like marrying a countess.—We, however, are the mothers-in-law.

---

Langevin is surely certain to come. Weiss, hopefully, too.[10]– Your theoretical optimism regarding the treatment of paramagnetism surprises me.[11]—Paramagnetism of gases, yes, fine. But *paramagnetism of solids?!*—Treating them as *gases* is surely nonsense.—It really is hard to believe that the "molecular magnets" [Elementarmagneten] can rotate *force-free* within a crystal (like molecules in a gas). So one would first have to explain

*why molecular magnets whose orientations are held by strong forces within a crystal lattice can nevertheless obey Curie's law:* $\chi = \frac{C}{T}$ [12] [gadolinium

$\chi$ ↓ (susceptibility)

sulfate follows this law very exactly up to 2° Kelvin!!][13]

So the mystery is: If molecular magnets are fixed in orientation by very firm crystal forces, then where does the great susceptibility[14] of a *crystal powder* at $T \cong 0$ (for *freely* rotating magnets, simply conceived) come from?

*Weiss* published (C. R. 1913) something incorrect on it.–[15]

I see only the two following possibilities toward solving the puzzle:

*Assumption A*. Each molecular magnet can assume *two opposite* orientations within the crystal with *equal* ease (pot. energy of ↑ = potent. energy of ↓ ) all divergent ones ↗, ↖ → →), with *much* difficulty (large potential energy). If one calculates (for crystal *powder* or a crystalline aggregate) the susceptibility in a magnetic field *one thus obtains exactly Langevin's formula*:[16] (Very similar to the derivation of the formula for osmot. pressure.)[17]

*Assumption B*. The orientation of each molecule in the crystal powder is *practically completely fixed*;

*however, an electron ring present within the molecule can execute a quantum mot[ion] of units* +1 *or* −1. Outside of the magnetic field, both motions occur equally frequently—*but within the magnetic field, not anymore*. If the susceptibility is calculated here,[18] [In order not to arrive at an incorrect result, one must set in the exponent of $e^{-\overline{KT}}$ not the *energy*, but "Routh's function," as was proved in the dissertation by van Leeuwen.] *one again obtains exactly Langevin's formula*.[19]

So far, this would be very nice indeed, but for the following problem: above *equilibria* furnish good susceptibility. *Yet how are these equilibria supposed to set in?* (particularly at those very low temperatures of 2°)—Quite extraordinary collisions are needed in order to make such an adjustment possible.–

*Perhaps the adjustment of molecular magnets to rapidly changing magnetic fields does not, in fact, occur so rapidly either* (for gadol. sulfate at 2°). I spoke with Onnes[20] about the problem of how such a lag, provided it exists, could be verified experimentally.–

The *reduction in susceptibility* with rising frequency seems to be difficult for him to demonstrate.

The *heat development* due to the lagging magnetization (I hope I am not mistaken that it must cause the powder to heat up) seems to be much more easily demonstrable to him.[21]

Because gadol. sulfate at 2° is already almost as strongly magnetizable as iron, something might be detectable.

Please give me a comment or two, on your part.

In any case, I really do not believe that it is permissible to treat paramagnetism of solids *seriously* like a gas. (See Smekal)–[22] It works here as with osmotic pressure.

---

*A barmy idea*: You know that W. J. de Haas[23] has been thinking a lot about how spontaneous magnetization of tiny crystals in iron could be proved directly.– Well, I considered whether there could be any use in the fact that in magnetic fields of a couple hundred gauss, superconductors are not superconductive anymore.—So:

*Assuming an Hg-layer of a few molecules thickness is still superconductive* (which would first have to be determined)—*and it remains so for very different base layers,*

then one would have to see whether it wouldn't lose its superconductivity *on iron*.

I think, *as is*, the thing is still nonsense—but maybe you can do something with it.

---

I would very much like it, if we could talk with Langevin and Weiss about *the nature of the inner field.*—Weiss has published on some very stimulating but, if I am not mistaken, very incorrect things (*Ann. de Physique,* vol. I [1914]) (also *Archives de Genève*, 1913 or 1914).[24]

—•—

Will you read all that I have scribbled down here, I wonder??

—•—

I am very, very depressed—partly because of perpetual (trivial!!!) financial worries, partly because I am not working at all. Whatever I am *able* to do is not science but making somewhat entertaining parlor and promenade conversation about the physics—of others.

=:=

In order to be able to understand Langevin's French somewhat, I recently took up reading in French, and specifically, I chose the philosopher Bergson.[25] He is a splendid fellow. You'll be astonished what a delight it is—when I show you a few selected pages from his books. My wife Tanya and I enjoy his work very much.[26]

How exquisitely happy I could be, if I weren't so jaded and *unproductively ambitious*. And it's all fully clear to me but there's no helping it—I can reap as much as I want from *any* pleasure—but it's like confection and marmalade—bread is the fruit of hard labor—and *that* is where I completely fail.

Sorry—you don't have to make the least reply to this whining drivel.

Expecting you with great impatience. But everyone expects you—everywhere,

Ehrenfest

Greetings from us to all of you.

Will there be a public scene at the scientists' congress? I very much fear—yes—to Germany's detriment.[27]

## 100. To Théophile de Donder

Berlin, 11 August 1920

Esteemed Colleague,

Many thanks for kindly sending me your papers.[1] I must admit, though, that I cannot understand many things in them, especially your treatment of energy. The operator

$$(1+\varepsilon_{\mu\nu})\,\Diamond^{\mu\nu}$$

carried out on an invariant (multiplied by $[\sqrt{-g}/\sqrt{-q}]$) does not furnish a quantity of the character of a tensor:

$\varepsilon_{\mu\nu}$ (better written as $\varepsilon_{\mu}^{\nu}$) is a mixed tensor;

$\Diamond^{\mu\nu}$ (better written as $\Diamond_{\mu\nu}$) also is a (covariant) tensor.

The product $\varepsilon_{\mu}^{\nu}\,\Diamond^{\mu\nu}$ does not have the character of a tensor,[2] however. Might you be thinking of the form

$$g_{\mu\nu}(g^{\alpha\beta}\Diamond_{\alpha\beta})\,?$$

Even in this case, though, I cannot make any clear sense of your equations. In my opinion, it would be very advisable if you kept the covariant forms clearly distinguished from the contravariant ones.

I am sending you the requested paper[3] under separate cover. It merely constitutes an attempt at showing, specifically, that electric charges are possible with the pseudospherical cosmological solution, without having to introduce new formal constructions in all the equations (cosmological constant, special term for the Hamiltonian function, which yields the cohesive forces required for constancy of the electron's spatial charges).

Whether the equations are derived from a law of variational principle or not seems unessential to me.[4] But I emphasize that the paper's content is just an attempt, which I do not, by any means, back with full scientific conviction.

In great respect,

A. Einstein.

# 101. From Erwin Freundlich

Heidelberg, 12 August 1920

Dear Mr. Einstein,

I have completed the first series of experiments with the oven and am leaving on vacation tomorrow.[1] The oven worked well and we have obtained good exposures of the cyanogen band at 3883 Å in *absorption* at a dispersion of 1 Å = 1.2 mm, thus sufficient for accurate measurement.[2] We are now having the oven modified a bit in order to extend the lifetime of the graphite tube. I am in suspense about how the line structure behaves and whether the results by Grebe and Bachem will be confirmed. In any event, I believe it will not be without interest sometime also to compare directly the solar *absorption* spectrum with a terrestrial *absorption* spectrum,[3] whose source type, as relates to pressure, temperature, Doppler effect, etc., is as clearly reproducible as with the oven. Important new tasks present themselves with this light source in other respects as well. Prof. Bosch,[4] who sends his regards, is offering us a member of the physics staff from his factory. This physicist will be officially employed by the Baden Aniline and Soda Factory and will work with us in Potsdam. I would like to suggest Dr. Stumpf for this position and assume that you will not have any objection to choosing him either. He is an experimental physicist & was trained specifically in optical methods in Göttingen as a student of Voigt.[5]

I learned enormously much here besides, while working together with Dr. Hochheim;[6] the work with gases, vacuum, etc., was something new that I first had to learn how to do.– I hear that the construction is making good progress. Siemens & A.E.G. have each donated 20,000 marks retroactively and, in addition to that, another 20,000 marks have been obtained from Dr. Kurt Albert, director of the H. Albert Chemical Factory.[7] I hope you are doing well. When are you going on vacation? My address until end of August is: *Dettendorf* (Aibling post office), Upper Bavaria. With greetings also to your family, yours,

Erwin Freundlich

# 102. To Paul Ehrenfest

[Berlin,] 13 August 1920

Dear Ehrenfest,

The inaugural lecture is almost like the school-leaving exam nightmare; it's an interminable little business that lasts as long as one lives and breathes.[1] I cannot be in Leyden by October 13 because I asked my boys, whom I have not seen for one and a half years already, to Benzingen near Sigmaringen on October 4. I suggest October 27. I do have a tail coat ⟨But it is just very inconvenient to drag it along to Kiel, [Bad] Nauheim, Hechingen, Benzingen, and Leyden. Perhaps a Leydener of about my dimensions can lend me a dress coat?—I will dare to leave mine here.⟩ and am going to bring it along. Please arrange for the passport visa very soon, as I am leaving Berlin on September 10 and am not going to return before Holland. I'll buy bow and case.[2] I do still owe you the 20,000 marks; calculation too complicated for my poor brain.[3] You don't have to pay anything more and, in my opinion, still have some chance of being allowed, legitimately, to feel rotten about it. Also don't forget that I still owe you for the stay; this problem is almost as difficult as the magnetic one. I did understand your account of it in the main.[4] If the gadol. sulf. satisfies Langevin's law right down to very low [temperatures], this proves that the orientational forces are extremely weak. Even lower, it really must degenerate, finally. The quantum treatment suffers from the problem that it is unknown whether quantum states are possible with the angle zero and $\pi$ between the magnetic axis and the field.[5] It would be very nice if Langevin and Weiss came.[6] I have taken particularly warmly to the former. Stop whining and don't be a self-flagellator. Surely *we*, too, may avail ourselves of the human right of becoming increasingly stupid and lazy with age. That way we do others the service of relieving their consciences . . .

Hearty greetings to all of you, from your

Einstein.

I know nothing as yet about any planned nonsense at Nauheim. You are making me curious.[7] It won't be the first, nor the last!

## 103. To Pieter Zeeman

15 August 1920

Highly esteemed Colleague,

Cordial thanks for the magnificent photograph[1] and for your friendly words. It is a special pleasure to me to have you as trustee ["curator"], and I think I can promise you that I will never subject you to any excruciating *cura*.[2] I can deliver the inaugural lecture only at the end of October, because I would like to spend the first half of October beforehand with my boys, who live in Zurich and only have vacation just then.

Wishing you and your family a nice vacation, I send you my warm regards, yours sincerely,

A. Einstein.

## 104. From Paul Ehrenfest

[Leyden,] 16 August 1920

Dear Einstein,–

Thanks for your postcard of Aug. 13.–[1]

1.) Your suggestion, October 27, I immediately conveyed to van Vollenhoven.[2]– As soon as I hear from him whether that works, I'll report to you (thus, a few days hence)—but in any event, for heaven's sake, do come here about that time and stay *long* with us—otherwise it will be quite impossible to coordinate everything (trustees, Langevin,[3] etc.) and *I'll go nuts* [*meschugge*].

2.) "Tails"—Well—you must decide that for yourself. Of course, plenty of people here will be willing to lend you their dress coat—but it is a geometrical problem whose solution remains uncertain whether one of them fits you.—I would say: simply let your wife decide.—Since (after the oration[4]) many people will want to "congratulate"[5] you, and the students will address you, you *must*, under all conditions, be wearing a dress coat.

3.) "Passport visa"—Enclosed letter by Prof. van Vollenhoven shows you that *shortly before August 12* the Minister *did* apply to the General Consul in Berlin to give you a visa for 6 months (valid for as many trips as you like through any border posts).– [6]

Please inquire at the Consulate *by letter*, with this letter by van Vollenhoven in enclosure, whether this "application" (!!—yes, not "instruction"!!!) by the Minister for Foreign Affairs has already arrived.—And please let me know the response *immediately*, so that I can pass it on to van Vollenhoven.

— • —

Don't get impatient with me—think of it, that I hop about among all you imposing creatures, harmlessly and helplessly, like a frog anxiously trying not to get squashed.

I neglected to do one thing: On July 16 a book arrived for you from Marie-Anne Cochet: *L'intuition et l'amour (Essai sur les rapports métaphys. de l'intuition et de l'instinct avec l'intellig. et la vie)*[7]—sender's address: Villa Fayet, Begunns sur Gland, Canton of Vaud, Switzerland. [The packing paper smells of perfume—the book itself, though, is 263 pages long.][8] Dedication (handwritten): "To the illustrious Profess. Einstein, a tribute by a metaphysicist Frenchwoman"—*To be forwarded to you or held here?!!*

*Furthermore*: The *students* from the Philos. Faculty at Utrecht invite you to give a speech—I would advise you to be cautious, for acceptance means: ditto in Amsterdam, Groningen, Delft, Wageningen, etc.—"Automatic settlement"?!–

— • —

Do you happen to know where Smekal[9] is now? I would like to write him.

— • —

This too: Van Vollenhoven informed you that the remuneration is always paid out on January 1,[10] but he told *me* that if, in the interval between talk and January 1, you should need part of the total, that you simply have to say so—then he will make the arrangements.

— • —

— • —

It does seem quite incredible that elementary magnets should be so freely rotatable in paramagnetic *crystals*.—Well, I am looking forward to being able to discuss that with you.– [11]

What I would give to accompany you to Copenhagen!!– I now have news of my physics friends in Petrograd and Moscow—everyone alive and more or less scientifically active or teaching. It is now a matter of providing them with offprints and books.[12]

Best regards to all of you.—Don't forget us completely in your *sentiments*, yours,

Ehrenfest.

# 105. From Théophile de Donder

Brussels, 11 Forestière St., 18 August 1920

Sir and highly esteemed Colleague,

All my thanks for your kind mailing.[1] At present, I am in Nismes (province of Namur); consequently, it was impossible for me to come into possession of the notice[2] that you were kind enough to advise me to consult. I am impatient to become acquainted with it upon my return to Brussels.

I am very sorry that my notation causes confusion. In my way of writing, the operation

$$\Diamond^{\mu\nu}$$

[contains] the *complete* derivatives;[3] hence, the presence [of the] numerical factor $(1 + \varepsilon_{\mu\nu})$. Thus, to fix the ideas, let us calculate:

$$\Diamond^{12}[(g^{11})^2 - g^{12}g^{21}] \equiv \frac{d}{dg^{12}}[(g^{11})^2 - g^{12}g^{21}]$$
$$\equiv -2g^{21}.$$

If, on the contrary, we prove the *partial* derivatives, one will obtain:

$$\Diamond_E^{12}[(g^{11})^2 - g^{12}g^{21}] \begin{cases} \equiv \dfrac{\partial}{\partial g^{12}}[(g^{11})^2 - g^{12}g^{21}] \\ \equiv -g^{21} \end{cases}$$

On the other hand:

$$\Diamond^{11}[(g^{11})^2 - g^{12}g^{21}] \equiv \frac{d}{dg^{11}}[(g^{11})^2 - g^{12}g^{21}]$$
$$\equiv 2g^{11}$$

is identical to:

$$\Diamond_E^{11}[\text{———}] \equiv \frac{\partial}{\partial g^{11}}[\text{———}]$$
$$\equiv 2g^{11}.$$

This numerical binomial is also present in the notices by Mr. Hilbert; he, however, informs the reader that *everything [is] understood*.[4]

All my calculations [in which] the Lagrangians appear *next to* $\Diamond^{\mu\nu}$ have furnished results in perfect harmony with yours.[5]

I hasten to add that I had already decided to adopt your very clear way of indicating the covariants, contravariants, symmetries, asymmetries, etc., in my future notices.

I hope that these elucidations will remove every obstacle in reading papers that I had the honor of submitting to you[6] and will permit you to form a definite opinion concerning their usefulness in elaborating your admirable theories.

I do not need to tell you how valuable it would be to me to have this opinion.

Please, Mr. Einstein, allow me to assure you of my utmost respect.

T. de Donder

P. S. If you wish, I [will send] you a proof [... of my] New Thesis of [Gravitation][7] in which I transcribe all my results *in your notation*, and in which I indicate *explicitly* the covariance of [diverse] physical quantities I had [encountered] there.

TDD

# 106. From Tullio Levi-Civita

Padua, 18 August 1920

Illustrious and dear Colleague,

Thank you very much for your extremely enjoyable letter of the 11th instant and for having satisfied my wishes regarding the two offprints with such thoughtful courtesy,[1] adding a third equally appreciated.

I immediately communicated your cordial assent to the Italian translation to Engineer Calisse with instructions to contact Vieweg for the editorial work.[2]

I join most fervidly in the wish that the true intellectuals very soon regain the sense of international solidarity, scientific and human, so sadly lost, even perverted, during the war.

It is true that in Italy the majority of colleagues have again become reasonable, but, unfortunately, there is still a not negligible minority in the world that thinks and works in a non-international way.

You warned me that the paper "Gravitationsfelder im Aufbau etc." should be considered only as a first attempt.[3] I understand, but meanwhile I have already read the paper with the most intense interest and I intend to study it as required by the exceptional importance of the topic, given that by now it is no longer acceptable to doubt the existence and stability of the electron.

Permit me, finally, to justify my reference to Newton,[4] relying on the authority of Lagrange for support. He wrote that Newton is unique, unique being the discovery of the system of the world. Now, from a speculative point of view, relativity has changed the scheme in such a profound way that we find ourselves unquestionably at the forefront of a new discovery of the world system. It is not, therefore, because of a proselyte's enthusiasm that I, alluding to the Newton of our day, offend your modesty.

Please accept with renewed thanks my most cordial and affectionate sentiments,
Your very devoted,

T. Levi-Civita

P.S. Since you have had the goodness to remember with interest the geometric complement that I had occasion to add to the theory of Riemannian curvature, I must draw your attention to two wonderful papers by Mr. J. Pérès (Professor at the University of Strasbourg) that are relevant: "Le parallélisme de M. Levi-Civita et la courbure riemannienne" and "A propos de la notion de parallélisme dans une variété quelconque" Rend. Dei Lincei, T. XXVIII (1.° semestre 1919), pp. 425–428, and T. XXXIX (1.° sem. 1920), pp. 134–138.[5] I will certainly write to Pérès asking him to send you the offprints of them.

## 107. From Adolf F. Lindemann[1]

Sidholme, Sidmouth, 18 August 1920

Esteemed Professor,

What must you think of me for not having answered your kind letter of 30 Nov. 1919? Unfortunately I have a very good excuse; I was compelled to undergo a serious operation on Christmas Eve, which tied me to my bed for many months and in a very weak state. I have only recently been able to take up my former intellectual work and was immediately overwhelmed with extremely important tasks.—Not until today have I had time to answer in peace your friendly lines & beg you to forgive my long silence. During this long period, your splendidly confirmed theory has been very much discussed in England, as you know, and my son also participated besides Eddington + Jeans.[2] I had intended to make use of your very kind information about yourself and your political views and for this purpose traveled to London on 10 Dec. for the meeting of the Astr. Society, but could [do] little.[3] Through my illness, which was discovered two days later, all my good intentions came to nothing and my son could act only indirectly and too late, as he stayed by my sickbed for some time. At that time, and more strongly than today, that incomprehensible, hateful mentality was, unfortunately, [prevailing] in many scientists' minds; otherwise you would surely have been awarded the Astr. Society's Gold Medal. I can tell you, in private, that it was *merely* a political issue and that, as you probably know, in the end, no Gold Metal was conferred for that year. It will probably interest you that my son was elected to the council of the Roy. A. S. in the year 1920. To return to your [valued] letter, I would just like to correct one thing, and that is that we absolutely do not claim to have been the cause of the solar eclipse expedition of 29 May 1919; rather, I just wanted to say that our report about stellar photography in daylight and the special purpose of perhaps being able to confirm your theory even *before* the total ⊙ eclipse of 29 May 1919 may possibly have supported the matter & in any event did draw attention to both.[4] As regards the redshift, as you know, Prof. Evershed of [Kodai]kanal Observatory likewise became thoroughly engaged with the problem & did not arrive at any decisive result.[5] He is curr. in England & I hope to be seeing him here.

Sir Norman Lockyer died here yesterday (16th) at the age of 84. He had not been very mentally astute for some years and was likewise no friend of German men of science.[6] My son left for France on the 13th to see a friend, Duke de Broglie, whom you also know, and will probably come to Berlin after his visit and will, I hope, have the pleasure of seeing you.[7] If you should see Prof. Nernst,[8] do please convey to him and his esteemed wife our cordial regards; and I remain, with amicable greetings, always yours sincerely,

A. F. Lindemann

# 108. From Arnold Berliner

Berlin W 9, 23/24 Link St., 19 August 1920

Dear Mr. Einstein,

At the Paris Academy on July 26, Deslandres presented a paper by Perot on the comparison of light wavelengths of one line in the cyanogen band in solar light and in a terrestrial light source.[1] According to these experiments, the wavelength of light in solar light is larger than the corresponding one in a terrestrial light source. The light-wavelength ratio lies close to the value required by the general theory of relativity. The terrestrial light source was an arc lamp. The solar light was taken from various locations on the disk for comparison.

Furthermore! Miss Ilse Schneider wishes to see her dissertation printed and published at Springer's.[2] I even promised her I would use my influence on Mr. Springer toward this end but do not trust my own judgment enough, after all, to want to motivate Mr. Springer definitely into doing so. Perhaps you will be so kind as to write me a line about it, such as how you evaluate the work by Ilse Schneider against the one by Hans Reichenbach.[3] I myself have a very good impression of it, as far as I have read the work, but I really do not feel qualified enough to pass final judgment by myself.

Finally, also the following: the issue of *Naturwissenschaften* that contains your dialog on objections to the theory of relativity[4] is completely sold out but is continually being reordered. Two years ago, when you wrote this dialog, you had the intention of writing a sequel.[5] Would it not be appropriate, in view of the many intrigues, to have this sequel appear now at some point? It might even be appropriate to expand this dialog further so that it could appear in two or even in three subsequent issues of *Naturwissenschaften*. One could then assemble in a pamphlet the part published two years ago together with the sequel to appear now. One would then have at hand an extremely suitable means of defense against the "Syndicate of German Scientists for the Preservation of Pure Science."[6] I find that Gehrcke and Gehrcke's comrades[7] are making superb publicity for relativity theory.

With best regards, yours very truly,

A. Berliner

## 109. To Edouard Guillaume

22 August 1920

Dear Guillaume,

I understand the beginning of your letter up to the formula[1]

$\Delta t = \Delta t' \beta(1 + \alpha \cos \varphi')$

This relates to the case[2]

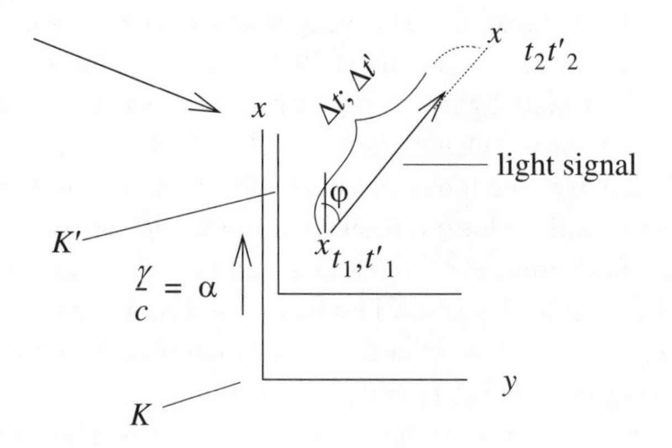

The time differences $\Delta t$ and $\Delta t'$, however, are not measurable by one clock (neither of the two are), rather only by means of a *system* of clocks that is synchronized with reference to $K$ and is at rest relative to $K$ or, respectively, a *system* of clocks that is synchronized with reference to $K'$ and is at rest relative to $K'$.[3] I absolutely cannot grasp the consideration that you say *I* would make. I absolutely do not see what I should understand by the clock frequencies $\nu$ and $\nu'$. Consequently, I also cannot attach any sense to the relations

$$\nu = \frac{\nu'}{\beta(1+\alpha\cos\varphi')}$$

$$\text{or } \nu = \nu'\beta(1+\alpha\cos\varphi')$$

either.[4] What is supposed to be the state of motion of a clock that measures $\Delta t$? A clock that is supposed to measure a system's time is set at rest relative to the system. Therefore, none of the rest of your letter is comprehensible to me. Discuss the matter with Besso sometime.[5] Maybe he can act as interpreter since he knows the relevant recesses of my brain inside out.[6]

With best regards, yours,

A. Einstein.

# 110. From Paul Ehrenfest

Leyden, 27 August 1920

Dear, dear Einstein!

Herewith I send you the fine diplomatic feat by my oldest brother[1] (he'll have a nice shock that I sent it to you!!)—Please, please arrange it so that you don't have to hurry away from Leyden again once you are here.– Date of your inaugural lecture not yet fixed—but highly probably Oct. 27 or Nov. 3.[2] As soon as it's fixed, I'll write you directly.—But please, please don't dash away from us—for only if you have *time* will you *relax* in Leyden instead of *tiring* yourself out.– We are all waiting impatiently for you!!–

Yesterday, received check for 336 guilders from *Methuen* as prepayment for the first edition[3]—but made out in your name, hence not cashable by me.—Okay like that?

Greet your wife and derivatives.– (In haste), yours,

Ehrenfest.

# 111. From Israel Malkin

Charlottenburg, 7 Wieland St., 27 August 1920

Highly esteemed Professor,

If I, as one of your innumerable auditors, dare to write you personally, this might perhaps be excused by the extraordinary occasion provided by the extremely deplorable spectacle made by two men at the Philharmonic last Tuesday.[1] As everyone else who has had the honor and the favor of becoming acquainted with you, highly esteemed Professor, as a teacher and a person, something every one of your pupils will remember with most pride from their period of study, the undersigned also felt scandalized, in the true sense of the word, by the shameless and inane manner by which attacks were launched against your character, quite apart from the spuriousness of the objections raised by the second speaker.[2] It scarcely has to be added that these feelings are shared by all of my colleagues from the University Course for Foreigners.[3]

At this opportunity I may perhaps mention that those of us who still have a glimmer of hope of matriculating, despite the unfavorable admissions situation at the university, were eagerly looking forward to the honor and pleasure of becoming your pupils and of being able to attend a systematic lecture of yours; having to relinquish this hope would be deeply painful to us, and that is why we would not like to believe that today's newspaper report about your intention to leave Berlin can be based on truth.[4]

Although these lines were caused by unfortunate events, I still warmly welcome the opportunity to express to you, highly esteemed Professor, my profound and sincere respect and loyalty.

Israel Malkin, Eng.

# 112. From Ernst Cassirer

Hamburg, 26 Blumen Street, 28 August 1920

Esteemed Colleague,

Only now, from your essay in the *Berliner Tageblatt*, have I found out about the attacks to which you and your theory have been subjected recently[1] and would not like to neglect to assure you in a few words, at least, of my most sincere and heartfelt compassion.[2] In matters of relativity theory I cannot claim to pass qualified judgment: but I number among those who have long followed this theory's development with profoundest intellectual engagement and I know how very much patient, persistent, and professional work and theoretical effort is concentrated and incorporated within it. That even this work, in which anyone with the slightest sense for intellectual merit could find pleasure and encouragement, has now been dragged down into the sphere of slander and political intrigue is deeply humiliating.[3] On the other hand, though, I am firmly convinced that all these things cannot touch you personally in any way and won't divert you from your path for a single moment. So neither do I believe the rumor, which was also represented in the last issue of the *Berliner Tageblatt*, that by reason of these last events you had taken the decision to leave Germany.[4] German science as a whole obviously has nothing to do with attacks by some political rabble-rousers and individual monomaniac scientists. If you let yourself be influenced by them in any way, it would come down to giving these attacks quite undeserving importance. The more conditions here worsen, the more we need the men who guide us about how to return to critical rationalness and staid, factual inquiry. Now I am glad that my little work on relativity theory is out of press and is just about to appear;[5]—thus I hope to be able to do my part in contributing at least a little toward checking the mental confusion about these things that still seems to exist in so many minds and seems to be deliberately exploited by some quarters.

In expressing my sincere respect, I am yours very truly,

Ernst Cassirer

# 113. From Ina Dickmann

Charlottenburg, 16 Eosander St., 28 August 1920

Esteemed Professor,

Following an inner compulsion, a layman would also like to write you a few words.

I listened to the talks organized by the "Syndicate of German Scientists,"[1] deeply ashamed of having to witness that the method of political election campaigns is now being used to bludgeon undesirable thinkers to death. I have never seen German representatives of science sink down so low as during that evening. How much I would have liked to shake your hand, but as a stranger I did not dare to approach.

The theory of relativity opened up for me a philosophical world of infinite breadth. In my mind's eye the constraints and boundaries of philosophical systems collapsed, and, for now, my thoughts wander about in the new world (—that is what it is to me) without a horizon in sight. Viewing relativity means to me the toppling of the absolute and the rise of another epistemological world.

But I also saw the importance of your research for generations still to come. It appears to me to be a treasure chest of new insights not just in theoretical science.

Perhaps there are many more among the lay who have had experiences similar to mine, who feel with me pain and outrage that in the already so humiliated Germany a major German personality be trampled in the dirt before the German people.

Allow me to express my utmost respect,

Ina Dickmann.

## 114. From Paul Ehrenfest

28 August 1920

Dear, dear Einstein!

There was a report today in our newspapers about the Weyland-Gehrcke campaign and your reply[1] (in excerpt) [what a pity that *you* answered for yourself, and not Planck, for inst.]–[2] I am terribly depressed that you have been dragged into the dirt.—Please, just spit upon *all* attacks. Unfortunately, Onnes is away (in Switzerland)[3] and I could not get hold of Lorentz yet. But I think I may assure you that all of those who love you *personally* so very, very sincerely here would manage to obtain for you, at least (to be safe) for 3 years, a full Dutch professor's remuneration (9,000, 2,000 + 7,000 guilders per year), in case you should still continue to be pestered in Berlin; provided only that then (without many more obligations) you would have to take up main residence in Holland and submit publications to the Amsterd[am] Academy. You do understand that I can say that right now on my own authority alone, because I know what Lorentz and Onnes and all the others are prepared to do for you.

You must understand me correctly: I know how you definitely rejected leaving Germany, defeated, a year ago, if only not to disappoint Planck.[4] And *really* (do please believe me), it's not that I might want to *entice* you away from Germany into Holland, making use of the "auspicious moment." But if they should sour your work there, then count on it that we here will set every force in motion in order to have you come over here!

Dear, dear Einstein—please don't write one more word in reply to these shabby newspapers—your friends must do that for you—primarily Planck–Lorentz. It suffices for you simply to inform me or Lorentz about any such attack that you would like to have answered. Then others will take care of it, all right!

With fond greetings also from my wife *and Tanya*,[5] yours,

Ehrenfest.

Can I receive your reply in the *Berliner Tageblatt*?!

## 115. From Kurt J. Grau[1]

29 August 1920

Dr. Kurt Joachim Grau feels impelled to express his undivided and sincere sympathy with the great scientist and person, in the face of the disgraceful and self-degrading attitude taken by a large segment of the German nation,[2] to which ice-cold, superior silence would be the appropriate response.[3]

The Jewish people are proud to have an Einstein and count him, alongside Spinoza and Moses Mendelssohn, among the most important men in their more recent history.[4]

## 116. From Moritz Schlick

Rostock, 23 Orléans St., 29 August 1920

Dear, highly esteemed Professor,

Various little motives are to blame for my once again having the pleasure of addressing a few lines to you. But I will keep it very short so as not to take up too much of your time.

At the beginning of this month, our local theoretical physicist died: R. H. Weber, a kindly man, to whom you also were introduced at our home, if I am not mistaken.[1] His chair (an extraordinary professorship) now has to be filled again, and you can imagine that, owing to the close relations that my philosophy has with physics, I am extremely interested in getting a very capable hand to join us here who is conversant in modern problems. As a private lecturer, I do not officially have the least to do with appointment affairs, of course, but this does not exclude my ability to nevertheless exert a little influence on the commission's resolutions through an occasional conversation with the gentlemen in charge, if in doing so I can quote outstanding authorities. That is why I would be exceedingly grateful if you could communicate to me in a line or two who the most suitable persons might be for this professorship. Maybe in this way I can contribute a little bit toward having theoretical physics at Rostock take off decently. I very much regretted the departure of E. Cohn to Freiburg.[2] In mathematics there is somewhat more life here since they obtained a second ordinary professorship for that subject in the spring (Haupt), a very nice person.[3]

Speaking of staff matters, I would not like to miss privately letting you know about a little rumor that reached my ears last month, the accuracy of which I cannot by any means verify, however. According to it, the German University in Prague supposedly has the intention of splitting the Philosophy Faculty and employing a philosophy specialist in the natural sciences section. They supposedly have even thought of me for it, already. That would be truly splendid! For with its geographical setting and intellectual life, Prague ought to be much superior to Rostock. But, as I said, we are dealing with a mere rumor, which I have not discussed with anyone else (not even with my wife, so as to spare her poss[ible] disappointment).

Now I have something else, without which I absolutely cannot imagine a letter to you, namely, thanks, warm, hearty thanks for what your benevolence once again gave me ample occasion to do. You were so kind as to recommend me to the *Berliner Tageblatt* for drafting an article on relativity theory for their almanac. I did, of course, immediately accept the offer by the editors[4] and in this way, from 8 days of labor, I earned a very pretty little sum that in prewar times would have been enough to take my whole family on a fine vacation trip. I just hope that you will be satisfied with the description as well. Most difficult was fulfilling the requirement of brevity while still remaining easily comprehensible.

More cordial thanks for your last amicable letter,[5] which once again was infinitely valuable to me! On the question of the causality of Newtonian space, it completely convinced me of your view, and it seems to me as if I really had been quite stupid; I did not consider the issue enough, physically. Nor can I adhere to the comment that I made about the question in *Space and Time* anymore; it must go, should the booklet happen to be granted another edition.[6]

What one has to put up with in Berlin nowadays, when the Philharmonic turns into a circus in which one clown follows the other onto the stage![7] I know that you stand, smilingly and completely unscathed, above the situation—but for that, magnanimity is required. I must admit that I did clench my fists a bit and feel heartily ashamed of Germans when I read about these things.

Will you be coming through Rostock or its vicinity sometime? It still pains us very much that we could not see you here last time.[8] At the very latest, I hope to see you in February if you are in Berlin then; for around that time I am supposed to give a talk there.[9] My wife, the children,[10] and I wish you and yours all the best. In offering best compliments to your esteemed wife and sending you my most cordial greetings, I am, in gratitude & admiration, yours truly,

M. Schlick

## 117. From Oscar Bie[1] et al.

Salzburg [30 August 1920]

In indignation about the pan-German incitement[2] against your exceptional character, we affirm to you, in truly international spirit, the sympathy of all liberal persons, who are proud to know you rank among them and number among the leaders of science worldwide.

Oscar Bie,[3]

Most cordially,

Joseph Chapiro, Werner Krauss, Andreas Latzko[4]

Warmly extending our hands in most cordial friendship,[5]

Alexander Moissi, Johanna Terwin, Helene Thimig,
Max Reinhardt, Stefan Zweig[6]

## 118. From Helmut Bloch

Berlin, 30 August 1920

Esteemed Prof. Einstein,

In greatest respect for you, I take occasion as a Jew—and consider it a pleasure to write you a few lines—as well as from the hum[an] point of view, to feel for you most warmly about the great injustice done to you![1] Although I am a layman of your science, I do gladly have great interest in becoming acquainted with a paper treating your work; and it should be a pleasure and an honor to receive such from you, either for reimbursement or I would demonstrate my most grateful appreciation by means of my above-mentioned articles (such as shaving instruments, etc.).[2]

I also have the *Tageblatt* article by you "My Response."[3] Likewise the article in the *Welt am Montag* by Mr. von Gerlach.[4] Even a shake of the head would be too much, and unworthy of you toward this category of people, whose newest symbolic badge is the swastika, and that means, according to its wearers, provocation, hatred, falsehood, and profanity! This degenerate type of human is, in my view, not normal![5]

The kaiser once said: Only a good Christian can be a good soldier;[6] that was too much said already, for now he sees how far he has taken it!

Genuinely good Christians do not hate, they have human compassion. One thing is incomprehensible to me, though, that the swastika, which has the effect of incitement, is not forbidden by the authorities, precisely because it is worn en masse with the intention of injuring the Jewish population; it does not hurt me, of course, because I carry with me the firm conviction that the swastika-person, or he without it, who carries falsehood and hatred within himself is mentally and morally infirm![7] Consequently, a *despicable* person. Yet that does not do away with the matter! Anti-Semites exist all over the world!

Yet I believe not so many as in Germany, for Germany's Prussians suffer under the pressure and burden of this sort of people, who could simply be eradicated through the full force of the law! However, even this sort is not spared profanity, and German prestige is only making itself more room in the other world, so the hatred that is based on it must necessarily rebound on itself! One can only be disgusted with the German Reich! So let us leave these vermin. This weed of nature. God created it too, and it will perish wretchedly as well. So, esteemed Prof., keep holding your head high, for one stands above the vile; and trust in God and in good, honorable people; one thus finds joy in existence and in one's life's work for oneself and for humanity! In thus begging for your esteemed advice, I voice my utmost respect,

Helmut Bloch

P.S. I beg k[in]d notice, when my visit is convenient for you; it would be a great pleasure for me.

## 119. From Fritz Haber[1]

Gastein, Strauburger Hotel, 30 August 1920

Dear Albert Einstein,

The *Neue Freie Presse* reports in two notices that you intend to leave Berlin and, therefore, Germany.[2] The reasons the newspaper mentions cannot have determined your decision—if you have really made it. If the anti-Semites assemble at

Philharmonic Hall in order to take a stand against you on the common ground of ignorance and antipathy,[3] to you this entente of mediocrity cannot appear to counterbalance the shared respect that all serious scientists have for you. Since Helmholtz,[4] no one in Germany has been ascribed by all those able to judge such a degree of leadership as you have been, and the annoyance by stupid fellows can cause an irascible neurotic like me to act *ab irato*, but not you, whose olympic equanimity could not be upset even by Wilamovitz's temperament.[5] That is why I believe that, if you are fed up with Berlin and want to go abroad, I presume to Holland, you will be expecting to find life richer in positive qualities elsewhere than Berlin. I think the world cannot offer you more than people and resources. That you find the people abroad more according to your requirements than in Berlin—that I can understand. It is not the pressure of fools that can drive you away, but the lack of stimulation. Everyone uses people for their own intellectual food and I understand that Berlin has become a platter eaten clean. If, however, it is not the people but finances that are the cause of your wanting to leave—*if* that is what you want—then I entreat you warmly not to decide anything before we have spoken with each other: the fact that your position has been completely neglected, economically speaking, while all other positions have been adjusted for the currency devaluation[6] was the subject of a detailed conversation I had with the government before my departure, and a change in this scandalous state of affairs will be sure to follow after the holidays.

The thought has often preoccupied me, these past months, whether I ought to stay in Berlin. What held me here, besides the institute I organized, was the intellectual environment, whose pinnacle is you. Your departure tears a hole that nothing else can fill, and a group of researchers whose numbers and motivation cannot easily be found elsewhere loses its direction and its focal point. I am egoistic in trying to stop you. Your needs and your family's feelings must take precedence. But in order to retain you I surely must say the little I can lay out before you, about how much we need you and how desolate it will be once you are no longer here.

Cordial greetings to your esteemed wife and daughters from your very unsettled

Fritz Haber

## 120. From Walther Meißner[1]

In the Berlin–Amsterdam train, 30 August 1920

Highly esteemed Professor,

I just read in the newspaper that you were thinking of leaving Germany.[2] During these vexing days, the most important German scientists have pled your cause,[3] so my feeble voice will count little. However, I would not like to refrain from telling you, nonetheless, how hard it would be for me, too, if your genius and your fine personality, which I personally was allowed the opportunity of becoming more closely acquainted with at the Warburgs',[4] were lost to Germany and particularly to Berlin physicists; and I think I may say that the majority of my colleagues at the Reichsanstalt also feel likewise and regret very much that one colleague has let himself be carried away into taking,[5] if only quite unofficially, such ludicrous measures against you.

In sincere admiration, most devotedly yours,

Walther Meißner.

## 121. From Toni Schrodt

Steglitz, Berlin, 54 Mommsen St. I, 30 August 1920

Esteemed Professor,

I have been following the notices published about you and your research in the papers and now also read the reports on the conflict.[1] I am neither a scholarly man nor a scholarly woman, but only a working girl from the middle class; but besides the compulsion to work for a living, there is within me a vital need for knowledge, for light. Obviously, I cannot penetrate your teachings and I lack the education for complete comprehension, but this much I do feel and grasp, that you sought and constructed something that will not pass away and that should fill the nation with pride and gratitude. This it surely does, for those who are now attempting to pull down your status [at] the public rostrum, with more—or less—pleasant sounding [words], are but a tiny fraction of the population, and in moral respects not among the most dignified, despite far superior academic learning. They weave no laurels for themselves and eclipse nothing of the image of you that has formed in our minds.

Now you want to leave Germany,[2] which precisely now, in the bleak situation it is in, needs a point from which it can draw energy and new hope and strengthen its self-confidence. *One* great person ennobles his entire surroundings, and from the glamor and renown that radiates from your name into all the lands, thousands of your humble fellow men are nourished. Please stay in Germany! Don't just listen to the slanderous yapping of a few dozen, but also feel the gratitude and admiration of all those who, even though not privileged to be taught by you and to take in the science in its full greatness, certainly do sense that they will be overtaken by shame if a noble, creative person is forced by jealousy and nasty reproaches to choose a foreign state as his fatherland. A mighty wall of friends and pupils and admirers stands around you; thus this struggle should give you new impetus for new creations whose proofs will turn the strident proclaimers into beaten, shamefaced mutes. Keep for us your efforts and your pacifistic ideas. As undulating fields are more fortuitous than the rattling of sabres and the glint of helmet spikes, thus also your intellectual seed and harvest will endure longer and be more beneficial for the German nation than the current effusions of minds clearly stuffed with scholarly learning but hobbling along without the slightest trace of "Christian charity." In joyful pride we could then say: Einstein is a German. They could take tools and the work of our hands away from us but not our minds.

Thousands agree with what I am writing and urge: stay with us; do not put a whole nation to shame for the sake of a couple of sinners!

Toni Schrodt

# 122. From Elsa Countess von Schweinitz und Krain

Bad Kissingen, Ölmühle Villa, 30 August 1920

To the founder of relativity theory!

Esteemed Professor,

I just read in the *B[erliner] T[ageblatt]*[1] about the outrageous personal attacks against you.

This is Germany's thanks, against *the one* person, who—now, when everything is sinking down in the crassest materialism—gave us illuminating values for eternity!

How pitifully small your opponent's plebeian hatred must seem to you, who carries the scale of infinity within himself!

I am familiar with the world and know: it forgives everything—save greatness—itself unable to ascend to greatness—it wants to drag the great down to itself. In this struggle lies the tragedy of genius.

But, believe me: in a few days, the baying of this pack will stop sounding . . . *Your* truth, however, will remain: *A solid rock to eternity!*

Yet—no matter how much they try to take from you: the best in you will remain far beyond reach:

That is, the memory of the hours in which *knowledge of your truth* was born of your mind! For this truth is yours alone.–

And this satisfaction is worth the labor of a lifetime!

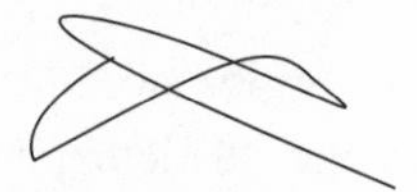

You are a man! A natural fighter! Nothing will break your intellectual blows—And yet—all knowledge notwithstanding, even the strongest is sometimes cast down—For, even the power of the best among us has its limits:

In such moments, send out your thoughts to those to whom you mean so many great things—and they are in the thousands!

My intellect cannot always follow yours (for I am a young woman and no learned mind); and yet I feel it in every nerve: your writings taught me, after a heavy blow of fate, to love the Earth again—for—they led me back from misanthropy and contempt of the world—to the silent majesty of eternal nature!

Just for that, I wanted to thank you herewith.

Greetings from a remote stranger,

Countess Elsa von Schweinitz and Krain

# 123. From Max Wolf[1]

Heidelberg, Königstuhl, 30 August 1920

Highly esteemed Sir,

To my horror I received today an announcement from Berlin about a lecture series against the gen. theory of relativity, on which I discover my name.[2] Thus I have unknowingly, and in an unwarranted manner, been connected to you and, consequently, you will surely excuse me if I disturb you with these few lines.

I want nothing more than to explain to you that I did not offer my name so as to act against you in this way, as seems to be going on in Berlin!

Mr. Weyland, previously unknown to me, visited me on Aug. 3[3] and coaxed out of me that I had reservations about the gen. theory of relativity, which is indeed the case owing to the time principle. At the same time, however, I told the gentleman that I took a neutral stance on the subject, and that, quite contrary to my wishes, measurements I had recently taken came out in favor of the solar eclipse data.[4] I did *not* promise Mr. Weyland a talk and am therefore appalled to find my name on the lecture listing.

I harshly condemn these whole proceedings; it reminds me terribly of the medieval councils.

I am writing to Mr. Weyland at the same time, in order to urge him to keep my name out of this affair.

I use this opportunity to assure you of my complete sympathy and respect. Most truly,

M. Wolf, Professor.

# 124. From Rütschke

Prösen n[ear] Elsterwerda, 31 August [1920]

In the name of numerous local pastors, I urge you please to remain in Berlin & spare Germany the shame that you went away.[1] Anti-Semites should not gain this triumph.[2]

Pastor Rütschke

# 125. From Matt Winteler[1]

[London, 31 August 1920]

"And yet, it moves!"[2] [*eppur si muove.*]

Matt Winteler

# 126. From Maja Winteler-Einstein

Lucerne, 1 September 1920

My Dears!

Your letter shook me up.[1] It is almost impossible for me to write now; I scarcely know why, myself. I thank you, dear Elsa, very much for the useful things you sent us through Mrs. Häfliger and both of you for the fine engraving of Carl Maria von Weber.[2] It pleases me greatly.–

The lectures in the Philharmonic seem to have degenerated into a slander campaign against you. Various papers published the announcement that you were giving up your Berlin job. Is that true?[3] In any case, this is a bad reward for your consideration toward the impoverished Berliners.[4] I'm earnestly sorry about it.–

What you are experiencing in the large arena I went through on a small scale. I also was attacked in the papers, supposedly because I had no talent as a teacher, place too much value on "finery and trappings" (!) and gave arbitrary marks.[5] In reality it was because the ultramontanes wanted to bring a nonacademic from Lucerne into the secondary school and the other colleagues were envious of our double earnings and of the schoolgirls' attachment to me.[6] Not a pleasant affair, either! A veritable witch-hunt was launched against me.

The money from the S[wiss] A[uer] S[tock Company] is anticipated for the end of October. I hope this doesn't cause you any inconvenience.[7]

Our Viennese lady is returning home the day after tomorrow. She gained 6 kg. A quite respectable result. I look forward to being alone with Paul again then as well.[8] Personally she was quite distant to us, even though she also ⟨is⟩ was a Jew. I'm going to accompany Mrs. Häfliger to Genoa and am inordinately pleased to be able to see Italy again after such a long time.[9] Mrs. Häfliger is traveling to America at the end of Sept. Will you still be in southern Germany at the beginning of Oct.?[10] I could possibly make a detour to where you are, in order to ⟨visit⟩ see all of you. I hope the stay by the lake did the children some good.[11] From Mrs. Häfliger I heard that poor little Margot was so ill again.[12] The poor child has to endure so much, and I knew nothing at all about it. Is she fully recovered again now?

We are socializing much with painters now. One of them, Einbeck, is, in our opinion, one of the very great ones.[13] He would interest you too, dear Albert. Sadly, the pianist-painter who became a close friend of ours has been in Paris since the beginning of July; he probably won't be coming back.[14]

Do write back soon so that we don't lose touch completely. All my love to you, the children, and Uncle and Aunt.[15]

Yours,

Maja.

# 127. From Paul Ehrenfest

[Leyden,] 2 September 1920

Dear, dear Einstein!

I finally managed to get the *Berliner Tageblatt* with "My Response."[1]—My wife and I read the whole text 4, 5 times over.

Forgive me—it is the first time that I am so bold as to meddle importunely into your affairs. I do it with great consternation, but I consider myself obligated to do so because we are so deeply fond of you.—If in the following any one of my comments should annoy you, or even offend you, do please *graphically* think of me, Van Aardenne,[2] and generally the whole ambience here in Leyden[3]—vividly imagine what you mean to us—then every last trace *must* vanish, then you *must* trust us.– So, with a heavy heart, I have the following to say:

*Such* is this piece, "My Response," that my wife and I absolutely cannot believe that *you personally* wrote down at least some of the turns of phrase. We don't forget for a moment that you were surely provoked in a particularly obscene way,[4] nor do we forget how abnormal the whole moral environment over there is in which you are living—but even so—there are totally un-Einsteinian persnickety points in this reply[5]—we could underscore each and every one of them in red. If you really did set it down in your own hand, then it means that these damned pigs succeeded in touching your soul, which is so very valuable to us.

You must understand me correctly: *I* might commit sins 100 times more serious—but it is not about me but about you—and this "My Response" simply doesn't suit *you*; it rather *reechoes* the filthy attacks on you.

—•—

What do I intend with my letter?

1. To tell you what my wife and I have on our minds.

2. To help prevent this poisonous filth from actually penetrating you.

3. At all events, to beg you very, very earnestly not to throw a single word more to that voracious beast, the "public."

—•—

And now—please—don't be cross with me. Whatever may happen, never forget how devotedly we all are attached to you here, from Pavlik to Lorentz![6] Yours,

Ehrenfest.

Today Onnes wrote me from Switzerland that Weiss will also be coming over while Langevin is here.[7]—I really do think that, if we all can come together for a week to discuss *!!!systematically!!!* first paramagn. of gases, then solid (nonconductive) crystals (Curie and anti-Curie behavior at low temperatures), liquid $O_2$, and *finally*, ferromagnetism, then something *will have* to come out of it.[8] I received a letter by Joffe from Petersburg[9]—they conducted *brilliant* physical exper.—among others, also a *very* clever one on the Einst[ein]–de Haas effect.[10] I am hoping to hear more in a couple of days; then I'll tell you about it. Apart from instruments, there is nothing they need more urgently than offprints. Despite repeated requests I cannot get any offprints from Franck or from Hertz[11]—I would like two copies: for *me* and in order to send one to the Russ[ian] physicists.

# 128. From Ludwig Hopf[1]

Munich, 22 Georgen Str. II, [2 September 1920]

Dear Mr. Einstein,

The wind blew onto my table the issue of the *Berliner Tageblatt* with your response to the "Co. Ltd."[2] Because I never had heard anything about the whole matter and the local papers took no notice of the whole uproar, I can only gather from your tone that you were annoyed. I'm sorry about that from the bottom of my heart; so please forgive me if I am so trivial as to express to you specially by letter that I—contrary to all fine custom—gladly share your annoyance and, like any decent physicist, will stand by you with all my might. It is a disgrace that political passions can exhibit such symptoms in our times; but I hope, on the other hand, that such phenomena will also open many eyes; anti-Semitism has now arrived at a level that no decent person can tolerate anymore.[3] Nor is it true that it is pervasive throughout the whole population; neither the masses nor the truly educated let themselves be dragged along, only the semieducated public, which is unfortunately drawing quite a lot of attention to itself at universities.[4] I consider it out of the question that even just *one* physicist of renown—no matter how far to the right he stands politically—will join the ranks of your opponents[5] and only hope that the newspaper report about your moving away is untrue[6] and that the voices of your thousand friends and admirers will weigh more heavily with you than this clamor, which you don't have to listen to at all. We Germans, who don't tarnish a true love for the fatherland with foolish nationalistic slogans, regard you as a piece of our cultural heritage that we do not want to lose. Please also remember that the rebuilding of new intellectual life, so direly needed by Germany now, is made more difficult if victory is made easy for the nationalist enemies of all stripes.

Otherwise, I hope you are physically healthy and mentally in high spirits, except for that small annoyed piece of your psyche, and I send you and your esteemed wife my best regards!

Yours ever truly and devotedly,

Ludwig Hopf.

# 129. From Willem H. Julius

Baarn, 26 Laan St., 2 September 1920

Dear Colleague,

What a strange and, if it were not so obnoxious, almost ludicrous smear campaign has been organized against you there![1] I did not think that Weyland's stupid plan would have any such success. At the beginning of August, I received a long letter from Mr. Weyland urging me to deliver a "talk against Einstein" somewhere, preferably in Berlin![2] A number of names of those whom they were hoping to add to the list were mentioned—who I could not imagine would accept, though. I was disgusted by the whole, obviously personally malevolent and *not* "purely professional" draft.[3]—Of course I declined. But this happened only about a week ago, because shortly after receipt of the letter I became ill, was not allowed to work at all, and do prefer to write a rather stern reply personally. [4]

Our summer has not been a happy one. My wife has been sick since the middle of June, was even absolutely bedridden for many weeks, precisely during our move to 201 Nieuwe Gracht, and thus had to be nursed in a hospital. Consequently, my daughters, who had overexerted themselves, did not fare too well, either; they are now in Noordwijk-by-the-Sea to recover.[5]— I hope we shall all be reunited at home again at the end of September.

Is it certain yet when you will be giving your inaugural speech in Leyden?[6] You really ought to stay in Holland for a longer period.

With cordial regards, yours very truly,

W. H. Julius.

## 130. From Hendrik A. Lorentz

Haarlem, 76 Zijlweg, 3 September 1920

Dear Colleague,

In the newspapers[1] I read about the unpleasantness that the so-called Syndicate of German Scientists has caused you and I feel the need to tell you how much I regret seeing you subjected to these kinds of attacks.[2] I do not understand what evil spirit has gotten into these people. Unfortunately I cannot ease the chagrin and vexation that these experiences must arouse in you, but there is one thing I truly can tell you (many others could do so, too). Namely: whoever knows your papers and has followed your research knows that you have been striving to seek the truth and to serve it in all modesty. Whoever knows you personally is doubly convinced of that. This applies to your friends in this country and fortunately also to the best among German physicists.

May the confidence that you will not be misunderstood by them make it easier for you to brush aside any such annoyance. Really, animosities like the ones I read about do not deserve your stooping to reply.[3]

Now I hope even more than before that you will be coming to see us soon after the holidays.[4]

With cordial greetings, also from my wife, yours faithfully,

H. A. Lorentz.

# 131. From Arnold Sommerfeld

Munich, 3 September 1920

Dear Einstein,

Sheer rage is what I have been feeling, personally and as chairman of the Phys. Soc., as I follow the Berlin stir against you.[1] A cautionary appeal to Wolf (Heidelberg) that he better keep his fingers out of it was superfluous. His name, as he has meanwhile written you, had simply been misused.[2] It is surely the same thing with Lenard. A refined lot, these Weyland-Gehrcke types![3]

Today I consulted with Planck about what to do at the Scientific Society. We want to give my colleague von Müller the cue to rebuff this "scientif." demagoguery sharply and to issue a declaration of trust in you. No official vote is supposed to be taken on it, but it is to be presented merely as an outburst of scientif. conscience.[4]

But you are not allowed to leave Germany! Your entire research is rooted in German (+ Dutch) science; nowhere will you find as much comprehension as in Germany. Abandoning Germ[any] now, when it is being so abominably abused from every side, would not look like you.[5] Another thing: with your views, in France, England, or America, you would certainly have been locked up during the war if you had protested, as I do not doubt you would have, against the Entente and its misinformation apparatus (comp. Jaurès, Russell, Caillaux, etc.).[6]

That you, of all people, should seriously have to defend yourself against charges of plagiarizing and evading criticism really is a mockery of all justice and reason.–

The *Süddeutsche Monatshefte* have requested an article by you and are very concerned about your response. You can also leave it to me, if you prefer. But we must have it as soon as possible because of the eventual reassignment. The *Süddeutsche* is widely read and is a reputable organ; you can also comment there, on the side, about the "bedbugs." I have not read your statement in the *Berliner Tageblatt*, but others have assessed it as not very felicitous and not quite like you.[7] The thing about the bedbugs was good, though.[8] The *B. T.* actually does not seem to me to be the right place to settle accounts with anti-Semite rowdies.[9] It would please us very much if you also did something in the *Südd.*

I hope you have in the meantime already refound your philosophical sense of humor, and pity for Germany whose sufferings are expressed, as everywhere, in pogroms. But desertion is completely out!

Warmly, yours,

A. Sommerfeld.

I asked Grebe to show his exposures in [Bad] Nauheim. He is going to do so. For the discussion, this issue seems to me to be the most important now. You are going to be coming to Nauheim for sure, aren't you?[10]

# 132. To Edouard Guillaume

4 September 1920

Dear Guillaume,

The conclusion about the rate of the moving clock cannot be derived from the formula[1]

$$\Delta t = \Delta t'\beta(1 + \alpha\cos\varphi') \;.....\; (1)$$

It rather results directly from the inverse Lorentz transformation:

$$t = \frac{t' + \frac{v}{c^2}x'}{\sqrt{1 - \frac{v^2}{c^2}}}$$

For the pointlike events that correspond to the ticking of a clock indicating seconds positioned at the origin of $K'$, is

$$t' = n \quad (= \text{whole number})$$

$$x' = 0.$$

The result is therefore

$$t = \frac{n}{\sqrt{1 - \frac{v^2}{c^2}}}.$$

From this, one may not by any means conclude that "the clock that indicates $t$ is running more slowly than the one that indicates $t'$." Rather, $t$ is indicated by many clocks and, to be precise, by each clock at rest relative to $K$ that just coincides with the $n$th tick of the clock positioned at the origin of $K'$. I see from this statement of yours that you have still not fully grasped the special theory of relativity, i.e., the theory of 1905, misunderstandings prevail instead. A *single* clock can only measure the time at a *single* location (of the frame of reference). In time evaluations in which many locations (with reference to the coordinate system) are involved, a *system* of set clocks is always necessary.

According to my conception, it is impossible to derive the Doppler principle from equation (1).[2] In order to define radiation as a *frequency* $\frac{\omega}{2\pi}$, I need a wave of the kind

$$\sin \omega \left( t - \frac{\alpha x + \beta y + \gamma z}{c} \right).$$

$\frac{\omega}{2\pi}$ is the number of waves that pass by a point in the coordinate system, while $\Delta t = 1$, i.e., while a seconds clock, arranged at rest relative to this point of the coordinate system, is making one tick. The Doppler principle results from the equation

$$\omega \left( t - \frac{\alpha x + \beta y + \gamma z}{c} \right) = \omega' \left( t' - \frac{\alpha' x' + \beta' y' + \gamma' z'}{c} \right),$$

whereby this equation must be made into an identity through the Lorentz transformation. I absolutely would not know how to indicate a simpler derivation of the Doppler principle. In any case, from equation (1) nothing at all can be concluded either about the rate of a moving clock or about the Doppler principle. You must think of it as if $\Delta t$ has the meaning

$\Delta t = t_2 - t_1$

$t_1 = A$ = time of emission of the light signal at $A$

$t_2 = B$ = time of arrival of the l[ight] s[ig]. at $B$.

$A$ ——————— $B$

⟶ light signal

There is no such thing as *one* clock that can indicate $\Delta t$.

Best regards, yours,

A. Einstein.

# 133. From Max Planck

Gmund on Lake Tegern, Grundner Farm, 5 September 1920

Dear Colleague,

On returning to Germany from Southern Tyrol, where no news reached me, I discovered the reports about the almost incredibly disgusting behavior that had occurred at the Berlin Philharmonic and about everything connected with it.[1] I am momentarily at a loss for an explanation of how such baseness can be possible among highly educated people. But of very much greater importance to me than this problem is the issue of what impression such machinations are apt to make on you personally and I am plagued by the thought of the possibility that you will lose your patience at last, after all, and could decide to take a step that would severely punish German science and your friends for what a despicable mentality has inflicted on you.[2] Adequate redress for you on the part of qualified authorities in science must and will not be lacking.[3] In [Bad] Nauheim I hope you will allay my grave fears. Until then, cordial greetings from yours faithfully,

M. Planck.

# 134. To Arnold Sommerfeld

[Berlin,] 6 September 1920

Dear Sommerfeld,

I did indeed ascribe too much importance to that undertaking by believing that a large proportion of our physicists were involved in it. So for two days I actually did think of "desertion," as you call it.[1] But soon my composure returned along with the realization that it would be wrong to abandon my circle of true friends. Perhaps I ought not to have written the article.[2] However, I did not want an enduring silence toward the criticisms and accusations, which are repeated systematically, to be interpreted as acquiescence. It is annoying that every statement of mine is capitalized on by journalists. I simply must close myself off.

I cannot possibly write the article for the *Südd[eutsche] Monatshefte*. I would be happy if I could manage my backlog of correspondence. A statement of that sort in [Bad] Nauheim may well be advisable for the image abroad, or generally for reasons of propriety.[3] For my sake, though, such a thing should not under any condition happen; for I am already content and at peace and read nothing that is printed about me besides matters of real fact.

Grebe's photograms are appearing soon in the *Zeitschr. für Physik*. They are truly convincing, that is, they refute the earlier findings about a nonexistence of the displacement effect. For a final decision on the question of the redshift, much thorough work is still necessary, though.[4] I am coming to Nauheim too and think it will turn out to be very interesting there.[5]

In thanking you earnestly for your friendly letter, I am, with cordial regards, yours,

Einstein.

# 135. From Konrad Haenisch

Berlin, 6 September 1920

Highly esteemed Professor,

With pangs of pain and embarrassment did I gather from the press that the theory you champion has been the object of malicious public attacks extending beyond the bounds of factual judgment and that even your professional integrity did not remain safe from disparagement and libel.[1] It is with special gratification for me that scholars of acknowledged reputation, among these even prominent members of the University of Berlin, are standing by you with regard to this affair, rejecting the baseless attacks against your person and pointing out that your research has guaranteed you a permanent[2] place in the history of our science.[3][4] As the best are supporting your cause, it will be that much easier for you not to give any more attention to such ugly activities.

Thus I may surely also express the determined hope that there is no truth to the rumor that due to these nasty attacks you wish to leave Berlin, which has always been proud, and always will remain proud, to count you, highly esteemed Professor, among the most brilliant jewels of its science.[5]

In voicing my very special esteem, I am yours very sincerely,

Haenisch[6]

# 136. From Isaak Meyer[1]

Stollberg, Erz Mts., 7 September 1920

Highly esteemed Professor,

Forgive me if I add another letter to the flood of correspondence you have been receiving in the last few days.

A request coming from the heart!

In case you really are leaving Berlin & Germany,[2] then there can be only one destination for you, the staunch Jew.—The reclaimed land of old[3]—Palestine. Jews from all over the world will lay teaching material at your disposal.[4]

Should my premonitions & yearnings be fulfilled, I hereby pledge in your honor, very esteemed Professor, to execute *the labor* completely for the entrance gateway in the finest wrought ironwork, entirely free of charge, for the newly established University of Jerusalem.

With most sincere best wishes for the New Year[5] to you, Professor, & your k[ind] family, also in the name of my d[ear] wife,[6] I am, respectfully yours very truly,

Isaak Meyer

# 137. To Konrad Haenisch

Berlin, 8 September 1920

To the Minister of Science, the Arts and Public Education, Mr. Haenisch, Berlin.[1]

Your Excellency's letter of the 6th inst. fills me with a sense of sincere gratitude.[2] Quite apart from the question of whether I deserve so much benevolence and high regard,[3] in these last few days I came to know that Berlin is the place in which I am most deeply rooted through personal and professional ties. I would follow a call outside of the country[4] [even to Switzerland] only in the case that external circumstances force me to do so.[5]

In utmost respect, Your Excellency's loyal servant,

A. Einstein.

P.S. I would like to use this opportunity to call to mind a letter that I directed to Your Excellency in favor of a budgetary appointment of the astronomer Prof. Buchholz (at the University of Halle).[6]

The ministry's communication: "Through the press, particularly the foreign press, alarming reports are repeatedly being made to the effect that Prof. Albert Einstein was thinking of leaving Berlin and Germany in the near future and of following a call to a foreign university. In order to knock the bottom out of these rumors, once and for all, which are being exploited in a biased way, particularly abroad, we communicate here Albert Einstein's reply to the publicized letter that Minister Haenisch had addressed to him a few weeks ago. Einstein writes:"

# 138. From Hedwig Born

9 Cronstetten St., 8 September 1920

Dear Mr. Einstein,

*When* are you traveling to [Bad] Nauheim, and which days will you give us? We shall tell *no one* about your being here; you are incognito here, if you wish. Paul Oppenheim, Jr., still seems to be away. Please send a postcard with your instructions.[1]

The vile squabblings that you are being harassed with sadden us deeply.[2] How injured you were is proved by the atypical step you gave way to in your more than justifiable irritation: the unfortunately very clumsy reply in the newspaper.[3] Those who know you are depressed by it, precisely because they can sense *how* affected you were by this notorious incitement, and they suffer with you. And those who do not know you get a false picture of you. This too is painful. Meanwhile, though, you are the old Diogenes again, I hope, and are laughing at the beasts driveling into your tub! It absolutely does not fit the image I have of you, which I have placed, among other venerated holy men, within the shrine of my heart, that people could still disappoint it or provoke it out of its tranquillity. You would not have withdrawn from the wild bustle of life into the still temple of science (see your Planck speech)[4] if you could have found in that bustle, in your fellow men, exactly *those* illusions, *that* happiness, and that peace as is in your temple. If the world's scummy floods are now lapping at your temple's steps, then just close the door and laugh! And say: it was not without reason that I went into the temple. Don't be angry! Stay the holy man in the temple and—stay in Germany![5] Scum exists everywhere, but not such enthusiastic smart[-ass] preachers as your quite pretentious

Hedi Born.

Darn it! Send us a word from you or your wife Elsa (to whom I give my warm regards) sometime again; otherwise I'll become a member of the Antirelativity Co. Ltd.,[6] or I'll found a competing outfit.

You absolutely must read *The Home and the World* by Rabindranath Tagore, it's the finest work (novel) I've read for a long time.[7]

## 139. To Paul Ehrenfest

[Berlin, before 9 September 1920][1]

Dear Ehrenfest,

I thank you heartily for your many letters. Ilse is returning from vacation only tonight.[2] She will find and send you the papers soon.[3] The two young Russian physicists[4] were here to see me already a day after submission of my petition; so it must have had an effect into the past (see relativity theory). You have probably already seen them now, too. The local Anti-Relativity Co. is already virtually bust,[5] and there is no reason to resort to extreme measures and "become a deserter," as Sommerfeld calls it.[6] But, in any case, it is touching that you should rush to my aid in my plight.[7] Thank you also for the prompt settlement of the consulate matter.[8] In case you still have something you want to write me about, I inform you that I am:

from 18–25 September at Born's (Cronstetten St., Frankfurt)[9]
from 4 October onwards at Father Brandhuber's, Benzingen[10]
(Hohenzollern-Sigmaringen).

My two boys are also going there. Now a big request.[11] It seems that you cannot cash the 700 guilders that arrived at your address.[12] But maybe you can manage, on the basis of the certificate that arrived, to send the money in francs to my ex-wife (Mileva Einstein, 59 Gloria St., Zurich). She is on the rocks with the children.[13]

Don't be cross if I write so rarely and so little. I am being so bombarded with letters that despite the best of intentions I cannot finish. So I'll save everything to tell you in person.

Warm greetings all round from your

Einstein.

Your letter about my article just arrived.[14] I wrote it one morning in one stroke, completely on my own. As my excuse, you must consider that nothing else was left to me than to defend myself in public against the accusations of deceptive propaganda, plagiarism, etc., which had been repeatedly made publicly against me.[15] had to do this if I wanted to stay in Berlin, where every child knows me from photographs. If one is a democrat, then one must grant the public this much right as well.

## 140. To Max and Hedwig Born

[Berlin,] 9 September 1920

Dear Borns,

Don't be so strict with me. Everyone has to offer his sacrifice on the altar of stupidity from time to time, for the amusement of God and man. And I did a thorough job of it with my article.[1] Exceedingly appreciative letters from all my dear friends prove it in that sense.[2]—A facetious acquaintance said recently: Everything is publicity to Einstein; his newest and slyest trick is the Weyland Co. Ltd.[3] That's how it is, or similar, at least. Like the man in the fairy tale, whose touch turned everything into gold, thus it is with me, with everything turning into banner-line news: *suum cuique*.[4]

During the first moment of onslaught, I really did consider flight.[5] But then I thought better of it and the old phlegm returned. Now I just think about purchasing a sailboat and a little cottage near Berlin by the water.[6]

I shall appear at your home somewhere around the 18th, if I can be of any use to you. If, however, it is necessary to live in [Bad] Nauheim during the scientists' convention, then please see to it, d[ear] Born, that we can be housed next to each other there. I am not reserving anything from here, because you can judge better how best to go about it. If possible, I would like to stay with you a little too, though, so that I can banter with the charming correspondent as well, since it does not work in writing with this annoying coagulation of my ink. Else is also coming but is staying with the Oppenheims.[7]

Then on the 28th we must be in Stuttgart, where I have to preach for the benefit of a public observatory.[8] After that we are traveling into Swabia, where I have arranged to meet my boys.[9]

Cordial greetings to you both, yours,

Einstein.

# 141. To Norwegian Students' Association

Berlin, 9 September 1920

Norwegian Students' Association, Christiania,[1]

In grateful reminiscence of the magnificent summer days I was allowed to spend as a guest of the Norwegian students in Christiania, I send you herewith a cordial greeting from my study in Berlin.[2] In you I became acquainted with a student body that holds high the ideal of culture common to us all, free from constricting nationalist tendencies.[3] It is a great pleasure for me, even though I am already quite high up there in my number of semesters, to have been welcomed among your cheerful circle, in which I felt so completely at home. Besides the impersonal matters of fact that unite the discerning people of all countries, for the recovery of international relations there is a need for warm, feeling free people of liberal views and good will.[4] You are such, and it is to be hoped that as students of a neutral country you will engender valuable contributions toward curing the European psyche.

With best wishes for a thriving Norwegian student body and research, yours.

# 142. From Marcel Grossmann

Zurich, 9 September 1920

Dear Albert,

In the same post I am sending you a paper by Mr. Charles Willigens from the *Archives des sciences physiques et naturelles*.[1] As you see, a cult is forming around Guillaume that thinks it must correct essential points of your concepts. Although this matter is unlikely to be of interest to you, I think it would be in the interest of relativity theory if you had a brief joust with Guillaume someday, such as in a short article for the *Archives*, for which I would gladly provide the translation, or simply in a letter to me, the scientific gist of which I could pass on, which would please our sympathetic colleague Guye very much.[2]

There is a danger that from the unchallenged appearances by Guillaume and his disciples—also in the dailies—dissemination of the fundamental ideas of the th. of rel. would suffer harm in the French-speaking region, which is always ready to claim Fren[ch] superiority on this issue as well. All the more so since the depraved campaign against you in Germany is also echoed here.[3] Thus I think I may ask you to let me know in brief outline for what reasons you reject Guillaume's ideas!

I very much hope all is well with you and yours. Both our boys,[4] who are in the same class at the Gymnasium, are already calculating with logarithms. We also are doing well, after my dear wife[5] withstood a nasty sepsis just a year ago that brought her to the brink of death. But now she is up and about again and more cheerful than ever. Are you still not ripe for Zurich yet?

With best wishes to you and your wife, I am your friend,

M. Grossmann

Cord. greetings from my wife.

## 143. From Felix Ehrenhaft

Vienna XIX, 70 Grinzinger Street, 10 September 1920

Esteemed Mr. Einstein,

I have been intending to write you for 3 weeks. At the end of August, I received the following letter, attached in carbon copy, which I initially considered a message from a crank.[1] Alerted by the newspapers about the extremely deplorable and, for our conditions, shameful events,[2] I remembered the letter and submitted it to our State Secretary for Public Health[3] at his wish. The latter sent the letter to *Die Freiheit* in Berlin, as he later informed me. From there it found its way into our *Arbeiterzeitung*, etc.[4] If you should be interested in the original, I shall be glad to send it to you, which at the present time is in the hands of the mentioned gentleman, or pass it on to you at the scientists' convention, where I hope to meet you.[5]

I cannot refrain from telling you how much I regretted those ugly proceedings. Although the whole affair must leave you completely untouched, I can perhaps grasp more than anyone else how much one suffers emotionally from such actions. It is my heartfelt wish that this be reduced to a minimum with you.

Certainly, the discord that I experienced, and still take part in, does not compare with yours; do not take this comparison as impertinence. I must tell you, though, that I for my part still had to suffer much this year, and still do, from the unscrupulous and even slanderous agitation staged against me by certain quarters.[6]

I am very much looking forward to seeing you again at the scientists' convention, where I myself intend to speak about radioactive emissions of isolated radioactive test particles of the order of magnitude of $10^{-5}$ cm.[7] You might find a free hour so that I can discuss this with you a little in [Bad] Nauheim, which I would appreciate very much.

Until the 17th, any messages from you can reach me in Vienna; in Nauheim I was assigned accommodations at the Carlton Hotel.

Sending you my most cordial regards, yours,

F. Ehrenhaft

[Enclosures][8]

1 Carbon-copy letter

1 *Arb. Zeitung*

1 offprint.

Paul Weyland's letter dated 23 July 1920:
"Esteemed Professor,

Now that unanimous agreement has been reached among *serious members of the exact sciences* about rejecting Einstein's research, we are planning also to present the educated lay public with counter-arguments, after it has long enough been fed to the point of *vomiting* with Einstein's ideas.

As *secretary of the Einstein opponents*, I inquire of you whether you are willing to participate in these lectures against Einstein and, on this condition, could provide you with more details upon receipt of your acceptance.

For reasons of urgency I ask you kindly to reply by wire. The business ought to yield a profit of about 10–15,000 marks for you.

In great respect, yours very truly, *Weyland*."

## 144. From Hendrik A. Lorentz

Haarlem, 76 Zijlweg, 10 September 1920

Dear Colleague,

Just today I happened to hear that, in the matter of your appointment by the Leyden University Council, all is now in order[1] and that you will deliver your inaugural speech at the end of October. I am very pleased about this and must tell you now again that I very much regretted your having to wait so long.

Have you read yet that now it is already possible to make money with relativity theory?[2] I enclose herewith what I read about this in a Dutch newspaper. It will interest you in any case.

You should really get the 5,000 dollars, and the *Sc[ientific] A[merican]* would have done better to ask you for an article and offer that sum as an honorarium. Now I wish that you had a short article ready and we were sure that the decision makers were reasonable enough to choose *it*. Then you would perhaps decide to descend into the arena yourself; it is not your fault that scientific achievements, such as those the world has to thank you for, do not free one from all financial worry. I could easily arrange for a translation into English here if that were to pose a problem in Berlin.

With cordial greetings, yours faithfully,

H. A. Lorentz.

# 145. To Willem and Betsy Julius

[Berlin,] 11 September 1920

Dear Colleague and dear Mrs. Julius,

I note from your letters[1] with deep regret that you have much misfortune and many unpleasant times behind you. Thus it is doubly nice of you to have such lively sympathy for the little troubles[2] regarding my public life. Actually, I was rather ashamed of myself for having allowed that little rumpus to upset me. I even let myself go so far as to write a reply in the newspaper, for which our dear Ehrenfest has already quite legitimately given me a good dressing down.[3] Now my opponents have even come up with the amusing idea of enlisting my dear friends in their enterprise.[4] You can imagine how much I laughed at this report of yours and how vividly I pictured the grim fury of my dear colleague Julius! I have long since found the comical side of the campaign and do not take it at all seriously anymore.

But I do feel terribly sorry that you went through such trying times, not only from the odious move but also from ill health. I keenly hope that all of you are healthy and happy again when this letter arrives and that at the end of next month we can spend worry-free and cheerful hours together again. (I am so unsuperstitious, you see, that I really and truly believe I will actually deliver my stale old sermon about the ether then.)[5] That disrespectful comment that you, d[ear] Mrs. Julius, made about your own piano playing thoroughly scandalized me; then I imagined the expression you were likely to be making as you wrote such an insult. But that was too hard for my power of imagination. Nonetheless, I did believe I would see an impish smile as opposed to any look of deep humility.–

Regardless of how everything else may be, I am heartily looking forward to seeing all four of you[6] again soon. In the meantime, sincere wishes for your health and well-being from your

Einstein.

# 146. From Paul Ehrenfest

11 September 1920

Dear Einstein,

Thanks very much for your letter.[1]

1. The two physicists are now here with me and we are currently making every effort to help my friends obtain books and instruments.[2] Lorentz and Onnes are helping famously![3]—Please beg for some books and *offprints* for us at the

scientists' convention.[4]—It would be a fine thing if Berliner[5] could give one copy each of the *Naturwissenschaften* for the two large cities (1918, 1919, or at least 1920). *Never* before have offprints been so thoroughly studied!

2. I cannot cash the check for 336 guilders from Methuen before you come (you slacker immediately wrote incorrectly: 700 guilders)—But in any case, on Monday, 1,000 francs are going out to your ex-wife (I already wrote her a letter today).[6]

My friends deserve every conceivable sympathy. They did some really exceedingly fine things in physics under incredibly difficult conditions.[7] I'll write you about that later when I can survey it all better.

—•—

I am very glad that, as you write, the agitation is subsiding.[8] Now, whatever may happen: 1. Do stay away as much as possible from that confounded newspaper wrangling. Because, God help me if I lie, your "My Response" is so chock-full of un-Einsteinisms that I could believe you had written it yourself only because it was unavoidable. So it simply proves that these darned wretches *really* did succeed in making you, for a very short time, "beside yourself."[9]—[In the meantime, I heard from Mrs. François, who was over here, how brutally vulgar the abuse toward you had been.][10]

2. At every moment, have rock-solid trust in your Dutch friends. You shall soon see how much you mean to us all.

Best regards, yours,

Ehrenfest

Your speech is almost certainly supposed to be set for 27 Oct. But only on 20 September shall I know for *sure*.[11]

# 147. From Arnold Sommerfeld

11 September 1920

Dear Einstein,

Many thanks for your very encouraging letter![1] From the letter by Lenard, overleaf, you see that, with Lenard's subjective and sensitive temperament, your case can have nasty consequences for our Phys. Soc.[2] I actually do not doubt that Lenard's name has been misused by Weyland, just as Wolf's and Kraus's.[3] Under this assumption, you might perhaps decide to write a placating word to L. that could work to the benefit of our [Bad] Nauheim negotiations.[4] In his newly reissued brochure *Rel., Äther, Gravit.*, he gave you very decent mention.[5] If you tell him that your defense was directed not at scholarly critics but only at the presumptive aggression of W.'s[6] comrade in arms and that upon his request you would state this in public, his anger will surely be soothed.[7]

Forgive me for trying to occupy you again with this distasteful slough. But you will approve of my motive.

See you soon! Yours,

A. Sommerfeld

## 148. To Marcel Grossmann

[Berlin,] 12 September 1920

Dear Grossmann,

This world is a strange madhouse. Currently, every coachman and every waiter is debating whether relativity theory is correct. Belief in this matter depends on political party affiliation.[1] Most amusing, though, is the Guillaume contest [*Guillaumiade*].[2] For in it, someone using scientific jargon has been serving the most pitiful nonsense to the illustrious experts in the field for years on end, and this with impunity, without being reprimanded.[3] *Thus one sees quite clearly how the judgments and values prevailing among the flock of scholarly sheep rest on the narrow foundation of a few discerning minds.* Refutation is not such an easy matter, though, when one is not even in a position to understand the other's assertions. I took every trouble: I thought about it, corresponded with Guillaume for a long time,[4] but met with nothing but mathematical symbols devoid of any sense. A *factual* sparring is absolutely unthinkable; rather, one can only state an opinion. I enclose one for the *Archives* with this letter.[5]

You ask me in your moody way: "Are you still not ripe for Zurich yet?"[6] This is how matters stand: on a personal level, it's wonderful for me here. My most immediate colleagues are genuinely welcoming and friendly. The Ministry attends to my needs. There is no lack of truly selfless friends, either. But it is exceedingly hard for me to support my family in Zurich; it would have long been impossible if unusual circumstances, which may not last particularly long, had not come to my aid.[7] I do not consider transplanting my children to Germany right.[8] So it could be that for these external reasons I must think of leaving my present position. I dread it, though, because desperate efforts will be made to keep me here, not so much because they want me personally as well as my brain, but more because I have become an idol due to the clamor in the press.[9] The role I play is similar to that of a saint's relics that a cathedral absolutely has to have. My departure would be perceived as a lost battle. It will be damnably difficult for me to summon the requisite hardness of heart, even when it does become necessary. I also think that, in an emergency, they will always drum up the necessary money. The tragedy of my situation is that I cannot muster even the tiniest fraction of self-esteem to play my role, which was allotted to me through no fault of my own, with "dignity."

I am deeply pleased that your wife is completely healthy and happy again[10] and that, generally, everything is going according to your wishes. It pleases me just as much that our boys are classmates, like we were.[11] Let's hope we can soon see each other again. This year I am having my boys come (in October) to Germany, because a trip to Switzerland is too costly for me.[12]

Cordial regards, yours,

Einstein.

For the *Archives*.[13]

In the past few years Mr. E. Guillaume has repeatedly stated his position about the theory of relativity in this journal and, specifically, attempted to introduce a new concept (universal time) into this theory. At the repeated prompting of the author himself as well as of other colleagues in the field, I consider it necessary to declare the following:

Despite taking the [greatest] trouble, I have not been able to attach any kind of clear sense to Guillaume's explications. Even by a lengthy exchange of correspondence conducted with utmost patience, I could come no closer to this goal. In particular, it has remained completely unclear to me what the author means by "universal time." My ability to understand does not even go far enough to be capable of a substantive rebuttal. I can only state my conviction that no clear chain of reasoning underlies Guillaume's explications.

---

Dear Grossmann, please ask the *Archives* to send the proofs to Guillaume.[14] The statement is hard, but I can find no other way; this nonsense has gone *too* far!

## 149. To Elsa Einstein

Kiel, Tuesday. [14 September 1920][1]

Dear Else,

Arrived after a successful, comfortable trip, was awaited by Mr. Anschütz at the platform.[2] I've rarely had it so nice—I say so not to magnify the tribulations of your trip but only to let you enjoy it with me in your imagination.[3] So we puttered away from the train station in Anschütz's motorboat up to a pier that belongs to the Anschützes' villa. It is set right near the water on a small knoll in the middle of a splendid garden. I was then led up to the attic of the villa, where there is an attractive little apartment to lodge visitors; it consists of two small, most tastefully furnished rooms with all the conveniences that the heart could desire and has a splendid view of the Kiel bay. Breakfast is also brought there, so I am surrounded by a matchless tranquillity and don't even notice that I am a guest. In addition, Mr. Anschütz and his wife[4] are quiet and content people who haven't the slightest notion about what it means to hurry and scurry about.

Yesterday evening I also accompanied Mrs. Anschütz to the Missa Solemnis by Beeth[oven].[5] Performance quite inadequate, composition magnificent, but not according to my ideal.[6] Mrs. Anschütz is still very young, pretty, more body than mind: she was very delighted with the presumption of having to mother me,[7] because the parental relationship would fit far better the other way round, although having such a well-endowed daughter with strawberry blond hair would always have a suspicious question mark attached. Today I am going to a talk by Becker on German educational issues and then—alas!—to an official dinner function.[8] (Nothing to be done about it!) Tomorrow morning is my talk.[9] Spengler's not coming.[10] Altogether, the Kiel week doesn't seem to me to grow beyond the local framework. Bourgeois public of civil servants, dull and simple-minded.[11] Not a single seat can be gotten for my sermon, but I think the little crowd will receive even less for their money than—I myself.

Think of the house and the sailboat.[12] We have to create a more human existence for ourselves as well, for all the rural simplicity. There is something fine about a life of meditation. This is most impressively set before my eyes now. Berlin is nerve-racking and deprives me of the possibility of quiet contemplation.

Kisses also to Ilse and Margot[13] from your

Albert.

Send my regards also to the grandparents and Anna, as well as to the minnesinger Moszkowski and spouse.[14]

Greetings from the Anschützes, who regret you declined. See you on Saturday. I am probably going to have to leave on Friday.

## 150. To the Association for Combating Anti-Semitism

Berlin, 14 September 1920

Dear Sir,

Prof. Einstein instructed me to inform you that in his opinion we Jews cannot contribute toward countering anti-Semitism through a direct campaign.[1] Since your view on this point deviates from that of Prof. Einstein, I respectfully ask you in his name kindly to refrain from your plan—of electing Mr. Einstein onto your association's board.

In great respect,

The Secretary

# 151. From Minna Cauer[1]

Berlin W. 62, 5 Wormser St., 19 September 1920

Highly esteemed Professor,

You probably don't remember me. During the raging war years, I met you a few times at conferences of the New Fatherland League. The impression you made on me is unforgettable.

Now the whole world is talking about you; naturally, the envious and the bickerers are also putting in their word.[2] That is human, all too human of those base elements who are dominating more than ever before. I was away for a long time and heard only now that there are men in Germany who do not bow in grateful acknowledgment of the work and knowledge of researchers and genius.[3] I perceive it as a humiliation affecting the whole nation.

You, highly esteemed Professor, are not touched by it. But we suffer painfully that even these most severe and difficult times we are living through do not have enough effect to suppress such base mentalities at all.

It will be a matter of indifference to you, highly esteemed Professor, that a woman, who marvels at the abundance of your knowledge, is writing you. Nevertheless, I cannot but thank you for having brought such a victory of science to mankind.

In expression of my special admiration,

Minna Cauer.

# 152. From Stefan Zweig

Salzburg, 22 September 1920

Highly esteemed Professor,

I think I did not go against your inner convictions by publishing the enclosed statement (in a Viennese paper). I had signed this telegram from Salzburg only with the explicit proviso that it be directed to you personally and not be published, which, however, evidently did happen before it reached you personally.[1] Undoubtedly, you will perceive such a public display of genuine sympathy as embarrassing as I do, and regard this explanation merely as clarification of the circumstances—onlookers could well have thought that the publication originated from you—.

I can imagine how much you must be suffering from the current importunity, as well as hostile malice: may calm, in which you are able to continue to develop your work, soon be restored to you. I welcome this opportunity to express to you my humble but very sincere respect. Most faithfully,

Stefan Zweig

# 153. To Ilse Einstein

[Bad] Nauheim. [On or before 23 September 1920]

Dear Ilse,

Thank you for everything. You are a diligent, good daughter.[1] Accept on my behalf at the federation of public observ[atories]. Topic: Physical foundations of the theory of relativity.[2] Tomorrow is the last day of the scientific convention here (for me).[3] It was magnificent in Kiel, less so here.[4] Mama also came here today.[5] Enjoy yourselves. Greetings to both of you, also to Anna[6] from your

Albert.

Forgive the postage surcharge. I'm in a hurry.

Greetings to the grandparents.[7] I'll write everyone when I have more peace.

# 154. To Ilse and Margot Einstein

[Bad] Nauheim [24 September 1920]

Dear Children,

Else sends her apologies. She's in bed with hemorrhages.[1] The nerves are also somewhat stressed, in part from the agitation about the L[enard] affair, which sets my colleagues in hefty motion.[2] Now she must have absolute quiet and be alone as much as possible. Then she's traveling to Stuttgart, if it doesn't improve quickly, to Ernst Levi's sanatorium.[3] When someone goes on traveling. . . .[4] I hope we can go to Hechingen as well.[5]

Kisses from your

Albert.

# 155. To Hendrik A. Lorentz

Hechingen (Southern Germany). [After 25 September 1920][1]

Highly esteemed Mr. Lorentz,

Your detailed letter[2] moved me very much.[3] Especially your offer regarding a translation into the English language. That American award was already known to me.[4] The founder sent me the announcement himself. I must confess, though, that I immediately decided not to take part in this competition. First, because I do not like to dance around the golden calf, and second, I have so little talent for this kind of dance that it would be hard for me to earn applause for it. Besides, I do not have any actual worries at all about a lack of money. If it ever becomes absolutely necessary, I can have my former wife resettle in Germany with the children; this has been avoided up to now.[5]

Lately, I had to tolerate various types of animosity, mainly from newspapers.[6] This is not regrettable, though, for it is an opportunity to distinguish my true friends from the unreliable ones. The strange thing is that, these days, every value judgment is made from a political point of view.[7]

The [Society of German Scientists and Physicians] conference had nothing new to offer of substantial importance,[8] but it did show that a very lively interest in pure science has remained, despite the war and the economic crisis. In other respects, many things have changed, of course. All these experiences have made minds that are so malleable that a prominent statesman could achieve great things; I am thinking of a union of European states.[9] It is a misfortune that now there is no farsighted, broad-minded leader at the top in France or England, who envisions more than his fatherland's limited, momentary material interests.[10]

For the next fortnight I shall be here in southern Germany together with my boys, a rare treat that has been filling me with joy for a long time now. Then I am coming to Holland, in order to deliver the inaugural lecture. I am very ashamed when I think of how much work you and Kamerlingh Onnes have done for my sake;[11] but I give way to the happy hope that the memory of it cannot disturb your affectionate attitude toward me.

In the hope of finding you and your wife[12] happy and healthy and being able to chat with you for an hour on physics matters (also about your allusions to Eddington's book),[13] I am, with cordial greetings, yours very truly,

A. Einstein.

# 156. From Eduard Hartmann[1]

Fulda, 26 September 1920

Highly esteemed Professor,

A meeting of the Görres Society is taking place at the beginning of the month of October. I have taken on a talk about the theory of rel. for the Natural Sciences Section, in which I intend to describe its high importance and its brilliant confirmation by experiment.[2] In order to be equipped for all arguments—the talk is being attended by a number of physicists and mathematicians—I ask you please to solve for me the following problem in a few words.

It involves the braking train carriage. According to the theory of rel., I have the right to use the carriage as an object of reference. *Before* the braking, I establish a uniform "backwards" motion of the railway embankment and the Earth. *During* the braking, a gravitational field forms in the "forwards" direction, under whose influence the backwards motion of the embankment and the Earth decreases.

Now, how can the observer inside the carriage explain the formation of the gravitational field and its maintenance during the braking process? In addition, each newly forming field must expand from the field-forming source at a finite propagation velocity. Here, however, we would have a field generated by the braking process, that, as it appears, would encompass the railway embankment, the Earth, and even the universe, all in one stroke, wouldn't we?[3]

In the hope that in your customary kindness you will remove this difficulty for me in a few words, I remain in exceptional gratitude and high appreciation, most devotedly yours,

Dr. E. Hartmann, Professor.

# 157. To Elisabeth Ney

[Stuttgart, 30 September 1920][1]

Dear Miss Ney,[2]

From Elsa I hear that you are not pleased, because you did not see Uncle Albert Einstein.[3] So I'll tell you what I look like: Pale face, long hair, and a modest bit of a paunch. Add to that an angular gait and a cigar in his mouth, when he has one, and a fountain pen in his pocket or in his hand. Bow legs and warts he has not, though, therefore quite good-looking; no hair on the hands as ugly men often have. So it really is a pity that you did not see me.

Affectionate greetings from your,

Albert Einstein.

# 158. From Society of German Scientists and Physicians

Leipzig, 48 Nürnberger Street I, 30 September 1920

Dear Colleague,

Re Scientif. Committee.

From the enclosed proceedings of the administrative session of our society on 22 Sept. 1920[1] please note that you have been elected a *member of the Scientific Committee* for the years 1921 to 1923[2] or, if the meetings only take place every 2 years, for the next 3 meetings.[3]

In notifying you upon instruction about this election I ask you please to announce your acceptance to us as soon as possible and sign respectfully,

The managing secretary of the
Society of German Scientists and Physicians,
B. Rassow.[4]

## 159. To Hedwig Born

[Hechingen, 1 October 1920]

Best regards from the most romantic point of our expedition. Here actual consciousness has not yet been discovered—so it seems; in any case, I'm dozing off.[1]

## 160. From Luther P. Eisenhart[1]

Paris, 14 Vendôme Place, 1 October 1920

[See documentary edition for English version.]

## 161. From Max and Hedwig Born

Frankfurt-am-M[ain], 2 October 1920

Dear Einsteins,

Judging from your postcard,[1] Hechingen must be a charming, sleepy haven, just right to let the excitement subside that you, to our distress, had to endure here and in [Bad] Nauheim.[2] Nor do we want to disturb your "dozing sensibility" with effusive letter writing; it is very healthy when even friends leave one's consciousness, and I have the feeling that *we* have to *disappear* like that just now. There is actually nothing more invasive than "sympathy"; it is an intrusion into a friend's life, an emotional denuding that embarrasses one later.

So, before we sink out of sight, like Punch and Judy on the stage, we approach you with two more requests, whose fulfillment I leave in your hands, dear Mrs. Elsa! That is, that you remind your husband sometime: *1*. That your husband write to Mrs. Hoff, 57 Güntersburg Allee. That really would not be a waste of precious time; for such people are few and far between.

2. My husband feels like slaughtering the American golden calf and, by lecturing there, earning the means to build himself a little house in Göttingen, according to his wishes. So, if you still have occasion to recommend someone for lectures there, please do name Max. He could go over in February, March, and April and thereby also satisfy his desire to revisit Broadway (although I do not understand this love, I do excuse it).

So, and now—without any further closing apotheosis—the two puppets,

Max and Hedi Born,

will duck out of sight, until you yourself think of the toy box again.

# 162. To Fritz Haber

Hechingen, [6 October 1920][1]

Dear Haber,

I would like to go to America but only next year, in the winter of 1921–22. This year does not work because of a few obligations I've already taken on. There is the difficulty that I have been asked twice unofficially already by ⟨Harvard⟩ Columbia University as well (once about 8 years ago already).[2] It might be possible to combine the two. Then I would also definitely have to go to Spain.[3]—Traveler in relativity.—Among the women I would just take *one* along, either Else or Ilse. The latter suits best because she is the healthiest and ⟨most skilled⟩ most practical.[4]

The undertaking with the filament holds no prospects, in my opinion. If molecules did not rotate, it probably would work.

If statistical equilibrium reigned with respect to rotation in a magnetic field, it would work only if

$$\frac{\text{[magn.] moment} \cdot \text{field strength}}{\kappa T}$$

were not all that small. The fraction of orientation would then be given by a number of this order, which under favorable conditions ought to be able to be increased up to 0.01. This would correspond to a paramagnetic orientation. Such, however, sets molecular interaction as a precondition for its onset, which does not exist in the molecular rays coming under consideration here.

The true behavior of dipolar molecules upon entering into the field is *adiabatic*, hence analogous to diamagnetism. The thing rotates more slowly against the field in momenta of larger potential energies than in momenta of smaller ones, whereby a *negative* time average results for the orientation from the outer field. That would correspond to a repulsion by the filament which, however, would be much too small to make it noticeable.

There is, secondly, also the Debye effect as a result of the polarizability of molecules;[5] polarizable things are drawn into the stronger field. This effect obviously occurs for every kind of molecule, not just for dipolar molecules. Yet this effect ought to be much too small to let itself be detected (I have not calculated it, but that is very easily done).

Thus I come to the result *that deflections probably are not detectable*. If no zero-point energy of rotation existed, there would be a certain fraction of nonrotating molecules that—provided they are of a dipolar nature—would probably have to exhibit a deflection. However, the specific heat function of $H_2$, as well as the Bjerrum spectrum of *HCl*, speaks *for* a zero-point energy of rotation.[6]

With cordial regards, yours,

Einstein

P.S. Forgive the hasty editing. I have to leave tomorrow morning[7]—I therefore have little time.

## 163. To Paul Ehrenfest

[Benzingen,] 7 October 1920

Dear Ehrenfest,

So, I am coming around the 22nd, with dress coat, to Leyden[1] and am then staying until Nov. 3, on which day I am giving a talk in Hannover.[2] Don't be indignant about the short length of my stay; I've been away from home for so long. I passed on the message to Berliner and hope that our Russian colleagues will receive the other desired things as well.[3] At [Bad] Nauheim, there was a ⟨bullfight⟩ cockfight of sorts, about relativity; Lenard figured, in particular, as my opponent.[4] To my knowledge, it did not come to any kind of manifestations of the sort you expected.[5] There was a very large attendance.[6]

To a happy reunion, yours,

Einstein.

Best regards to all of you, also from my boys.

Many thanks for procuring the money for my wife.[7] Also think of the reimbursement for my last stay. I now have accumulated hefty debts with you; but it will soon be set right again.

# 164. To Elsa Einstein

[Benzingen,] Thursday [7 October 1920]

Dear Else,

After unhappy farewells and a happy voyage (with Rudolf Levi), arrived in Sigmaringen, where, after about one hour, the boys arrived, who are very nice and cheerful.[1] At the station below we happened to meet the priest[2] who came up with us. We picked up the luggage today by cart. None of the villagers go out for fear of foot-and-mouth disease, which is rampant here. I hope you are well; endure the 2 weeks in Hechingen.[3] The children brought food and cigarette butts along; I'm smoking one right now after a festive lunch of chicken.[4] Think of the gift for Mrs. Brandhuber.[5] The Zürchers are requesting a picture of me.[6] Take good care of yourself & be careful.

Heartfelt greetings from your

Albert.

# 165. To Ilse Einstein

[Benzingen,] Thursday. [7 October 1920]

Dear Ilse,

Yesterday my boys arrived in Sigmaringen and I arrived in good shape with them in Benzingen. The priest sends his greetings.[1] *Please send two of my children's books over here immediately*,[2] one for the priest, one for my Albert. I will go to Hannover if the people give me 1,500 marks;[3] I'll write myself. I shall probably take you along to Spain and America, but only in the winter of 1921.[4] Mother has a bladder infection; it goes away by dieting and makes itself noticeable again with each lapse.[5] It's a tedious affair until it's *completely* over. I advised her to stay for another fortnight in Hechingen, but I don't know whether she can stand it for so long.[6] We were in Haigerloch the day before yesterday.[7] The landscape there is gorgeous. I liked the Hohenemsers *very* much.[8]

Best regards from your

Albert.

Greetings to Margot, Anna, & the grandparents.[9]

Dear Ilse,

Your letter just arrived. According to both passport visas, the time of entry is not restricted.[10] So you don't have to do anything. I'll deal with the rest of the things myself.[11]

It's very nice with the boys. Enjoy yourselves for the rest of your solitude.

# 166. From Hedwig Born

Leipzig, 7 October 1920

Just for *you* alone to read!

Dear Mr. Einstein,

Today a very serious, friendly word to you. I so much would have liked to grant you a restful vacation,[1] but too severe consequences are involved which, since Nauheim, have been upsetting your friends' peace of mind.[2]

You *must withdraw* the permission given to Moszkowski to publish the book *Einstein im Gespräch* [*Conversations with Einstein*],[3] and to be precise, *immediately* and by *registered mail.* Nor should it be allowed to appear *abroad either.*

I wish I had the persuasiveness of an angel in order to make the consequences clear to you.

I happened upon Moszk.'s *Freibad der Musen*[4] [*Swimming Pool of the Muses*] here; the level of this book disgusts me so much that I wrote the enclosed nasty remarks, which—this I swear to you—I will publish if you do not immediately withdraw your permission.[5] And I have much more venom [to] spray if it is a matter of saving the honor and moral standing of a friend. I am not painting a too gloomy picture: first of all, the preliminary announcement of "The Works of Moszkowski" (available to read on advertising pillars, newspapers, etc.)

*Die Unsterbliche Kiste, 1000 der besten jüdischen Witze*[6] [*The Immortal Box, 1000 of the Best Jewish Jokes*]

*Freibad der Musen*

*Einstein im Gespräch*

*Sokrates der Idiot.*[7]

This already makes such a very good impression!

Then to the *content.* That man doesn't have the slightest inkling about the [essence] of your character, wherefrom follows what is important and valuable to you, and to us about you. Otherwise—if he understood, or even had a glimmer of respect and love for you—he would neither have written this book nor wrung this permission out of your good nature. Thus your "conversations" will be conducted at a very low level indeed. Narrated with Jewish snottiness and extreme superficiality. Every "pen pusher" will rise and draw a very bad picture of you for his readers. Thereafter you will be quoted everywhere, your own jokes will be smirkingly flung back at you as proof that they know the book. Couplets about you will be written, an entirely new, awful smear campaign will be let loose, not just in Germany, no, *everywhere, and your revulsion of it will choke you.*

And *we*, your good friends, how should we defend you then? "I beg your pardon—Mr. Einstein, your 'humble' friend—gave the permission himself, you know." Then it would be no use [for us] anymore to protest that you gave permission out of weakness, out of good nature. *No one will believe that* (my father, who studied with Moszkowski and has told me many things about him, also tells me so).[8] The *fact* will then simply remain, *that a man, still in his early forties, thus early in his life, gave permission to one of the most despicable German writers to record his conversations.*—If I did not know *you*, I would not concede to a single other living soul, to whom the above fact applied, innocence. I would definitely believe it was vanity. For everyone, except for about 4–5 friends of yours, this book would constitute your moral death sentence. Additionally, it would be the *best confirmation of the accusations of publicity for yourself.*[9]

We friends are profoundly alarmed by this prospect. The book—should it appear *anywhere*—will be the grave of your tranquillity anywhere and for good.

Now I also see very clearly why Moszk. always imposed himself upon you. *He caught wind of the goldmine*. For every one mark spent on each egg that he thrust upon you during your illness,[10] what a well-invested speculation: for each m[ark], he now earns a thousand.

If Moszk. had even only a trace of real heartfelt interest in you, he would be the first—especially after the recent vilification[11]—voluntarily to forgo publication of the book. That he does not do so—even at the request of friends (Freundlich–Max)[12]—gives you the *right* to be *hard.*

Please reassure us directly about this worry of ours, which pursues us day and night. Max just wrote me today: "An express letter by Freundlich just came in with Moszkowski's reply, which is negative, of course, and reveals a vain old donkey. I don't know what I should do yet. I would so much like to discuss it with you; I am so worried every day."

Please, dear friend, quickly dispel our worries and don't rebuff advice and pleas. I am never going to tell anyone about this whole story, for I have heard more than enough about how abhorrent it is to you when women meddle into your affairs.[13] "Women are only there to cook [*Kochen*], of course," but sometimes they can boil *over* [*überkochen*] as well.

Yours,

Hedi Born.

I have also discussed with Father the legal consequences of your retraction,[14] since Max just informed me by telephone that Moszk. wrote: "he has already signed his rights over to the publisher."[15] (That sly dodger!) So now the situation is that the publisher will demand a compensation amount and that it could come to legal proceedings. It is possible, says Father, that the judge will then either consider a rel. *low* sum (for the already incurred costs, printing, paper, etc.), or, on the other hand, that he also—in case printing hasn't started yet—won't demand any damages for the publisher, since your retraction came *in consequence of the new situation resulting from the smear campaign*. That is what you must be prepared for with your withdrawal.

But whatever happens: *You must* (Father thinks so too) *now simply retract permission*.

Even if the book should *nevertheless* appear, because one is powerless against such crafty types, then *you* or your *friends* can present the situation in the papers and, depending on the circumstances, energetically attack Moszk. *You then are in a different moral position*. *We* all accomplish nothing with Moszk.; with us he condescendingly invokes his friendship with you. We too find ourselves now in a distasteful position toward that fellow that can change only if you coldly, without much fanfare, withdraw your permission and henceforth do not grant anymore discussion, any more contact. He shouldn't be allowed to exploit your "friendship."–

Please do not show this letter to your wife; her nerves do seem to be very shaken as a result of the Weyland-Lenardiana affair, and her so direly needed recovery really should not be upset by such troubles.[16]

## 167. From Fritz Haber

Dahlem, Berlin, 4–6 Faraday Way, 7 October 1920

Dear friend Einstein,

In the attached you will find a large amount of printed paper, and I would like it very much if you could look favorably upon it. It involves a matter that you, based on your way of thinking, ought to have originated yourself, had it not already existed. You surely share my opinion that it is a cheerless state of affairs when, with all these wage disputes, hospitals can no longer be heated because they cannot obtain any coal, children no longer get milk, and water and gas are lacking and therefore no more meals can be cooked, and more of the like.[1] I believe it is the binding duty and obligation of all decent people to make sure that vital services continue to operate while managers and workers are otherwise engaged in screaming bloody murder at one another.[2] The institution existing in Germany that achieves and secures this maintenance of vital services is the Emergency Technical Aid

[*Technische Nothilfe*], which this enclosed stack of paper addresses.[3] The man who called this Emergency Technical Aid to life is a friend of mine named Lummitzsch; and I find that you and he are the only two people to have created anything of note and of international significance in Germany in recent years.[4] That is why I would like to get you interested in helping Mr. Lummitzsch in Holland.[5] The help would consist in finding a man in Holland who would take an enthusiastic and avidly profound interest in the cause of this Emergency Technical Aid and allow himself to be associated with this enterprise in Holland, as is the case in Denmark, Sweden, and Norway, following the German model. The Emergency Technical Aid over here would then turn to this man in order to get in touch with Dutch circles and to prepare the way for international communication in this area, which is just as essential as the Red Cross was and still is now.[6] Your daughter told me[7] that you are going to Holland right now, and if you find someone there who meets these qualifications, then please write one of your postcards, which always particularly delight me, and do not be indignant that I want something again from you.[8] I do think, though, that you could help; for what, ultimately, do you do with all that relativity, when in the wintertime Lummitzsch does not prevent an absolute shortage of gas, water, and electricity for *you* in Berlin, which no gravitational potential can alleviate, or even any arbitrary curvature of space can improve?

With hearty greetings to you and your dear wife, your friend,

Haber

# 168. From Arnold Sommerfeld

[Munich, 7 October 1920][1]

Dear Einstein,

Yesterday I was talking to Geiger, a philosopher colleague[2] who is very close to me. He had received and declined an invitation by a "Working Association 1920" to give a speech; they want to ask you for a lecture as well. Whether the university will unlock its halls for this is doubtful. This syndicate involves younger literati (Jewish, as Geiger stresses, who himself is a Frankfurt Jew) of a kind of Bohemian type; one of these, of the name of Holländer, I believe, is *possibly* the same one who distinguished himself in the [Bavarian] Soviet Republic.[3] Hence this lecture cycle does smack of sociopolitics, contrary to Mr. Weyland's direction, of course.[4] Geiger had the quite correct feeling that if you spoke in Munich, you would want a lecture without such a tinge, on a purely scientific podium, which all of us would most heartily welcome. He declined for himself because he does not favor the fusion of science with the trends of the day.

I am writing you all I know about this case and told Geiger that stopping you from a lecture would be much easier than motivating you to give a talk. I now proceed pertaining to the latter:

You know that, at the first suggestion from you, you will always be received with open arms here as a speaker, for example, at the local chapter of the Physical Society. I have the funds from the Anschütz Foundation to reimburse you properly for the trip, etc.[5] I would very much like to invite a broader range of students, i.e., have your lecture take place in the main auditorium. I do not believe that the Berlin scenes could repeat themselves here[6] and will employ all means to try to prevent it by limiting the ticket distribution to certain categories of students.

But I do not want to bother you with this invitation. You should not reply to me now, either, rather only when you deem the point in time has come when you feel inclined toward a Munich talk. What I would like to achieve with this letter is merely the following: if you speak in Munich, then speak under our auspices and not before the literary "Working Association 1920."

With cordial greetings, yours,

A. Sommerfeld

## 169. From Erich Wende[1]

Berlin W 8, Wilhelmstreet, 8 October 1920

[Not selected for translation.]

## 170. To Elsa Einstein

[Benzingen,] Saturday [9 October 1920]

Dear Else,

It's very nice here.[1] We even managed to get hold of a violin and played from Haydn's "Creation."[2] The priest is somewhat overworked and his heart is not quite in order; his official business in Sigmaringen took an additional toll on him.[3] But I think the change we are bringing him is doing him quite a lot of good. Fidelia's[4] roof is now finally being fixed, so there'll be an end to her being in the trenches. The Borns now also want me to recommend him to America so that he can earn the money from my theory for a house in Göttingen.[5] That seems a bit unconscionable [*ausgeschämt*] even to me. Stay in Hechingen until the 20th so that you can recover thoroughly.[6] Greetings & kisses from your

Albert.

# 171. From Moritz Schlick

Rostock, 23 Orléans St., 9 October 1920

Dear, highly esteemed Professor,

For the last few days I have been reading with greatest enjoyment the booklet by Reichenbach about relativity theory and *a priori*[1] knowledge. The work really does appear to me to be a very outstanding contribution to the theory's axiomatics and to physical knowledge in general. You surely were also very pleased with the logical probity. Obviously, on a few points I cannot concede Reichenbach to be completely right; I hope to be able to arrive at an agreement with him about that by letter, for this matter really does lie very close to my heart. I would have liked to ask for your opinion, but in writing that would be too inconvenient; perhaps I may be allowed to come back to it in conversation, for I fervently hope that I shall be granted the favor of seeing you again sometime in the winter. Reichenbach does not seem to me to have done justice to Poincaré's theory of conventions; what he calls *a priori* correspondence principles and rightly distinguishes from empirical correspondence principles seems to me to be completely identical to Poincaré's "conventions" and not to have any meaning extending beyond that.[2] R.'s reliance on Kant seems to me, upon closer examination, to be merely terminological. I would later like to ask your opinion also with respect to a passage in the magnificent book by Born on the theory of rel.,[3] of which I saw the correction proofs. It concerns the juxtaposition of matter and field (in the last section of chapter V). I exchanged correspondence with Born about this, and although his reply fully placated me with regard to the passage itself, questions did arise in connection with it which, owing to their philosophical importance, I would like to present to you once in person. I heard some nice things about [Bad] Nauheim,[4] and I would heartily like to have been there, but the trip seemed just too long to me from here. What trip doesn't seem too long nowadays?

I would again like to shake your hand in sincere gratitude. For, I have perceived from various quarters that, in the interim, you have been solicitously mindful of me again. Through your recommendation, I received invitations to deliver talks in Danzig and Harburg;[5] furthermore, to write articles for the journal *The Monist* and for the paper *Berliner Tageblatt*.[6] Nothing came of the Danzig talks, because the treasury of the local scientific society was unable to promise me sufficient travel reimbursement, but I am going to speak in Harburg. The article for the *Tageblatt* turned out badly because of the obligatory, exaggerated brevity; by contrast, the one that I was permitted to write up for Mosse's *Almanac*, also as a consequence of your generous recommendation, does seem to me to be better. Neither of the two have appeared yet. Writing for *The Monist* gives me much pleasure, and this contact with England is extremely valuable to me as well. It was truly refreshing to read what has been written in England about you and the Germans' treatment of you. Berliners deserve to hear the truth told: I still shudder at the thought of the possibility of them succeeding in making it too loathsome for you to stay in Berlin! A short while ago Mr. Bröse, who translated *Space and Time* as well as Freundlich's book, visited us here from Oxford.[7] A very nice person and most highly talented musically. Did I already tell you that I entered myself in the *Scientific American* contest for a popular account of your theory?[8] That cost me much labor, because of the 3,000-word limit, but the prize is so enormously high (5,000 dollars) that I believed I ought to try, even with the extremely low chances: my family would be helped along for a number of years then. Incidentally, there is at the moment a vacant philosophy professorship in Erlangen, and I am dutifully informing you in case you should have any connections there. I do think, though, that only a historian of philosophy would come into question. Forgive me please for having bothered you in my last letter[9] with the question about physicists who might come into consideration as a successor to our deceased R. H. Weber.[10] It would not have been of any use, after all, even if I could have recommended a couple of names, for the candidate list was completed here in all speed, without making numerous inquiries. In the meantime, Lenz in Munich received the call but will probably not come. Ewald and Kossel are in second and third place.[11] The Kant Society wants to offer a prize for a paper on the relation between rel. theory and modern philosophy; I am going to be one of the judges. As the physics judge, Vaihinger definitely wants Wiener (Leipzig). Do you consider him suited for it, at all?[12]

Now I have come to the bottom of this paper and I have scarcely told you anything of interest. But I am glad to be able to express to you once again my profound admiration. My whole family heartily wishes you health and well-being. Won't you be making another trip northwards again sometime that leads you through Rostock? Best compliments to your esteemed wife. Your grateful[13]

M. Schlick.

## 172. From Hermann Anschütz-Kaempfe

Kiel, 24 Bismarck Avenue, 10 October 1920

Esteemed Professor,

I owe you a letter about the metal spheres.[1] Initially I obtained quite good results with 3 magnets and one copper hemisphere of 3 m/m wall thickness. Then I tried the same with a whole aluminum sphere of 1 m/m wall thickness and, to be precise, with two rings of magnets of 10 magnets each.

A sketch is attached:

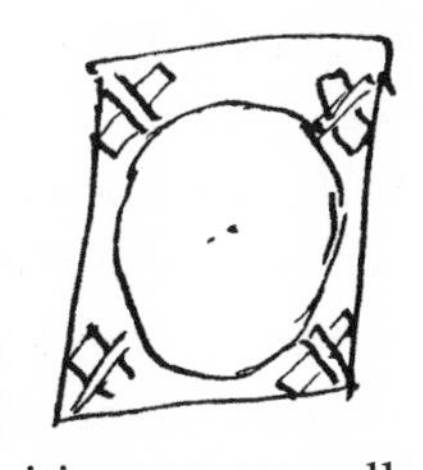

Distance between the magnets and the sphere 5 $^{m}/_{m}$.

It is thereby revealed that the sphere has preferred positions that are caused by irregular thicknesses of the sphere wall or by varying conductivity of the aluminum sheet. A repetition of the experiment with an Al sphere, carefully machined inside and out, of 2.5 $^{m}/_{m}$ wall thickness was very successful. Preferred positions very small. The sphere's settings between the rings of magnets very clear.

I am now having a new experiment prepared with a sphere of 220 $^{m}/_{m}$ ∅ and built-in gyroscopes, hoping that it manages to catch the lines of force in such a way that the 3 gyroscope bodies do not cause preferred positions; for that purpose, I have arranged two more iron sheets inside the sphere.

In any event, the thing looks very hopeless. As of now, I have not registered any patent yet; still want to wait and see what other surprises and problems the new apparatus still has to show for itself. Then the firm A[nschütz] and Co. will contact you and send you the patent proposal and solicit your approval.

The second trial with the heated copper cylinder has yet to be done, because I am still waiting for a pure copper rod of 50 $^{m}/_{m}$ diameter, which has already been ordered.[2] I first provided for frictional heat, by the simplest arrangement, since I then have no worries about power connections [on both sides] or the like. I shall report to you about the experiment that my cousin Schuler is going to conduct during my absence.[3]

We are in the act of dismantling our tents and moving southward; the cold is chasing us away. It would be grand if your path led you to Munich and brought us the pleasure of your visit. And we are quite certainly counting on a hopefully very long visit from you and your wife next summer. Then I am going to demonstrate to you the hopping sphere *in natura*.

With most cordial greetings, yours,

Anschütz.

I too am looking forward to next summer, which promises to bring a visit by you and your wife. Your surrogate mother,

Reta Anschütz.

## 173. From Ilse Einstein

[Berlin,] 10 October 1920

Dear Albert,

Yesterday evening I received your postcard;[1] thanks very much indeed. I would have liked to write you properly, but now it's 11 o'clock at night (Sunday) and I've been answering letters for you all afternoon until now, so I'm almost fed up to the teeth with that job.[2] These godforsaken people are really out of their minds to write you so much. Each time a mail delivery is dropped off, I let out a powerful curse, but that doesn't stop them at all from their senseless business.

I am sending you as enclosures a few things that you might perhaps answer, considering all the time you have in that cozy nest of yours.[3] I am glad that you are having such a nice time with your boys. The weather is, of course, splendid, too.

In my free moments I am constantly singing "Beautiful Spain, far off to the south" in preparation for our trip.[4]

Many affect. greetings from your today particularly crazy [*meschuggene*],

Ilse.

## 174. To Max Born

[Benzingen,] 11 October 1920

Dear Born,

Your wife wrote me an incensed letter[1] about the book by Mr. M[oszkowksi].[2] *Objectively*, she is right, although not in her severe judgment of M. *I have informed the latter in a registered letter that his magnificent opus is not allowed to be printed.*

With warm greetings to both of you, yours,

Einstein.

My hearty thanks to your wife.

## 175. From Max Born

Frankfurt-am-M[ain,] 13 October 1920

Dear Einstein,

The accompanying sheet from the booksellers' periodical *Buchhändler-Börsenblatt* reached me from various quarters.[1] Commentary superfluous. It seems that you are less upset about it than your friends.[2] My wife already wrote you what I think of the matter.[3] (She does already regret, though, having otherwise also wanted to turn your name into gold by sending me to America;[4] poor women must bear the brunt of existence and grasp at any relief.) You *must* shake off Moszkowski, otherwise Weyland will have won the whole way, Lenard and Gehrcke will be triumphant.[5]

According to the experts consulted, the best course is this: You write emphatically to Moszkowski that, after having been accused of seeking "publicity,"[6] you cannot authorize publication of the *Conversations*, especially since the advertisement in the *Buchhändler-Börsenblatt* is offering your opponents more ammunition. If, as is to be expected, M. refuses, you obtain from the public prosecutor a preliminary injunction against the appearance of the book and see that it gets into the papers (or else we arrange that). I am going to inform you about the precise details about where to file the petition. Experts have determined that, just as it is forbidden for anyone to publish a picture of someone else without his consent, no more may the conversational thoughts of another be printed either. This course is more appropriate than having the printer's proofs sent to you to read through because then you have *absolutely no* responsibility for the book. Otherwise, if in the foreword it is stated that you had read and approved the correction proofs, all the dirt that is flung about as a result of the book will land on you. I *beseech* you, do as I write. Otherwise: Farewell Einstein! Then your Jewish "friends" will have achieved what the anti-Semitic gang were unable to do.[7]

Forgive the insistence of my letter, but it concerns all that is dear to me (and Planck, Laue,[8] et al.). You don't understand; in such things you are a little child. You are loved and you must obey; that is, perspicacious people (not your wife).[9]

If you don't want anything more to do with this business, then give me full authority *in writing*. I shall drive to Berlin, if necessary, or to the North Pole.[10]

Yours,

Born.

## 176. To Lucien Chavan and Jeanne Chavan-Perrin

[Benzingen, 15 October 1920]

Dear Chavans,

I am sitting in this isolated hamlet with my boys[1] and reminiscing about you and the times in Bern.[2] Albert likes telling me about all the nice things he experienced with both of you.[3] A few years ago I wanted to visit you at the office;[4] but you were out of town.

Good luck and *auf Wiedersehen* (hopefully soon)! Yours,

A. Einstein.

## 177. From Vilhelm Bjerknes[1]

Bergen, 18 October 1920

Esteemed Colleague,

I very much regret that I had to travel to the extreme north of Norway—on official business—during the time that you were holding your lectures in Christiania.[2]

Now, however, I learned during a visit there, whence I recently returned, that it might perhaps still come to pass that we later meet there, after all. For according to long-standing agreements, I am, in all probability, going to be moving [over] there again as soon as my current "geophysical" period has ended, for which there is prospect in the not too distant future[3]—and now I learn that the issue of your being called to Christiania is being seriously considered.

Whether something will come of what to many is an "exotic" plan, I dare not say. Naturally, you would find the University of Christiania intellectually poor. Nonetheless, it is not completely without tradition, and particularly not so in your own field. No one before yourself has made a more energetic attempt than my father[4] to trace gravity (and other forces of nature) back to the effect of inertia, although not to such an effect of the gravitational masses themselves within curved space—rather that effect of free, space-filling masses within Euclidean space.

I take the liberty of sending you the most important publications with regard to this—all of them unfortunately by me;[5] he himself never sat down to writing, probably partly an effect of his isolation in Christiania.

I myself long to return to the area of research into which I had been introduced already as a boy.[6] I would very much like to know, before I stop, *why* these extended analogies exist. Are they the play of a—very improbable—coincidence? Or do deeper formal or real reasons lie behind it, and if so, which ones? I would like to have a chance to discuss such questions with you all the more when I take up this work again.

Yours very truly,

V. Bjerknes

## 178. From Zionist Student Association of Eastern Galicia

Lemberg [Lwów], 18 October 1920

Your Eminence,

Highly esteemed Sir and Master,

Moved by the outrageous hate campaign that is being conducted by German students and pseudoscientists against your person as a Jew and a scholar,[1] we, the representatives of the Zionist university scholars of East Galicia in Lemberg, feel moved to express to Your Eminence our sentiments of profound respect and devotion.

We Jews are most highly proud of the honor and good fortune of being able to count you, highly esteemed Professor, among our own and hope that in the not distant future the Hebrew University in Jerusalem will offer Your Eminence the possibility of educating a student body worthy of this distinction.–[2]

We can just recall the famous words by Disraeli to Bismarck: "We stood on the pinnacle of civilization and had people at the height of greatness while you were still crawling about in bearskins in the forests."[3]

We have the honor of undersigning respectfully as Your Eminence's deeply devoted "Histadrut Akademim Zionim,"

Dr. Abraham Schwarz

## 179. To Elsa Einstein

[Sigmaringen,] Tuesday, 19 October [1920]

Dear Else,

Now I'm back in Sigmaringen, where I'm picking up Brandhuber for Benzingen. The trip with the boys through the Danube valley was magnificent.[1] We then also climbed up the Hohentwil near Singen.[2] Finally, we drove to Constance, where I bought Albert a few musical scores and Tete a book by Mark Twain.[3] They then traveled to their mother in Donaueschingen, from where they are continuing homewards.[4] Today I saw the resplendent art treasures at the local castle.[5] The boys have developed grandly; Albert is purely practical, disregarding his passion for music.[6] He'll make his own way; he wants to go to South America after completing his studies. The boy, for all his childishness, is very far advanced. He read to me *Budge & Toddie* by Habberton, his favorite book.[7] Deeper human perception seems to be inaccessible to them both—alas. We were very chummy, no more of the tensions from earlier;[8] but I cannot see them as my temporal successors. They have large chubby hands and for all their intelligence something indefinably four-footed about them. Tomorrow or the day after it's onwards to Holland[9] and then—thank heavens—soon back home again. Think of the house purchase![10]

Fond greetings to all of you. Yours,

Albert.

Console Moszkowski. His book about me must *not* appear. It would be catastrophic![11]

# 179a. To Elsa Einstein

[Leyden], Friday. [22 October 1920]

Dear Else!

Yesterday 12 o'clock after happy night trip (thanks to Ilse) arrived safely in Utrecht. There I stayed (with Julius, who suffers of heart ailments, as does his wife) until the evening and then traveled to Leyden, where I surprised the Ehrenfests. It was a hearty reunion. Since the two Russians, who were also in Berlin, are with him, colleague Onnes has invited me and I am indeed staying with him, in a very beautiful room. He and his wife are very kind. I am now coming on the 7th of November to Berlin, because Ehrenfest and Kam[erlingh] Onnes have convinced me through a letter they sent to Benzingen that I move the talk in Hannover to the 6th of November.

Warm greetings to all of you, your

Albert.

[...]

# 180. From Bertha Moszkowski

22 October 1920

Dear, esteemed Professor,

Although you will already have learned from your wife, Elsa, what I am writing you about here, I really cannot help repeating it.

On that terrible Friday when your letter arrived, Alex immediately dropped everything to fulfill your wish.[1] At 10 o'clock the publisher was over here with us and, as he saw the state my husband was in and heard his earnest pleas, he initially granted him his request and telegraphed right away to Leipzig to stop the printing of the book, which is almost finished. He also promised, since he is much indebted to my husband, to assume the costs arising out of it and to convince his co-publisher—a new publishing house was founded for the book: Fontane, Hoffmann & Co.—to take the same position.

That was on Friday.

On Saturday and Sunday we heard nothing. On Monday, however, at a meeting both publishers declared that it was impossible for them to keep their word, the business could not be prevented anymore, they could not expose themselves to this disgrace before the many booksellers who have already placed their orders, it would place them in disrepute and they would never again receive a response to their sales offers. It is already evident now that the sacrifice would be too great.

All my husband's suggestions were to no avail, he was ready to sacrifice anything. He even wanted to relinquish everything for himself, but the publishers absolutely do not want to give up the large profits that they are anticipating.

What else am I to say to you? I am exhausted from the excitement of the last few days.

In the foreword of the work it will be added that you did not read the book and that my husband is solely responsible for the authorship.[2]

And now I imagine that today, after having spent such a nice time with your friend and the boys, you are a bit more removed from all those unjustified notions. Who or what can hurt you? How few of your friends and counselors are completely free from envy? Don't they all want something from you? Don't they all use your name to their advantage? And that my Alex in particular should be denied this? Would it silence your enviers? Great men are a good target, easy to hit, but the arrows must bounce off you.

I hope and think that, with your sense of justice, you will not allow a shadow to fall on our friendship. My husband is innocent, and time will tell how much he was prepared to give up.

You would cheer us up if you sent us a word of reassurance, if only through Else.

How happy my husband was at first that the book was to appear. Now he has been deprived of his joy by this unfortunate incident, and he, as I and Richard, will be able to be happy again only when we know for certain that your friend Born's persuasiveness does not withstand the facts.

Born's biography of you is certainly enthusiastic and could never hurt you.[3]

X rays penetrate through everything and thus it is also with you and your fame; you and your enviers must come to terms with that.

In respect and most sincere friendship, yours,

Bertha Moszkowski

# 181. From Edgar Wöhlisch[1]

Kiel, Med. Univ. Clinic, 25 October 1920

Esteemed Professor,

Occupied by a paper on the connection between the molecular volume and energy content (heat of combustion) of liquid organic compounds,[2] I take the liberty of applying most respectfully to you for your advice on a question in theoret. physics.

In the literature I have perused so far, the generally held view is that, given various isomeric compounds, those with larger combustion heats must also have larger (true) molecular volumes. Yet, the rough survey I have made of the available material does not always seem to confirm this view, and I would therefore very much like to know whether the mentioned parallelism between molecular volume (as is measured, for example, by Van der Waals's *b* constant) and combustion heat is at all a postulate of our modern views on molecular theory, or whether it is also possible to conceive of cases in which relations are reversed. Is there any literature in theoretical physics on this question yet?

In giving my most obliging thanks in advance for the kindness of eventual consideration of my problem, I am, in great respect, yours very truly,

Edgar Wöhlisch

# 182. To Max Born

[Leyden, 26 October 1920]

Dear Born,

I categorically forbade appearance of the book by M[oszkowski].[1] Ehrenfest and Lorentz advise against court action, which would only heighten the scandal.[1] The whole business is a matter of indifference to me, along with the clamor and opinion of *all* persons. So nothing can happen to me. In any event, I applied the strongest means available outside of the courts, threatening in particular with a break of relations. By the way, M. really is preferable to me than Lenard & Wien. For the latter cause problems for love of making a stink,[3] and the former only in order to earn money (which really is more reasonable and better). I shall live through all that awaits me like an uninvolved spectator and not let myself be agitated anymore as in [Bad] Nauheim.[4] It is completely incomprehensible to me that I could have so utterly lost my sense of humor in such bad company. Yesterday Lorentz spoke about your lattice equilibriums in his lecture;[5] I was also in there! Such an admirable person!

Cordial regards to you and your wife.

I am having a nice time here in Leyden. Weiss & Langevin are here.[6]

# 183. To Elsa Einstein

[Leyden,] Tuesday. [26 October 1920]

Dear Else,

It's grand here.[1] I've been living all week with Kamerlingh Onnes's brother, an excellent painter who lives in a fabulously furnished home, has a son, who is also a heaven-blessed painter, two very nice daughters and, it goes without saying, a dear, considerate wife.[2] This displacement is because the two Russians are living at Ehrenfest's, and Weiss and Langevin at K.-O.[3] The latter [Langevin], whom

I'm very particularly looking forward to meeting, is arriving this afternoon. Tomorrow is the famous inaugural lecture that has been casting its shadow so far in advance.[4]

Yesterday I received a letter from Born with a singularly tasteless booksellers' advertisement for Moszkowski's book.[4] I wrote ⟨him⟩ the latter that in the event of the appearance of the book all personal contact with him will be cut off.[6] I refrain from legal prosecution (which the Borns are advising), because this would only increase the scandal.[7] This affair is much more bitter than the Co. Ltd.[8] If it gets too unsettling in Berlin, I'll simply absent myself.[9] Then it's all the same to me.

Kisses to you, the children, and the parents.[10] Yours,

Albert.

## 184. To Elsa Einstein

[Leyden,] 28 October [1920]

Dear Else,

Yesterday was my inaugural lecture with very great attendance.[1] It all went well and pleasantly. That evening was an invitation at K[amerlingh] Onnes.[2] Langevin is a darling fellow.[3] We kissed heartily upon seeing each other again. Saturday is Academy.[4] I'll be visiting Lorentz one more day.[5] The Moszkowski affair is worse than you think. But we'll get over it, if we don't read anything and don't react to anything that arouses the people's anger.[6] What is certain is that M. has done me more harm than Lenard and Gehrcke put together.[7] You're right that you pressed to have the title changed.[8] But little use.

Greetings & a kiss to all of you, yours,

Albert.

Take care of the house purchase.[9] Your provisional acceptance for America for next winter was correct.[10] I don't have time to deal with correspondence here.

## 185. From Max Born

Frankfurt-am-M[ain], 28 October 1920

Dear Einstein,

I am very glad that you acted forcefully against that book by M[oszkowski].[1] The future will tell whether it is enough to prevent a big stink. The main thing is that you are determined not to let your peace of mind be disturbed anymore. But, ultimately, you are not the only one involved; rather we, who venture to call ourselves your friends, are likewise enveloped in the stench and I fear we are not going to be able simply to hold our noses as you intend. You can just slip away to Holland, but we are stuck here in the land of Weyland, Lenard, Wien, and their cohorts.[2]

I am hurrying to write you in Holland because I would like to know the address of Mr. Fokker. He sent me a fine paper in which he absolves me from a sin of my youth; the address was also on the envelope, but since I was sick and in bed with asthma,[3] I could not pay attention to it and so my children destroyed the envelope. I would so much like to thank Mr. Fokker; Ehrenfest will know where he lives.

Do also have Ehrenfest show you the transcription of a letter by Boguslavsky,[4] which I sent him, and think about how that poor person can be saved. Planck wrote me he was very ready to help personally but thought nothing could be done officially in Berlin. Now I am negotiating with Hilbert to have him invite Boguslavsky through the Wolfskehl Foundation.

I am glad that all is going so well for you in Holland. But you must not be annoyed with me if, after the last incidents, I doubt your knowledge of human nature so strongly as not to share your admiration for Lorentz.[5] You see Lenard and Wien as devils and Lorentz as an angel. Neither is quite accurate. The former suffer from a political sickness[6] that is widespread in our famished country and is not at all based on innate malice. While I was in Göttingen just now, I saw Runge emaciated to a skeleton and correspondingly embittered and changed.[7] Only then did it become clear to me what is happening around here. By contrast, Lorentz: he refused to write something for Planck's 60th birthday, you know.[8] I hold that *very* much against him. You are welcome to tell him so. One can, of course, have a different opinion from Planck, but his honest, noble character can only be doubted by those lacking in such qualities. Lorentz evidently fears a loss of his contented Entente friends more than he values justice. The fact that he lectures about my lattice calculations in his course does not charm me. Besides, that is not the only thing I have against him, but I am not writing just to complain; rather, I frankly confess that when I know you are with Lorentz, Ehrenfest, Weiss, and Langevin,[9] I find that much more welcome than socializing with the author of *Freibad der Musen*.[10] You have probably also seen Mr. Chulanovsky[11] there from Russia; ask him about Mr. G. Krutkow,[12] who sent me a paper on adiabatic invariants that seems to me excellent. He must be an outstanding theoretician; I never heard of him before.– My wife sends her best regards,[13] she is slaving away because our cook had to be removed a few weeks ago for theft and deceit (in countless instances). Added to that, I lay in a miserable state in bed until yesterday with asthma and had to be "nursed." The children are well.[14]

With heartiest greetings, yours,

Max Born.

# 186. From Paul Hertz[1]

Göttingen, 34 Riemann St., 28 October 1920

Esteemed Mr. Einstein,

I received a letter from a Hungarian acquaintance in which she asked for my intervention for a Hungarian engineer.[2] He is supposedly an exceptional person; he was a public commissar, and that is why he is now in danger of being sentenced to death. As the lady writes, he never did any harm to anyone.

In jail he wrote some scientific papers and my acquaintance thinks if they were very favorably reviewed, this could perhaps move the Hungarian government to a reprieve. I cannot send you the papers because they are not here at the institute. So you probably can easily obtain them yourself (József Kelen, Budapest, issues 20 and 30 of *Zeitschrift für Elektrotechnik und Maschinenbau*, 1920).[3] Now, it is unlikely to cause much of a stir if I wrote something about it; an article by you, however, could perhaps make an impression.

Another way would possibly be to get important physicists and engineers to launch a public appeal. This solution was successfully taken, in its day, for Lukács (Harnack also participated).[4] As I hear, Philipp Frank (Vienna) wants to attempt something.[5]

It is a pity, of course, that I cannot give you any details at all. I wrote the lady to kindly provide you with information right away (character, prehistory). Because of the extraordinary urgency, I am writing you now and it must be left to you whether you would like to wait for more information or be able to do something immediately, possibly with the help of our Dutch colleagues.[6] I wrote the lady to convey the necessary information directly to you.

Excuse me for causing you so much trouble. But it does involve saving a person's life. With best regards, yours,

P. Hertz

# 187. From Bertha Moszkowski

28 October 1920

Esteemed Professor,

The heavy, unfair blow that you delivered in the letter my husband received today lands on a completely innocent person.[1]

At all events, he *Buchhändler [Börsen]blatt* that was sent to you is not meant for the general public and came into your hands only through the indiscretion or malice of some publisher.[2] Not even newspapers take these advertisements, and you must believe me that neither my husband nor I have ever seen that paper nor know about the advertisement.

My husband gave the book to the publisher; he has never concerned himself with marketing. When Else informed me about the advertisement, I immediately arranged that she personally discuss it with the publisher, since she knew the exact wording.

She will confirm what I tell you; she could also hear from him about the impossibility of withdrawing the work.

In the meantime, you have probably learned that the title has been changed, hence that your personal collaboration is completely eliminated; you likewise now know that much more that could have in any way exposed you to attack has been removed from the book.[3]

And now I must point out that anything my husband ever wrote about you was never "publicity."[4] In the *Book of 1000 Wonders*[5] he placed you in as elevated a place as befits you and which your enemies are jealous of. The work has now had a printing of $40^{m}$, has thus been read widely, and never was it mentioned when your enemies brought forward evidence of "publicity."

Max Born opens E[instein]'s theory of relativity with *a portrait of Einstein*.[6]
Freundlich, the foundations of E.'s theory of gravitation

*With a foreword by Einstein.*

—You know, it is 10 lines.—

So strictly scientific thinkers would have had to make do without publicity for their book. You, however, surely know very well how much more valuable your book is with the aid of a portrait and foreword. You believe in the most absolute disinterestedness of these friends and you surely have reasons to do so.

But more loyal friends than we are to you, you cannot have, and your hard words could not have hit anyone as heavily as they hit us.

The dreadful advertisement appeared *once* in the *Buchhändlerblatt*; after my husband learned of it he, like your wife, extracted from the publisher the promise that nothing appear, even there, that he had not seen and authorized.

My husband is incapable of writing at the moment; that is why I have taken it on.

We have only one wish; that you gain the firm belief that my husband neglected nothing in complying with your wish, but failed before the impossible, and that when you know all the accompanying facts you will take back your hard words.

Yours,

Bertha Moszkowski

# 188. To Elsa Einstein

[Leyden,] Sunday. [31 October 1920]

Dear Else,

Yesterday I was at the Academy together with Langevin & Weiss.[1] That evening I was invited by Zeeman while the other two were visiting Lorentz.[2] The Frenchmen are soon leaving and I would do so, too, if I hadn't rescheduled my talk in Hannover to the 6th.[3] Now it'll take a whole week more until I'm finally back with all of you. In the interim, work is being accomplished and otherwise the living is nice. I have become frightfully fat. Ehrenfest is suffering somewhat from a stomach problem, a little similar to how it had been with me.[4] He couldn't come along yesterday to Amsterdam.

Kisses to all of you, yours,

Albert.

# 189. From Adriaan D. Fokker

Delft, 2, Rotterdamsche Way, 2 November 1920

Dear Professor,

One always wants to ask you more questions. Now the following problem has entered my head.

In a Euclidean-Minkowskian space-time $(-1, -1, -1, c^2)$, a gyroscope is made to describe a plane circle at a certain velocity. If the rotational axis is not set vertically to the circle's plane, will the gyroscope exhibit no precession?

According to the Lorentz contraction, the "center of inertia" of the gyroscope masses do not lie along the axis. This could be a reason for precessions.

This problem occurs to me because it might be able to explain the discrepancy between Schouten's calculation and mine.–[1] One would expect, though, that such a precession would be dependent upon the magnitude of the velocity of rotation (because at a faster rotation the center of inertia shifts farther away from the axis of rotation) and this would make the explanation for the discrepancy illusory.

Did you ever worry about this idea? What should one believe?– Tomorrow we are going to be at Ehrenfest's around 11 until half past 11 to say goodbye.[2] If you would like to telephone Ehrenfest, he could give me your response.

So you are traveling on Thursday?[3] Most cordial greetings from both of us,[4] and so long! Yours most sincerely,

A. D. Fokker.

# 190. From Willem de Sitter

Arosa,[1] 4 November 1920

Dear Einstein,

It was with *very* great enjoyment that I read your Leyden inaugural speech.[2] What pleases me particularly in it is that you emphasized with such finality the untenability of a purely mechanical explanation of nature. Even as a student (so, around the time of 1894), I always bristled when someone explained matter by the ether or by electricity, only to turn around again and seek material explanations for the ether! That always seemed preposterous to me. Now you have decided to call the $g_{\mu\nu}$-field the "ether," and you show convincingly that this ether is just as good as "matter," if not better, as a physical primal substance [*Ur-ding*]. In my opinion, there is consequently *no* reason left to look for a *material* carrier of inertia. Mach's requirement also seems to me simply to be a residue of the quest for a *mechanical* explanation of nature (on the basis of action at a distance).[3] The *ether* is the carrier of inertia. The material points are just discontinuities in the ether, i.e., in the $g_{\mu\nu}$-field; the field itself is what is real.

From this point of view, though, it also seems to me that the ether of your system ($ds^2 = -d\sigma^2 + c^2dt^2$) does not have any more advantages over mine

($ds^2 = -d\sigma^2 + \cos^2\chi c^2 dt^2$);

$$d\sigma = \text{spatial line element} =$$

$$d\sigma^2 = dr^2 + R^2\sin^2\chi[d\psi^2 + \sin^2\psi d\vartheta^2] \qquad \chi = \frac{r}{R}.$$

My system has the advantage that it avoids the inconvenience of closed space in that it makes the "journey around the space" impossible.[4] In your theory, there are ghosts of the Sun that are *visible* there (but *not materially*) where the Sun once was, for inst., 500,000,000, 1,000,000,000, 1,500,000,000, etc., years ago. Astronomically (and geologically) these are *short* periods. A substantial portion of the objects appearing to us as stars would thus just be ghosts. From that it would follow that there would have to be many more young (apparent) stars than old ones. In fact, there are many more old stars than young ones, however. [This can be interpreted as indicating either that the creation of stars is essentially already coming to an end, hence that only a few new ones are being added, or that the young (i.e., "giant") stages are passed through much faster. The second interpretation seems to be the true one.][5] It would be possible to get rid of these ghosts by assuming absorption. But I do not believe that. The loss of light, according to Rayleigh's law, along a path that is traveled in 500,000,000 years with the appropriate density for "world matter" would, according to your theory, be only about $^1/_{100}$.—There is, of course, a gravitational ghost of the Sun as well. But this one does not coincide with the light ghost. I have not calculated where it could possibly be found—it would not be so easy, either, I fear. Maybe it coincides with the current position of the Sun, in which case it would not be dangerous. Since the Sun's velocity has certainly not been uniform during the period of 500,000,000 years, however, I do not know whether one can assert anything about the gravity ghost without complicated calculations. There is no such thing as gravity absorption.[6] I showed (from the motion of the Moon) that the absorption coefficient of gravity (cgs units) is certainly less than $4 \cdot 10^{-16}$ for a journey around the universe that would yield an absorption of 1/10,000,000.

Nevertheless, it is not merely and not primarily a fear of ghosts that makes your theory somewhat unpalatable to me; most [predominant] for me is that you make time absolute again. Your hypothesis violates the principle of special relativity. A Lorentz transformation is not permitted in your world. We have often quarreled about this already, and it ultimately remains a matter of taste, which system one wants to consider most probable.

On p. 13 of your talk you say that even the tiniest positive mean density of matter in the world must necessarily lead to the assumption of a spatially closed world. I think that this can be upheld only if one makes the additional hypothesis that the world is in (statistical) *equilibrium*. I elaborated a bit more on this in a short article that I sent to Lorentz to submit to the Amsterdam Academy.[7] I hope you have nothing against that.

With cordial greetings, yours sincerely,

W. de Sitter.

Give Ehrenfest my regards!

## 191. From Paul Ehrenfest

Sunday, Maarn, 7 November 1920

Dear Einstein,

Forgive me if I hurriedly write you now, without being asked, about the Wisconsin business[1]—it happens on the basis of a very serious conversation with Mr. de Ridder (Kernhem) [*father* of Carl de Ridder].[2]

—•—

We beg you earnestly to follow this advice *strictly*:

1. You calmly hear out Smedeman[3] (Wisconsin agent) and ask him to summarize the main points of *his* proposal briefly *in writing* as well.

2. You do not agree to anything either in writing or *orally!!*[4] [Not even any "contingent consent," such as, e.g., the following: "In case I go to Columbia Univers., I shall come and see you too.] *Absolutely* no consent!! Instead, you say:

3. I am going to reply to you in about four weeks *in writing*, *after consultation with my friends and after I have calmly reviewed my currently outstanding obligations*. Specific advice of Mr. de Ridder![5]

4. You inform me briefly but *precisely*

α) What Smedeman has offered

β) What other offers from America exist (precisely!!)

5. You do absolutely *nothing* more about the America business, either regarding Wisconsin or with any other offer—nothing that binds you in the slightest way before you have a reply from us (from me; Mr. de Ridder).

—•—

*Reasons*:

1. An America trip by you is justifiable by no other single aspect than that you can thereby finally resolve your Swiss worries.[6] = Net gain of 20,000 dollars.[7]

[2.] The standing or interests of German science will be promoted by your visit to America only if you are invited by the 2–3 of the *most prestigious* universities in America. If, however, you go to the 5th, the 6th, or the 7th best university, you then *damage* this standing exactly like many a German privy councillor has damaged the interests of German scholars in Holland through ineptness (of a different sort).[8]

3. Your attitude toward the America matter up to now convinces me that you will achieve neither purpose 1, nor purpose 2 unless

people with very much *tact* (like: [. . .])[9]

and *experience* (Mr. de Ridder)

settle this business

*for you*.

Van Aardenne naturally hesitates to force his help upon you.[10] And has, moreover, an unjustifiably great confidence in *your* judgment of such things. (*I* don't: that you consider it irrelevant, for inst., what ranking the university inviting you has, means that you will damage the interests of all the younger physicists who will be coming to America *after* you.)

Mr. de Ridder is so interested in your obtaining the money and respect that you deserve that he has (spontaneously) declared his willingness to *help you in every way*, e.g., he is even ready to come to Berlin or to send a suitable friend to Berlin, in case that's necessary.

But I want to repeat briefly what the main thing is:

1. Commit yourself neither orally nor in writing in any sense before we here can assess everything exactly.

2. Inform us briefly and precisely about everything that is being offered to you.

— • —

Pardon the meddling. Very hearty greetings from the Kernhemers and from Maarn,[11] yours,

P. Ehrenfest.

Greetings to wife, Ilse, and Margot.

Van Aardenne wants me to say explicitly that he finds this letter very superfluous and that he does not think such patronizing of you necessary!

# 192. From Edgar Meyer

Zurich 1, 69 Rämi St., 7 November 1920

Dear Mr. Einstein,

As you know, the professorship for theoretical physics over here is still not filled. You yourself declined last year and some time ago were so kind as to give us your advice about the appointment.[1] Well, today I again approach you with a request regarding this matter, and a big request at that. I don't want any more advice, rather I would like, or better put, *we* would like to try once more to get you to come yourself. Dear Mr. Einstein, I still know everything that you said to me at the time, but hasn't the situation, through the mental derangement of the times, possibly taken on a form that you then described as a condition for your coming to Zurich? I heard that you made a statement about wanting to leave Berlin[2]—and so you must forgive me if I come again with my old request. And what can we offer you here? There is no Planck or Warburg, of course; but Debye is here and, first and foremost, our Weyl![3] And then one more thing that became so completely clear and valuable to me after my [Bad] Nauheim stay: here we have a democracy that is understood by every one![4] I do not need to tell you on behalf of Weyl, Debye, myself, or even the rest of our colleagues *how* delighted we all would be if you wanted to fulfill our deepest wish. And you can also be assured that the material side and everything else that you should desire will be entirely as you like. I am really at a loss for words to express the happy sensation that I feel at the thought of your being here.—Dear Mr. Einstein, I dare to approach you with my request only because you did authorize me to do so at the time. You said then: "If circumstances become such that . . ." And circumstances in Berlin really have become less attractive!

Dear Mr. Einstein, do fare very well, give my best to your dear wife, and offer some hope to your optimistic and grateful

Edgar Meyer

## 193. To Edgar Wöhlisch

[after 7 November 1920][1]

By no means can it be expected that [this] theorem be generally valid, even though, under otherwise identical conditions, the closer together a molecule's atoms are, the smaller must be its potential energy. The reason why the theorem could not be generally valid is that the *nature* of atoms surely must play a role, along with *the atoms' more exact relative geometric orientations*. Although it is very probable today that all molecular forces are of an electr. nature, we are far too badly informed about the constitution of atoms and molecules to be able to address this problem theor. in a quantitative manner.[2]

## 194. To Jolán Kelen-Fried[1]

Berlin, 8 November 1920

Through many colleagues[2] I have been informed about the distressing situation in which the engineer Kelen finds himself in Budapest. I would like to undertake a drive together with Viennese colleagues in the profession to help him. A very favorable opinion by the professor of electrical engineering in Delft exists as well.[3]

I herewith declare myself ready to come to this valuable man's defense everywhere in written form. For now you may make free use of this letter, if you have occasion to do so for this purpose.

It is one of the most important obligations of our day to protect good and valuable people like Kelen against political persecution [*Parteiwut*]. His papers demonstrate to me that he is a discerning scholar.[4] That is why I feel justified and obliged to stand up for him.

With greetings,

A. Einstein.

## 195. To Carl Runge

[Berlin,] 8 November 1920

Esteemed Colleague,

You are completely right with your correction. This error does indeed originally appear in my *Annalen* paper.[1] It has already been rectified, however, in Teubner's collected edition *Das Relativitätsprinzip*.[2] The correct formula

$$B = \frac{\kappa M}{2\pi\Delta}$$

furnishes the observed light deflection.

With best regards, yours,

Einstein.

# 196. From Friedrich Adler

[Vienna, 9 November 1920]

Dear Einstein,

There isn't any particular need to convince you of such an action. Just one thing, the matter is *urgent*![1] Cordial regards to you, yours,

Friedrich Adler

# 197. From Mário Basto Wagner[1]

Lisbon, 108 R. do Seculo I, 9 November 1920

Esteemed Professor,

I take the liberty of forwarding to you in the enclosed the first three parts of my "Thermodynamik der Mischungen" [Thermodynamics of Mixtures].[2] If you would inform me of your opinion of it, I would be very indebted to you.

At the same time, I would like to beg another big favor of you. As Dr. Cassel[3] informed me a while ago, just as did Professor Planck only a few days ago, to whom the line of argument of my thesis seems quite fruitful, incidentally, objections have recently been raised against Planck's extension of Nernst's heat theorem for solutions,[4] which, as it seems, are justified. Dr. Cassel let me know now that you have published some things on this question.[5] Since living here in Portugal one is almost wholly isolated from the scientific world, even with regard to professional journals, as they are not kept here at our universities at all, I would be very grateful if you would send me an offprint of your papers pertaining to this. I hope that you will not hold this request against me.

Heartily thanking you in advance for your efforts, I sign in great respect,

Maria Basti Wagner.

# 198. To Stefan Zweig

Berlin, 10 November 1920

Dear Mr. Zweig,

The friendly declaration by the artists from Salzburg heartily pleased me at the time.[1] With special satisfaction do I see from your kind letter[2] and the enclosed newspaper note that a statement to me personally, not to the public, had really been intended. I gladly take this opportunity to express my cordial thanks to you and your friends.

With kind regards,

A. Einstein.

# 199. From Georg Count von Arco

Tempelhof, Berlin, 49/50 Albrecht St., 11 November 1920

Esteemed Professor,

Through your daughter Ilse I have already requested that you grant me an interview about a patent issue.[1] It might suit you better, though, if I present it to you in writing:

You know that the cathode tube, with all its numerous applications, is the latest advance in wireless telegraphy. Our company owns an apparently very important patent in this area by Dr. A. Meissner, according to which the tube can be used for generating vibrations. This patent is regarded as a pioneering patent outside of Germany as well, e.g., in publications of the Marconi Company.[2] Owing to the great electrical advantages of this system, other companies are naturally also endeavoring to skim off the cream. We are going to try to obtain clarity about the rights, by instituting a series of legal proceedings, initially within Germany, but then probably also in other countries as well. The matter is therefore of paramount importance to us. There are very few experts, however, and among these, even fewer who are completely impartial and who, at the same time, know about patent law and can assess a patent.

What my request is driving at, now, is whether you would allow us to suggest you as our expert before the courts.[3] I would not dare to direct this request to you if I did not know that you are particularly well versed in patent law through prior occupation and, moreover, that you have already figured as a legal expert for patent matters in Germany, and this in so clear and factual a manner that it attracted the greatest admiration of all concerned. The work for you that would result from this is, I believe, not so very considerable, since the patent literature is relatively sparse and the individual patent specifications are not particularly expansive, especially also because it does not involve tricks in design but exclusively issues of principle.

Would you be so good as to leave word by telephone or, should you be in doubt about whether it might, after all, be too troublesome to take on this responsibility, have me come in to see you?–

On Sunday, when I heard that Mrs. Sklarek[4] would be spending the evening with you, I asked her by telephone to take along the book I had lent her, *Das Gesetz der Serie* [*The Law of Series*] by Dr. Kammerer,[5] Vienna, and to ask you to read through the chapter on the "hypothesis of inertia" and the subsection "physics." I must say that, while reading this book, I often had the impression that in this area Kammerer absolutely could not be taken seriously. I would be very interested to hear from you sometime about this as well.

With best compliments, yours very truly,

Arco.

## 200. From Paul Hertz

34 Riemann St., 11 November 1920

Dear Professor Einstein,

I thank you very cordially for the steps you have taken in the Kelen case.[1] I imagine that all that has happened will eventually help. Kármán published an article in the *Frankfurter Zeitung* (albeit anonymously), who passed the article on to the Hungarian government and legation.[2] Arco also took steps.[3]

Since you inquire about the patent office issue, I must tell you that I have been delaying the decisive step for a little while longer.[4] I could not decide so easily now to cut myself off from purely theoretical work before I had tried to find out if there wasn't another way. (Besides, I heard that the starting salary is very low, so in view of the higher Berlin prices I would be in a worse position in Berlin than without supplementary income in G[öttingen]. So I will see whether it isn't possible to stay in Göttingen and to do theoretical work for firms from there.[5] I wrote to Siemens-Schuckert, who advertised for a mathematician about this.[6] If they do not accept that I work for them here (and only come to Berlin occasionally), I would perhaps decide on a relocation to Berlin, after all; if that does not work, then the patent office would still be an option. But I would first like to try staying in G. I wrote to S.-Schuckert that they can make inquiries about me with you. I assumed you wouldn't be annoyed about it because you had granted me permission in [Bad] Nauheim to do the same with regard to the patent office.

With best regards, yours sincerely,

P. Hertz.

# 201. To Vilhelm Bjerknes

Berlin, 12 November 1920

Dear Colleague,

Cordial thanks for the transmittal summarizing your father's life's work[1] and for your friendly letter.[2] I must admit that I am no longer an adherent of mechanical analogies to explain action-at-a-distance, for the reason that the electromagnetic field has turned out to be more fundamental than ponderable mass.[3] I think that[4] the formal analogies are primarily *geometrically* determined, such as, e.g., the analogy between Newton's law and the photometric distance law for a point-shaped light source. Also, the fact that energy can be described mostly as a homogeneous quadratic function certainly plays an essential role in some[5] analogies.

As far as the possible call to Christiania is concerned, it is more natural if *you* as a Norwegian received the appointment. That is what the colleagues in Christiania are also thinking, and are right. I personally believe I can and should stick it out in Berlin, even if it isn't always very easy. In the end, a piece of one's own life's purpose is left behind in each place that one leaves after a longer stay. Let's hope, though,[6] that independently of this we shall soon have occasion to converse with each other about the physical problems of mutual interest to us.[7]

With best regards, yours,

A. Einstein.

# 202. From Jolán Kelen-Fried

Vienna XVII, 33 Jörger St. III. 20. c/o Göde, 12 November 1920

Esteemed Master,

Your warm words of Nov. 8th in regard to my husband[1] sincerely moved me and I am infinitely grateful to you for it. You spent your valuable time in the interest of a man who deserves this offering of yours.

I have published your warm letter, in hope of your retroactive consent, with a little alteration as though it had not come to me but to a professor here at the local university.[2] This was necessary so that the matter looks more informal and not staged by me.

I have another favor to ask of you, since for my husband's sake I have to be presumptuous. It would be very useful and might be of infinitely much help if you would address an appeal directly to the Hungarian government[3]—in part[icular], to the Prime Minister, Count Teleky himself—which was signed either by you alone or along with a few others, even if they are less prominent yet impressive personalities. In order that the whole action not be hushed up by the current Hungarian masters, I would also like to receive a copy for publication. This business is extremely urgent, since the verdict is expected at the end of November, latest beginning of Dec.[4]

My husband is bearing his fate with the calmness and circumspection of an earnest person, continues his scientific work as long as he is able, and regards the whole thing as a natural course of history. He has the conscience that for deeper reasons he could not have done otherwise and is taking the consequences of his honest actions with a simplicity that contains neither a martyr's pose nor remorsefulness. For economic reasons he is an advocate of organized production, and for love of mankind an adherent of a fairer system of distribution. At the Charlottenburg polytechnic one will still recall the excellent student Josef Klein (Kelen is the Magyarized form); he also sought to transmit his knowledge at the free school for workers in Berlin, to those who were in a lesser position to attend university.[4] He always lived very humbly and worked with dedication and selflessly in science and society, without considering the consequences. He did not particularly distinguish himself politically during the dictatorship.

I tell you all this in order to give you as intimate an impression as possible of my husband's character.

With sincere thanks and gratitude, in full awareness of the greatness of the honor granted us,

Jolán Kelen-Fried

# 203. To John G. Hibben

Berlin, 14 November 1920

Highly esteemed Mr. President,

After returning from a longer trip I find your friendly invitation to Princeton University, which our colleague Mr. Eisenhart conveyed to me in your name.[1] I am inclined, in principle, to follow this invitation, but not until fall 1921, since I cannot make myself free before then. I would then embark for America in September.

It would be very enjoyable for me to become acquainted with your economically and politically highly developed country, whose importance in world development is constantly on the rise.

I suggest that I deliver throughout two months a related series of lectures on relativity theory, three lectures per week. As honorarium I suggest 15,000 dollars (fifteen thousand).[2]

I must inform you that I have received an identical invitation by another North American university, to which I have conveyed a similar proposal.[3]

From the letter by our colleague Mr. Eisenhart I perceive with great pleasure how great is the interest in relativity theory at your university, so I enthusiastically look forward to scientific interaction with local colleagues and students there.

In utmost respect,

## 204. To Hugo Lieber[1]

Berlin, W 30, 5 Haberland St., 14 November 1920

Esteemed Doctor Lieber,

After arriving home, I reviewed my earlier correspondence with Princeton University. Thereby I became convinced that following the foregoing exchanges I am obligated to negotiate with that university myself,[2] especially considering that it had not availed itself of any mediating person either.

In thanking you cordially for your kind willingness to negotiate for me in America, I thus ask you please not to take any steps for the time being, until that correspondence with Princeton has been settled.

It goes against my grain to prompt invitations directly or indirectly to any university myself. If the two universities that have directly invited me do not come to an agreement, then I would temporarily have to abandon the trip to America. The initiative must not come from me.

In any event, I thank you heartily for your readiness to support me and I am pleased to have made the acquaintance of a person who offers his energy and experience in such a selfless way for the healthy recovery of international, and especially also of German research efforts. In begging you not to hold me to blame for taking, upon closer inspection of the situation, a somewhat changed stance from that of our discussion today, I am very respectfully yours,

## 205. To [Minna Cauer][1]

[Berlin,] 19 November 1920

Esteemed Madam,

Your letter really heartened me,[2] because it comes from a person whose genuine sentiment I could thoroughly sense in the few minutes of our passing acquaintance. In thanks I warmly shake your hand. I know very well, though, that as a person and an intellect I fall far behind the image that you have of me. But what of it? There have to be people to serve as bearers of others' illusions, such have I become, peculiarly enough. As long as one is clearly aware of this situation, no harm is done by it.

The various attacks[3] do not burden me, I even understand them. They are a natural reaction to the excessive adulation, which helps to some extent in restoring the natural balance.

In the hope of soon meeting you again, with friendly greetings and all due respect, yours,

A. Einstein.

## 206. From Marcel Grossmann

[Zurich, 20 November 1920][1]

Dear Albert,

Many thanks for your postcard. I received your letter about Guillaume's work[2] and forwarded it at the time and know from Prof. Guye, whom I met recently, that your notice is going to appear soon in the *Archives*.[3] I, for my part, added a short math. statement that attempts to grasp the math. nonsense, as it is very much more overt than the physics, where everything is obscure shades of gray.[4]

With kd. regards from all of us to all of you, yours,

M. Grossmann

## 207. Augustus Trowbridge to Heike Kamerlingh Onnes[1]

[Washington, 22 November 1920]

= PLEASE COMMUNICATE FOLLOWING TO EINSTEIN RESEARCH COUNCIL NATIONAL ACADEM[Y] AUTHORIZES ME EXTEND INVITATION TO LECTURE AT SEVERAL LEADING AMERICAN UNIVERSITIES TO POSTGRADUATE STUDENTS MATHEMATICAL PHYSICAL GROUP PROBABLE DU[RA]TION OF TRIP TEN WEEKS PREFERABLE BEGINNING IN JAN[UAR]Y HONORARIUM THREE THOUSAND DOLLARS AND FULL TRAVELLING EXPENSES WILL WRITE DETAILS ON RECEIPT REPLY STOP CABLE ANSWER COLLECT TO NATIONAL RESEARCH COUNCIL WASHINGTON = TROWBRIDGE.

# 208. From Heike Kamerlingh Onnes

Leyden, Huize ter Wetering Haagweg, 23 November 1920

Dear Colleague and Friend,

Most cordial thanks, also in my wife's name, for your kind letter, as well as to your esteemed wife for the kind letter to my wife. I must now restrict myself to a single word of friendship, because I just received from our coll. Trowbridge the following invitation for you, which I believe is best dispatched via registered express mail.[1]

I am very willing, of course, to telegraph the reply to our coll. Trowbridge, as telegraphing from Germany is perhaps not so secure either, and Trowbridge turned to me for this reason as well. But perhaps just because he suspected that you were here and your address is in any case known to be over here with me. At any rate, by choosing me as spokesman, he has done me a great favor.

Very warm greetings. I want to get this letter out into the post as quickly as possible. Cordially and sincerely yours,

H. Kamerlingh Onnes

# 209. To Paul Ehrenfest

[Berlin,] 26 November 1920

Dear Ehrenfest,

You are a man of sterling worth and you are right. I asked for 15,000 dollars each from the Universities of Princeton and Wisconsin.[1] That will probably scare them off, so I will be able to stay contentedly at home. If they do bite into the big sour apple, I actually will buy myself economic freedom, which is nothing to scoff at.

I really do believe that the Hall effect works like that and am happy that you agree.

Either the field is gradually compensated by eddy currents *in such a way* that the electromotive force in the whole superconductor can be derived from a potential, or the Hall force generates currents that grow until the respective local superconductivity is destroyed. Experiments speak for the first possibility.[2] Experiment: Gradual shielding off a magnetic field permeating a current-carrying plate. Process could be done as slowly as you like, in accordance with the chosen current intensity. I hope Kamerlingh Onnes does these experiments.

I am going to decline the invitation sent by Kamerlingh Onnes,[3] for the time being, in consideration of the pending negotiations with the two universities. What's the news on your America plans? It really would be nice if all three of us could set off together. I don't know yet whether I would take my wife along. Probably nothing will come of the matter at all.

The stink from my dear Mosz[kowski]'s pot of trouble still hasn't dissipated yet, thanks to the slowness of the printing presses.[4] It's a pity that I can't demonstrate it to you *in natura*; you have a sense for such things. Mrs. Born's crescendo finally grew into such an awful fortissimo penetrante[5] that I had to silence her with a very slight hint of irony. A little Dutch breeze would do them good.

I am coming no closer to my pet research goal (overdetermination and unification . . .[6]).

Hearty greetings also to your wife and children from your

Einstein.

Best regards to the two Russian youngsters [*Maltschiks*].[7]

# 210. To Augustus Trowbridge

Berlin, 27 November 1920

Dear Colleague,

I cordially thank you and the gentlemen of the Research Council of the National Academy for the honorable invitation.[1] The invitation fills me with gladness not only because it confirms the general interest in the theory of relativity but also, above all, because it is a sign of the renewed beginnings of international exchanges between scholars.

Unfortunately I cannot accept your invitation for the time being, because I am already negotiating with two American universities.[2] Therefore, I cannot think about committing myself otherwise before these negotiations are completed.

In great respect,

# 211. To Edgar Meyer

[Berlin,] 28 November 1920

Dear Mr. Meyer,

I hesitated so long with this letter because refusal—one more time—is so hard for me.[1] It is perhaps even rash of me to stay in Berlin, but I simply cannot leave. Not that I do not value the scientific milieu of Zurich just as highly as that of Berlin. Nor do I underestimate the physical advantages and security in any way. I know that we would fit together perfectly and would accomplish much of profit. I love Zurich and would have my dear sons nearby, which means very much to me. However! You too will turn your back on this provincial university when you have the chance. Scientists are simply valued quite differently here than in Zurich. I am independent in an entirely different way and my will counts for something. And the deafening outcry if I went away from here, and justifiably so![2] I hardly dare to think of it. It would be reckoned as a humiliation for Germany (as ironic as this is upon sober reflection) if I were to leave.[3]

I am now being continually berated by pan-German papers and professors.[4] Even so, this does not bother me, because it is a part of "fine politicking."[5] All the friendlier are my closer colleagues here to me, and my Jewish friends. It was only here that I lost the painful feeling of isolation under which I had always been suffering in Berne and Zurich. So in short: I cannot take myself away, stupid and unpractical though it may be!

If tomorrow morning I were in Zurich, without any of my doing, and came upstairs to you in the morning at the institute, I would certainly be content then. But I cannot extricate myself from here . . . I see you laughing, not unkindly, but rather in that good-natured way of yours.

I hope I can come to Zurich again next year (just for fun). Until then, cordial regards to you, yours,

Einstein.

Friendly greetings to your wife and the children also from my wife.

# 212. From Hans Albert Einstein

Zurich, 28 November 1920

Dear Papa,

How are you? A short while ago we received the long-hoped-for money.[1] In our cash box we still had, *nota bene*, 10 francs . . .

We thank you kindly for it. You also write we should come to Germany.[2] Well, we discussed it and also asked the Zürchers for advice[3] and just came to the conclusion that it would be irresponsible if I had to interrupt my schooling now, because I would, in any case, lose at least one year and for the time being we do mainly have to see to it that I, at least, become self-sufficient as soon as possible.[4] This is our only prospect for improving our circumstances, you know.

That is why we ask you to make it possible that I at least graduate without disturbance. I do have very little of you and so you really could do this for me.[5]

Mama also thinks that, as she saw, nutrition really isn't that good in Germany, especially for foreigners who don't know their way around the stores, and that would just have the result that Teddy's and her hard-earned health would go to the devil in the process.[6]

You'd have to say all this, yourself, if you thought about it more carefully and that's why you shouldn't always write it again and again, especially because the main effect is just agitation for Mama, linked with sleepless nights, etc. You did write us and the Zürchers, you know, that you could continue the transfers for the next year.[7] Why do you write something else now, yet again?

I'm having no great success with my hunt for patents because such a detail isn't patented yet and I can't gather from anything how the current organs are built, for if a firm is manufacturing it now already, then it can't be patented, can it?

Otherwise we are doing quite well, I hope you are, too. Now Christmas will be here soon, and vacation.

Many greetings from

Adu.

P.S. I hope you don't hold this letter against me again. It's not meant to be nasty in any way and because I'm not yet used to writing such letters you will surely understand when sometimes it doesn't come out as I want it to.

## 213. From Walther Nernst

Berlin, 26a Am Karlsbad, 28 November 1920

Dear Colleague,

Enclosed I'm sending you the two $H_2$-papers by me and by Dr. Günther.[1] You will find my special theory of gas degeneration in my monograph on the heat theorem,[2] p. 167, general discussion pp. 157–162; specifically, I show p. 161, fig. 21, how extremely simply and transparently the liquid-vapor balance results, *without* any chemical constant, from variation in the spec. heats of gases at low temperatures.[3] Certain unclear points which, e.g., are also attached to Stern's first paper—1913, *Physik. Zeitschr.*[4]—are thereby probably eliminated. E.g., the zero-point energy that Stern spoke about has nothing to do with these thermodynamic considerations.

It might interest you that Tammann has established isomerism in mixed crystals, applying various arrangements in many sorts of experiments.[5] Below the melting point it disappears quite rapidly. If $E$ is the energy content at equilibrium of the mixed crystal modification and E′ is that of any other modification, and if at the high temperature $T$ the compensation happens very rapidly, then the first law of thermodynamics yields

$$(1)\ E'_T - E_T = Q_0 = \int_0^T (c' - c)dT$$

($c'$ and $c$ the molecular heat). At $T$, naturally, $c' = c$.

The second law of thermodynamics probably does not offer anything here, but my heat theorem *possibly* furnishes (because $Q_0 = A_0$)[6]

$$(2)\ Q_0 = A_0 = T\int_0^T \frac{E' - E}{T^2} dT.$$

Checking (2) won't be easy but it is probably feasible. In any case, my heat theorem does provide a possible application here as well. Whether it is quite certain I still don't want to say at the moment, because heating in order to determine $c'$ is an irreversible process, at high temperatures at least. What is certain, of course, if we measure $c'$ so quickly, is that the heating remains reversible:

$$A_0 - A_T = T\int_0^T \frac{\int_0^T c'dT - E}{T^2} dT.$$

$\int_0^T c'dT$ is now, of course, different from $E'_T$ .

So, see you on Tuesday!
With greetings, yours,

W. Nernst

# 214. From Willem de Sitter

Arosa, Wald sanatorium, 29 November 1920

Dear Colleague,

I thank you very much for your letter.[1]

About the stability of the Milky Way galaxy and the problem about the extent to which it is held together by its own gravitational influence, Eddington has done some fine analyses,[2] whose results unfortunately are not available to me here. As far as I can recall, though, he has come to the conclusion that with the given stellar velocity, it is very well possible that it is kept together by gravitation according to Newton's law. But in any case: a value for $\lambda$ derived from the Milky Way's *mass* would make the universe *much too small*; and the $\lambda$ value that corresponds to the *density* of matter in the vicinity of the Sun makes the universe so *large* that the Milky Way galaxy could not hold itself together in it better than without $\lambda$. For the problem of the clustering of the Milky Way galaxy, in my view, the $\lambda$-term cannot be of any help.[3]

You say: "Would it not be more satisfying to assume $\lambda = 0$ there, if one lays no significance on the existence of a mean density of matter anyway nor on interpreting inertia as an interaction between bodies?"

It certainly is so, and I also have always been very inclined *not* to assume the λ-term. However . . . the apparent repulsive force that results out of my interpretation of the universe $\left(\text{with } g_{\mu\nu} = \cos^2\frac{r}{R}\right)$ really does seem to exist! When I wrote about it in 1917, the radial velocities of only 3 spiral nebulae had been measured by then.[4] Today, *25* have. And, with 3 exceptions, all are *positive*. The average, if one excludes the two brightest and thus probably the closest ones, is +631 km/sec. The largest observed velocity is 1,200 km/sec. The velocities are *radial*; the nebulae are distributed unevenly throughout the whole sky.[5]

You say that the universe is too inhomogeneous (optically [speaking?]) to allow ghost suns to come into any focus.[6] Yet the universe is incredibly *empty.* It is an *observational result*, whose accuracy can scarcely be doubted, that outside of the Milky Way up to a distance of about 100,000 light-years there is still *no* sign of absorption or dispersion of light—thus *certainly* less than 1/20.—It is not *excluded*, of course, that over much larger distances (of 100,000,000 light-years) absorption and dispersion would become noticeable. But we still know nothing about that at present.

Yours sincerely,

W. de Sitter.

## 215. From Wander de Haas

Delft, December 1920

[Not selected for translation.]

## 216. To Hans Mühsam

[Berlin,] 1 December 1920

Esteemed Doctor,

By instruction of Prof. Einstein I enclose for you a short handwritten manuscript by Prof. Einstein for the benefit of a charitable Jewish cause.[1]

In great respect,

The Secretary,
Ilse Einstein.

# 217. From Arnold Berliner

Berlin W 9, 23/24 Link St., 1 December 1920

Dear Mr. Einstein,

Your letter, in which you wish to come to the aid of Mr. Harry Schmidt, who has been so roughly handled by Reichenbach,[1] would, if I were willing to publish it—I am not going to do so, as I would like to mention at the very outset, because I consider it uncalled for—force me onto a path that would lead to unforeseeable consequences and even to grotesque consequences. A short while ago I had to publish the review of a zoological dictionary that I also might describe as bilious, as you did Reichenbach's review, but because the reviewer was known to me as a knowledgeable and reliable man, I obviously did not make the least objection to it, and likewise with many, many other reviews that I had to publish during the course of the eight years of the existence of our journal. I would have to allow a defender to put in his word for *every* one of these reviews if I did so even in a *single* case. Criticism is not merely a matter of judgment but also a matter of taste. Upon careful inspection of the book reviews in the *Naturwissenschaften*, you will find that no unauthorized reviewer is granted a say. And that is what is decisive. But criticism is additionally also a matter of taste. Now—it is generally known that there is no arguing about taste. It is possible that Reichenbach hit upon a slightly inappropriate tone, but after a short while this review is forgotten and I do not believe that it will cause Mr. Schmidt any mentionable harm, and the *status quo ante* will, in my view, then be only imperceptibly changed by it. It is entirely different, however, when Mr. Einstein now comes forward for injured Harry Schmidt in the *Naturwissenschaften*, as "avenger of his honor." Everyone would regard Einstein's advocacy of a book about the theory of relativity as an advertisement of enormous weight and this letter would have a temporal and spatial effect of unimagined extent, and this for the glory of one of many popularizing expositions of the theory of relativity, which distinguish themselves little from one another and are announced, one and all, in the *Buchhändler-Börsenblatt* with some favorable, more or less veiled allusion to Mr. Einstein's praise. ⟨This time⟩ In this case, though, just by your book dealer! Such recommendations in the *Buchhändler-Börsenblatt* are not objectionable, since they become known only to publishers and very occasional readers of this paper. And then such praises originate only indirectly from the founder of the theory of relativity, hence can only be exploited very incompletely as publicity before the public at large. It is entirely different, however, when the founder of the theory of relativity puts in his personal word for an exposition of the theory of relativity in a journal that has already acquired a certain reputation. I don't want to carry my point to the end and would just like to indicate that guilt and atonement, if one may use these two words for such a harmless proceeding, must approximately balance each other out. And now Mr. Reichenbach is standing on one side with perhaps scathing but by no means devastating criticism, and on the other side we have (you forbade me so often from complimenting you that I will suppress any qualifier) Mr. Einstein with a recommendation that looks like an advertisement. You know that I always completely naturally comply with any wish you express, but here I have to say no for the first time and I hope you will not be cross with me for it on the basis of the arguments I just made.

With best regards, yours very truly,

A. Berliner.

## 218. From Paul Winteler

Lucerne, 1 December 1920

Dear Albert,

The promised letter was a bit late in coming, but now I can communicate to you in more detail what I have to say.

The money from Zurich will be remitted to me in the coming days. That is why I'm asking you what I should do with it; I'm still waiting for your instructions.[1]

As far as we are concerned, we have the intention of moving away from Lucerne next spring; it concerns no more and no less than my retirement (with c. 42% of my salary); and specifically I am *very content*, owing to [ugly] personal relations that have arisen, because through this unusual fact at my age I would be a free person who can pitch his tents wherever living is cheap or where additional earnings can be made.[2] We are naturally making all kinds of plans and are not thinking of loafing about; on the contrary, now the stimulus to work is aroused more than under the ambitious opportunists whom I would ultimately have had to serve as a shoeshine. Maja is participating insofar as in the event of my death she retains her half of the pension for life.[3] It was only for the sake of this pension that I stayed in this job for so long![4]

Because my thoughts are now wandering to various options, you can also see that I'm thinking of you in particular, as well. I have long thought that you have not drawn the material fruits of your scientific labors that are quite naturally your due. It really must be somewhat uncomfortable for you to have me request advances in Zurich for you; furthermore, to have worries that it isn't always possible for you to meet your obligations toward your first wife in time,[5] spec[ifically], whereas everything could have been organized much more favorably for you and your peace of mind. In the end you do have 2 families to think of and that is no easy matter.[6]

For, I thought, you never remembered that you have a lawyer in the family who can put in his share toward lending your finance department some kind of order. Don't be offended if I make you suggestions about this. Your publisher of the *Popular Account*[7] has cheated you flagrantly. You helped a *rich and shrewd* publisher (all publishers are shrewd nowadays! and how!) make a fortune, or he raked it into his big strongbox, for a pittance [*Butzenstiel*]![8] 100 free copies per printing run that you can't even verify, *including translation rights (!)*, as I hear. Do you really believe that this booklet would be sold for a single penny less because you gave your little opus, so to speak, for free? Your motive was touchingly good, but the consequence simply was that you made a rich publisher even richer without making the purchase cheaper for a single reader!

I imagine the business as follows: little will be alterable in the rel[evant] contract. In this regard publishers are not so soft as to let go of the profit they had nabbed for themselves. But I hope you didn't give away your *future* productions as well, too. If not, I suggest the following to you: pull yourself together and write a *textbook*,[9] a summarizing work on relativity theory; it doesn't need to be comprehensive, perhaps a short histor[ical] introduction with a detailed exposition of your theory as part II, the mathematical exposition as part III, all in concentrated form so that both nonspecialists and specialists in the field, but notably students, find everything they wish to know. The whole thing would thus be in part "popular," in part addressed to specialists.

Vieweg will gladly deal with such a book, *but any other publisher you like would as well. But then you have to draw up a contract that rewards your work thus far*; indeed, I hope your exceedingly unfavorable agreement with Vieweg will even be rectifiable this way. As a lawyer I would gladly concern myself with it and then your families will be provided for! It is something else when you can actually live off this work instead of being dependent on the Zurich business!

Don't tell yourself that your theory is most quickly turned into common property if you make such gifts to publishers while you have financial worries! Besides, the psychology of the public is such that it only values what it pays for. I can imagine your scruples and do find that you are on the wrong track, if not on a rocky road of tribulation.

Up to now I've been speaking as a "materialist," if you will, but I'm thinking of you, of your family, of the whole disagreeable existence of a man who must constantly scrutinize his pockets for holes and never escapes the unease about how he can drum up enough Swiss francs in order to provide for the necessities of his families in time.

And now the lawyer in me. My existence as an agent of this honorable profession is, according to definition, to act on behalf of the *material* interests of other people. How much more eagerly could I act for you and your own people! For me personally—I hardly need to say so, I don't want anything for it, I just find that the lawyer in the family saves measures taken with other people, and this lawyer has to look on passively while other people fatten themselves on the money they made off you and are still making off you.

I would like to hear what you think of this matter. Of course I understand if these ideas could appear unsavory to you at first, but the conditions up to now were surely even more so! They should and must be remedied and can be as well, as soon as you have created the preconditions for it and give a reliable family lawyer the opportunity to work for you instead of other people.

With warmest regards, yours,

Pauli

P.S. Could you inform me—in the event you do detect some salt in my long speech—about *everything* you agreed on with Vieweg? For I only know about vague outlines and nothing specific and would like to whittle it down to the legal core. Publishers' contracts are, for all their sanctimonious appearance, confounded legal masterpieces!!, at the expense of intellectual workers.

## 219. To Harry Schmidt

Berlin, 2 December 1920

Dear Colleague,

In the enclosed I send you a letter by Dr. Berliner.[1] The arguments by Dr. Berliner did in fact convince me. Thus I am not in a position to write the promised defense and must request that you make no use in public of my earlier statements about your booklet.

Very respectfully,

## 220. From Maja Winteler-Einstein

Lucerne, 6 December 1920

My Dears,

The letters from both of you drew me away from agonizing worries. I didn't know what was to become of all of you after the unpleasant experiences in Berlin and [Bad] Nauheim.[1] Now you all seem to be continuing your lives and therefore I can be content with that, since it was always harmonious and nice. Dear Albert, I thank you especially for thinking of my birthday. (I actually ought to thank you, dear Elsa, you must have prodded him.)[2] It delighted me hugely to hear from you directly again.

We need the well wishes particularly now, because we are embarking on a quite uncertain future. Pauli has lately become so irritated in his life in the civil service that his nerves are worn very thin and he urgently needs to recuperate for a longer time.[3] Because health is ultimately more important than a large income, we decided, after mature reflection, on retiring.[4] This step alone already calmed Pauli down a bit. We rented out our apartment for a year and are now going (around Dec. 20th) for 4–6 weeks into the Vorarlberg so that Pauli can fortify his nerves in the mountain air. We are then thinking of going to Paris, where Pauli wants to work as a legal consultant for Swiss firms and as a journalist. I have prospects through friends of working in the art trade. These are vague plans, but the solid basis is the pension, with which we can get by with a very modest lifestyle. With ⟨this⟩ the annual sum now due from our share of the S[wiss] A[uer] S[tock Company], we are completely liberated from our debts, thank heavens and praise be! I'm not at all attached to Lucerne; I'm looking forward to getting away.[5]

Dear Else, your long letter pleased me particularly, because it reported about so many relatives about whom I otherwise no longer hear a thing. I always felt sorry for poor Marie but what can be done for her?[6] She was never very clever but neither did she seem downright foolish. The worries and the misery are surely going to affect her very much. Now it'll soon be a year since we brought poor Mama to Berlin.[7] You can imagine that I also constantly have to think about that. How much the poor dear had to suffer! But in Berlin she did have a couple of moments of real joy from Albert and the love of so many family members.

What are Aunt and Uncle doing?[8] I often think of the two of them as well. I have quite a few pieces of the gray material of which Aunt Fanny would like to have her share, but I don't know how I should send them.

Dear Ilse, I thank you for your good wishes.[9] Are you going out a lot again this year? Be sure to have a good time. Is Margot completely well again?[10] How's the music doing? Has she composed anything else? I always like to use the notebook she made for me and always think of her warmly whenever I use it.

Now all of you certainly have a long letter. Please don't let me wait so long again without news. It is so depressing not to know anything about one's closest and dearest.

Heartfelt greetings and kisses from your

Maja.

# 221. From Paul Mühsam[1]

Görlitz, 4 Bismarck St., 7 December 1920

Highly esteemed Professor,

When a few days ago I was in Berlin for a short time with my cousin Hans Mühsam,[2] he told me—what I already knew in any case—that you were a convinced pacifist and asked me to forward to you my little work *Aus dem Schicksalsbuch der Menschheit* [From the Book of Mankind's Destiny], in which I as a similarly keen pacifist discussed the madness of war in [poetic] form.[3]

In fulfilling my cousin's request I meet at the same time a wish and heartfelt need of my own and am pleased by the fortunate opportunity allowing me to convey to you, highly esteemed Professor, most [truly] and sincerely my respect and admiration.

It would fill me with pride and joy to know that the words I use in my book to lament the fate of self-destructive mankind also [finds] resonance in your heart.

Yours very truly,

Dr. Paul Mühsam

# 222. From Heinrich Zangger

[before 8 December 1920][1]

Dear friend Einstein,

Even your [Hans] Albert has no news and Besso also inquired.[2] I said consolingly: It's always like that when one is sitting full of hope on a golden egg and from experience is justified to hope. I can be nice, too. Furthermore, I managed to obtain a great improvement in the negotiations for Weyl—above all: time off for the coming *winters* for his lung cure in the mountains; added to that, I can arrange that a fund be mobilized so that he can use the vacation for the cures without worries, so that his world line isn't jittery and can even neglect redshift, independent of c.[3]

(When[4] I was feeling bad about people and healthwise, I added a few extra jitters; bad joke.) Weyl has to take care of himself. Huguenin's parting concern for Weyl was: on the last two days before leaving he called after me: I should watch out for Weyl, because he is sick and so valuable, which is what he thought. I know that you think as highly of Huguenin as I.[5]

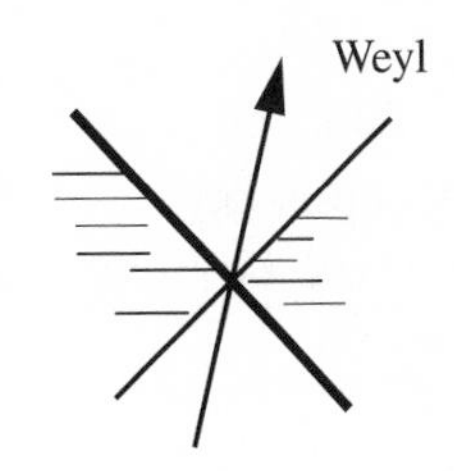

Debye, like Born, should write a book, for himself and others, a fabulous lucid mind with simple conceptions.[6] I've been reading Born's book these past 2 days in bed:[7] The Divine Comedy Einstein, where it also says every time at the end of the song: *e amor e le altre stellae* ["love, which moves the sun and the other stars"].[8] In the next edition he should be a little more detailed, spec[ifically], *you* really ought to comment a bit more on the development of the ideas on [pp.] 34ff., 54f., 92f., 118ff., and on the end *yourself*, in the way you made many things clear to me: But [it is] surely the most comprehensible book—despite Freundlich[9]—much, much better, more contained, more logical than his "Structure of Matter," which was too hastily drafted.[10] Whereas Sommerfeld is somewhat difficult to read,[11] because he retains the authors' nomenclature—thus leaving one to labor (on one's own even ν), and the requisite compilation for self-ordering [via proofs] becomes tiresome.

By the way: I don't understand: how for the nucleus of H, 1/2000 of the electron should be = 1/2,000,000 of H. The cohesion would be the energy-mass.

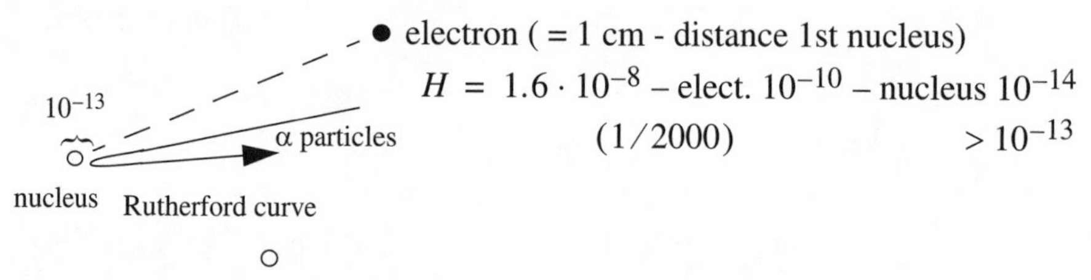


o

Where is this leading to?

We want to try to have industry contribute a fund so that we can invite Rutherford, Aston, etc., to Zurich to give lectures,[12] because this train of thought is more difficult for me to access and yet quite suggestive (as earlier Kelvin and Faraday helped me very much).[13] Langevin is also coming.[14] You too? You old Swiss.

Besides that, I'm in the middle of the boundary surface conditions; I'm hoping that my first love, the membranes, their structure acquire new life (despite stress and tiredness).[15] The papers by Kossel—who is going to come to Zurich as well—are also valuable to me as a hint, but not yet thought through enough for my taste.[16]

You see, I am full of lively physics hopes and delivered a talk on spectra and spectroscopy.

With best regards,

Zangger

# 223. To Max M. Warburg[1]

Berlin, W 30, 5 Haberland St., 8 December 1920

Dear Mr. Warburg,

Upon closer consideration, the favor that my wife recently asked of you on my behalf does appear to me a bit presumptuous. I would like to ask you earnestly whether you really want to keep the promise, so disconcertedly given, of approaching your brother.[2] If you could interest another person with sufficient experience, tact, and influence, in the negotiations to be made in America, instead of your busy brother, I would still be very gratefully obliged to you.

Until now I received official invitations from the Universities of Princeton and Wisconsin (Madison) and from the National Academy in Washington; Princeton, represented by President Hibben,[3] proposed that I go there immediately and lecture throughout the whole semester; respecting the business side, they remarked: "that the matter would be settled later according to my wishes." Wisconsin negotiated orally through the American envoy in Christiania, Schmedemann,[4] who visited me here.

I informed both of these universities that I was prepared to lecture there three times a week for two months each; to be precise, in the period between October 1921 and March 1922 (altogether 4 months). In payment, I asked for 15,000 dollars from each university; however, I demanded no compensation for travel or accommodations.[5]

I wrote to decline the Academy in Washington, which offered me 3,000 dollars in payment for the whole America trip (with free travel and free accommodations), on the grounds that I was negotiating with the two mentioned universities.[6] I have heard nothing more from Wisconsin University since my reply (6 weeks). From Princeton I got no answer, either; but it's been only three weeks. I do have the impression, though, that my demands are too high for both these universities.

I am determined to take upon myself the loss of time and exertions of such a tour only if I can thereby gain a certain financial independence. This I can achieve only if I can lay aside at least 20–25,000 dollars, that is, remaining after subtraction of the travel costs. If this is not achievable, I prefer to stay here.

Under the circumstances described, nothing can happen along the official route, of course. However, an influential person in America surely could assess whether the goal I am striving for is attainable by a suitable division of the time among many locations.–

The evening we spent together recently was very stimulating for me. I am pleased to have made the acquaintance of such a farsighted and strong personality as you.

With amicable greetings also from my wife, I am yours,

## 224. From Max Born

[Frankfurt,] 8 December 1920

Dear Einstein,

I'm sending you in the enclosed the circular of the *Math. Annalen*. I never received a submission for this journal before, nor do I know anything about it and therefore added no comment.[1]

In addition I'm attaching a copy of a letter from Russia, from my student and friend Boguslavsky. The letter arrived some time ago already.[2] The content might interest you. From it one sees that something has to be done to try and invite the poor man (he has a lung problem besides) to Germany so that he doesn't starve. I tried all sorts of things, first with Planck, then with Klein and Hilbert at Göttingen,[3] whom I asked to prompt the academies to send B. an invitation of some kind. But they all declined; they don't want to have anything to do with "foreign policy," as Hilbert put it. Maybe you will think of a way. What Boguslavsky writes about his research is, in part, evidently nonsense; but this probably is explained by his deplorable condition; he is a smart, fine person. A mutual friend, Dr. Bolza in Würzburg,[4] incidentally also made an attempt to send Boguslavsky something through the Red Cross; whether with success I do not know. To stay on the same topic: some time ago I sent you a letter by Epstein, who was requesting help. An answer has meanwhile come in from G. N. Lewis in America, whom I had written about this matter. He has created a position at the university at Berkeley in California for Epstein and offered it to him.[5] But I haven't heard anything from Epstein, whether he wants to accept; the Swiss might be keeping him.[6] An attempt to bring him here as my successor fell flat against the faculty's resistance. Nor could I get Stern into 1st place, because Wachsmuth wanted Madelung; but Stern is placed 2nd, Kossel placed 3rd.[7]

In science I tried many things without getting enthusiastic about anything. What attracts me most is a decent theory of irreversible processes in crystals, as Debye once suggested;[8] but I haven't arrived at reasonable general propositions. The measurements at the institute on mean free path are going quite nicely; the main thing was to keep the gas pressure constant during the ½-hour-long vaporization of the silver; we are now getting it to 5%. On the other hand, clean measurement of the silver deposit thicknesses is still not accomplished, because we first have to gather together the various optical apparatuses gradually.– Landé,[9] who was recently in Heidelberg at the colloquium, told me yesterday that Ramsauer (alias Lenard)[10] had raised serious charges against my relativity book, because in it I had acted as though Maxwell's suggestion (to determine the absolute motion of the solar system from the eclipses of Jupiter's satellites) had in fact been carried out with a negative result.[11] I find that the charge is not unjustified and therefore expect a resounding attack by Lenard or one of his cronies. Healthwise I was not doing well for weeks, which may well have been recognizable from the bilious tone of my last letter to Holland.[12] Now I am feeling quite well again, just the political situation depresses me more than I would care to admit to myself. Cordial greetings, yours,

Born.

## 225. From Paul Ehrenfest

[Leyden,] 8 December 1920

Dear Einstein,

Many thanks for your dear letter.[1]

I am sending you herewith two letters and two newspaper clippings.—The brochure by Fr. Bauer, which is the subject of the letter by him, is going out to you at the same time with a package of offprints by *Millikan*.[2]–

I have just received letters from the physicists in Petrograd again. It is very probable that my friends Profs. Joffe and Rozhdjestvensky will come to Holland for a few weeks.[3] *Perhaps* I will be in need of your help then for a visa for Germany. I guarantee you that their trip serves purely scientif. purposes.

I am very glad that your stay here left a cheerful, calm impression on you despite the endless debating.[4] You can hardly imagine how warmly all the people in the painter Onnes's home speak about you.[5] Particularly father Onnes does not hesitate a moment to say that dealing with you makes dealing with almost all other people (this "almost" is very diplomatic!) seem very boring to him. I have a couple more fine fellows in reserve to show you the next time you come.

Find an opportunity to write a few kind words to Julius and see to it that the people who are working on redshift take note of his papers *as is their due*.–[6]

I am *reading up* on the research about resonance-ioniz[ation] potential—it is musically beautiful—e.g., Franck-Knipping on helium.[7]– The papers on band sp[ectra] I cannot quite grasp, but I want to understand them![8] Cordial greetings to all! Yours,

Ehrenfest.

## 226. From Harm H. Kamerlingh Onnes

[Oegstgeest] 8 December 1920

[Not selected for translation.]

## 227. To Paul Ehrenfest (?)

[ca. 9 December 1920]

[Not selected for translation.]

## 228. From Rudolf Goldscheid[1]

Vienna, 13 December 1920

[Not selected for translation.]

## 229. From Albert G. Schmedeman

Christiania ...13 December 1920, 12 o'clock 20 min.

regents university wisconsin just informed me[1] that fifteen thousand dollars beyond their means[2] they wish to know if i[t] is possible to arrange series of lectures at wisconsin prince[ton] and one other university for one or two weeks each for the coming summer stop on what terms will you consider such a proposition telegraph me reply = schmedeman[3]

## 230. To George B. Jeffery

Berlin, 14 December 1920

Dear Colleague,

Your extremely friendly letter of October 14th filled me with extraordinary joy and gratitude; I welcome it not just from the personal point of view but particularly as a sign of the renewal of friendly relations with fellows in the profession across the Channel. On my side, certainly nothing should be left undone that can alleviate the hardships of the past and the present.[1]

First, as far as the publication of papers by me on the relativity problem is concerned, a collection of the more important ones already exists that appeared with Teubner in German, which I am having forwarded to you at the same time.[2] Of course I have no objection to a translation by you and Dr. Perett.[3] Might it not seem appropriate to you to translate the papers by Lorentz and Minkowski as well? As regards the finances, I would be satisfied if I received one shilling per copy.

Your offer to lend journals to me and send them to me regularly demonstrates a sense of good will rarely encountered in this world.[4] I warmly shake your hand for it. However, I am not in a position to make use of your great kindness because I am anyway not in a position to study so many papers myself; I learn about most of them from oral reports at our colloquium,[5] during which the papers by English physicists are frequently and thoroughly discussed. Recently, the papers by Aston, especially, found enthusiastic reception among us.[6]

In thanking you heartily again for your kind letter, I am with amicable regards, yours,

# 231. From Erwin Freundlich

[Potsdam,] 14 December 1920

Dear Einstein,

Below are the requested densities: the formula for the spatial density $\rho$ in star clusters is set in the following manner:[1]

$\rho = \varphi(r) = \frac{3}{4}\frac{N}{a}\left(1+\frac{r^2}{a^2}\right)^{-5/2}$ the corresponding apparent density projected onto a plane is:

$$f(r) = N\left(1+\frac{r^2}{a^2}\right)^{-2}$$

$N$ & $a$ are constants; $N$= number of stars per unit area in the center of the cluster. The total number of stars

$N_\alpha = \pi \cdot N \cdot a^2$; $a$ thus means the radius of the corresponding homogeneous cluster with the same total number of stars and a mean density equal to the central density in the real star cluster.

According to this formula the drop in density in the Messier 3 cluster (N[ew] G[eneral] C[atalogue] 5272) for a count of all the stars up to a limiting size class, without taking types into consideration:[2]

| Messier 3 | | | | Hercules cluster Messier 13 | | |
|---|---|---|---|---|---|---|
| $r$ | $n_{obs}$ | $n_{rep}$ | $n_{rep} = c \cdot r^{-3}$ | $r$ | $n_{obs}$ | $n_{rep}$ |
| 0 | (143.4) | 100 | | 0 | 132.6- | 110.9 |
| 1 | 104.0 | 87.4 | | 1 | 110.0- | 102.0 |
| 2 | 62.0 | 61.0 | | 2 | 76.0- | 81.1 |
| 3 | 37.5 | 37.6 | | : | 58.5- | 58.2 |
| 4 | 23.0 | 22.3 | | : | 45.5- | 39.6 |
| 5 | 14.0 | 13.3 | | : | 33.5- | 26.3 |
| 6 | 8.0 | 8.1 | | | 23.5- | 17.6 |
| 7 | 4.6 | 5.1 | | | 15.5- | 11.8 |
| 8 | 3.3 | 3.3 | | | 9.0- | 8.1 |
| 9 | 2.5 | 2.3 | | | 5.6- | 5.7 |
| 10 | 2.0 | 1.3 | | | 3.9- | 4.1 |
| 11 | 1.7 | 1.1 | 1.8 | | 3.0- | 3.0 |
| 12 | 1.4 | 0.8 | 1.4 | | 2.3- | 2.2 |
| 13 | 1.2 | 0.6 | 1.1 | | 1.8- | 1.7 |
| 14 | 1.0 | 0.5 | 0.9 | | 1.4- | 1.3 |
| 15 | 0.8 | 0.4 | 0.73 | | 1.0- | 1.0 |
| 16 | 0.6 | 0.3 | 0.60 | | 0.8- | 0.8 |
| 17 | 0.5 | 0.2 | 0.50 | : | 0.6- | 0.6 |
| 18 | 0.4 | 0.2 | 0.42 | : | 0.4- | 0.5 |
| 19 | 0.3 | 0.2 | 0.35 | 19 | 0.2 | 0.4 |
| 20 | 0.2 | 0.1 | 0.31 | 20 | 0.2 | 0.3 |

$a = 113^0$ $N_\infty = 1122$ $a = 146^0$ $N_\infty = 2074$

Jeans contends, in particular, that describing it as an adiabatic gaseous sphere doesn't work at the edges of the cluster and that $c \cdot r^{-3}$dm reflects the drop in density better.

Generally, the following applies for the conversion from the surface distribution into the spatial one:

$R$ = outer radius of the cluster

$F(r)$ = surface density; $f(r)$ = spatial density

$$F(r) = \int_{-\sqrt{R^2-r^2}}^{+\sqrt{R^2-r^2}} f(\sqrt{h^2+r^2})dh = 2\int_r^R f(\rho)\rho\frac{d\rho}{\sqrt{\rho^2-r^2}}.$$

If one sets

$$h = R^2 - r^2,\ x = R^2 - \rho^2 \qquad F(r) = \varphi(h)\ \ f(\rho) = \frac{d\psi}{dx} = \psi'(x),$$

then one obtains the integral equation treated by Abel,

$\varphi(h) = \int_0^h \frac{\psi'(x)dx}{\sqrt{h-x}}$, for which Abel found the solution:

$$\psi(x) = \frac{1}{\pi}\int_0^x \frac{\varphi(h)dh}{\sqrt{x-h}}.$$

After partial integration and slight reformulation one finds

$$f(\rho) = \frac{1}{\pi}\int_\rho^R \sqrt{r^2-\rho^2}\frac{d}{dr}\left(\frac{1}{r}\frac{dF}{dr}\right)dr.$$

I hope this information suffices for you; otherwise please telephone me. Yours,

E. Freundlich

P. S. I almost forgot to give you the requested details concerning the American donors. So, Mr. Pagenstecher[3] is father-in-law of Mr. von Estroff in Potsdam, where I spoke to him during the summer. He is 82 years old and is very interested in all new research. His daughter, Miss Pagenstecher, who also sent 100,000 marks, is naturally more deeply interested than her elderly father. Do please write a letter to Miss Pagenstecher that is directed at the father at the same time, and enclose a copy of your booklet on relativity theory with a brief dedication. Please send the letter and booklet to me over here; I will assemble everything for the mailing, along with [my] letter.

Best regards, yours,

E. F.

# 232. To Hans Albert and Eduard Einstein

[Berlin,] 15 December 1920

My Dears,

In the near future, some more money will be reaching you, which I was able to turn into ready cash.[1] I did receive your letter, d[ear] Albert. It's a pity, though, that all of you are being so badly advised.[2] In my opinion, you definitely should go to Darmstadt. There is a good polytechnic there, and not only could you all live much better there than in Zurich but also make substantial savings, whereas now virtually all of my income is being spent on providing a meager living for you in Zurich.[3] Every expert with whom I discuss this finds this situation a screaming disgrace and me foolish. In this way not a single penny is being saved up and if I die there will be *nothing*. I even had to take out loans in Swiss currency so that I can give you the bare necessities.

I would prefer most to look for a furnished apartment for you all in Darmstadt that you could maintain for yourselves and you would rent out your Zurich apartment[4] furnished as well. That would be the most advantageous way. As soon as you are decided, I'll travel there to look for something suitable. Albert[5] could perhaps also go there so that everything is the way you yourselves want. Think it over quietly. In Germany one gets more than twice as much for the franc than in Switzerland and you will get just as much in francs as in Switzerland as long as I am able.[6] Don't think that living happily is only possible in Zurich. You would have it much nicer and more comfortable there, quite disregarding that Albert and Tete[7] would get a lot more out of me if we weren't living so terribly far apart.[8] Do speak with people sometime who aren't so fanatically obsessed with Switzerland.[9] Your resistance to the change is so unnatural that I've been accused here many times already of [neglecting] my responsibility by giving in. Why does divorced Mrs. [Max] Wohlwend live here with her child[10] (and many others from Switzerland)? Just because she can live here much better than in Zurich; and added to that, Darmstadt is undoubtedly considerably cheaper and more pleasant than Berlin.[11] She chose Berlin because she wants to study music here.

In the summer I am probably going to take a half-year lecturing trip to North America.[12] How nice it would be if we could see each other beforehand. Everything would be better if you all lived closer by.

Food is available here for you as much as you want for less money than in Switzerland.[13] Why else are there export controls?[14] With the money that you have available here you could live perfectly well.

So don't be so obstinate, instead, write soon in agreement to your

Papa.

# 233. To Edouard Guillaume

Berlin, 16 December 1920

Dear Guillaume,

I have so much obligatory work to do at present that I cannot think of writing a longer paper.[1] Thus I am unfortunately not in a position to accept the friendly challenge. You might write Mr. Xavier Léon that he could address himself to Langevin, who is an outstanding expert in the theory.[2]

Grossmann recently asked me for an assessment of your papers in the area of relativity theory because it was supposedly necessary to take an official position on it, finally.[3] I asserted that despite diligent attempts I was unable to make any progress toward comprehension and that I personally was convinced that there is no clear theoretical idea behind it. Don't be cross with me; it was no longer appropriate to keep silent about my opinion on this point.[4] It is impossible to assign, in any meaningful way, a universal time to the totality of inertial systems.

Amicable greetings to you and your wife, yours,

A. Einstein.

# 234. To Albert G. Schmedeman[1]

Berlin, W 30, 5 Haberland St., 16 December 1920

Highly esteemed Mr. Minister,

Thank you very much for your detailed telegram.[2] I thoroughly agree with your suggestion and will try to arrange things so that I can lecture at many places for shorter periods. It is naturally quite impossible, however, to conduct the complicated negotiations required for this from here. I therefore applied to the banker Paul Warburg in New York (former shareholder of the company Kuhn, Loeb & Co.),[3] who will either conduct the negotiations himself or delegate them to a suitable middleman.

With hearty thanks for your efforts until now, I am, very respectfully,

# 235. From Arnold Sommerfeld

Munich, 18 December 1920

Dear Einstein,

I come to you today as an irksome supplicant. As I hear, you are delivering a lecture in Vienna on January 10.[1] The return route passes through Munich. We have long been wanting you here one day at the Physical Society and at the university, generally speaking. Indeed, I even think that we have a greater right to you than other places because your relativity gospel found root here earlier and more firmly than elsewhere.[2] Furthermore, Dr. Anschütz has commissioned me to invite you as his guest.[3] Hence you will spend particularly pleasant days here—the Munich Anschütz residence is a temple of the arts without equal!–

I would like to ask you to present a "popular" lecture here on Friday the 14th of Jan. in the main auditorium. Audience: 1. Physical Society and colloquium. 2. Students with a special interest in physics and philosophy. 3. All lecturers of the university. 4. The local engineers association (to whom I would otherwise have to deliver a talk, which I would like to avoid). 5. About 500 students; total seating and standing capacity 1,200.

So as to reduce the throngs, we would charge an entrance fee to benefit the student residence (support association). You yourself would receive 1,000 marks from grant funds as travel compensation (hence not from the entrance money); if your financial circumstances are bad, as I suspect, we could also go higher.

Yesterday I spoke with the General Student Committee in the presence of the rector. The committee enthusiastically welcomes a guest lecture by you. If you prefer, we could also organize two such lectures on succeeding days, such as on the 13th and 14th. Best time, ¼ past 6 until about ½ past 7. After the lecture Anschütz wants to invite the faculty to his home.

Now, I could imagine that you are slowly getting fed up with popular lectures. But I do hope that you will not treat Munich worse than other places. In consideration of the student body, a popular lecture would be most particularly preferable to me. If, however, your disinclination should be too great, we would naturally also be grateful for a specialty lecture in a more restricted colloquium setting. It is unlikely, though, that we could offer such a high honorarium for the latter (only about 500 marks).

Delight me with a rapid acceptance, if possible before Christmas. I still have many preparations to make. You will also get to hear a very interesting thing from Herzfeld about the Einstein–de Haas effect (explanation for the ½ factor!!).[4]

Give my regards to your wife, whom I thank kindly for her postcard from November. I count on her not foiling our hopes through dissuasion.

I presume you haven't read my relat. article in the *Süddeutsche Monatshefte*; if you have, I hope you did not take offense at my indiscretion (communication of letter excerpts).[5] It was governed by the [style]. Yours,

A. Sommerfeld

# 236. To Arnold Sommerfeld

[between 18 and 28 December 1920][1]

Dear Sommerfeld,

My heart is pounding perceptibly at the thought of not being able to follow your kind invitation.[2] But it is completely impossible. On the 13th I have to give a talk in Vienna and on the 17th in Dresden, which I had promised the Polytechnic students half a year ago already.[3] So it would be too disquieting and exhausting to slip in one more stop in Munich; we do have to make the pitiful bundle of nerves that Nature endowed us with for our sojourn last a whole lifetime.

I liked your article in the *Südd. Monatshefte* very much,[4] as with all that you write, incidentally. Your book, too, is wonderfully transparent and fine, you know.[5] The glowing personal goodwill of the article cheered me additionally.

I am very curious how Herzfeld explains the factor $\frac{1}{2}$.[6] Recently, Bohr, whose atomic structure has again made bold advances, was here to see me.[7] His intuition is very much to be admired.

Heartily wishing you happy holidays and a good 1921, yours,

A. Einstein.

# 237. From Hermann Anschütz-Kaempfe

Munich, 6 Leopold St., 19 December 1920

Esteemed, dear Professor,

As I gather from Sommerfeld, there is hope of seeing you here.[1] So it is surely self-evident that I cordially repeat my invitation again, also in my wife's name.[2]

For egotistical reasons I would be especially pleased if I could speak with you in the course of January; I would like to report to you the experiences made so far with the metal sphere and the rotational heated cylinder.[3]

And besides that, I consider it proper to discuss with you the obligation by the firm Anschütz & Co. to pay tribute.[4] You must allow me that, along with representing your own interests.

It would be very particularly generous of you if you would not cut your stay here too short and if you would take your wife along.

Your room and the music room with the organ are waiting for you.

With most cordial greetings from me and my wife and best compliments from both of us to your spouse, yours,

Anschütz-Kaempfe

# 238. To Jewish Community of Berlin

Berlin, 22 December 1920

After careful consideration, I cannot resolve to join the Jewish religious community.[1] As much as I feel myself to be a Jew, I stand aloof from the traditional religious rites.[2] Nevertheless, in order to show how close Jewish issues are to my heart, I am very ready to give a fitting annual contribution to Jewish charity.

In great respect,

# 239. To A. J. Reingold

Berlin, 22 December 1920

Dear Sir,

Your statement of sympathy honestly pleased me; I thank you cordially for it.[1]

The attacks directed against me[2] were, incidentally, not by any means as bad as it might appear from abroad.[3] On this occasion I experienced more goodwill than animosity from our colleagues in the field and other people. It should also be taken into consideration that Germans currently deserve forbearance in their severe economic and political situation.

In great respect,

# 240. From Frederick A. Lindemann

Sidholme, Sidmouth, 22 December 1920

Highly esteemed Privy Councillor,

I would not like to miss extending to you and your esteemed wife my best wishes for a Merry Christmas and a Happy New Year. The delightful hours that I was permitted to spend with you in September will always remain for me one of the finest memories of the year 1920.[1] I hope that you will allow me to repeat this pleasure in the coming year, although I cannot deny that you may perhaps say that pleasure is also an observer-dependent concept.– I take the liberty of enclosing a couple of offprints and hope that they will meet with your approval. The essay on relativity caused general irritation at the philosophy conference in Oxford.[2]

I would be extremely grateful if you could have your offprints forwarded to me. While in Berlin, I had stayed so long with you that I forgot to make this request.–

I hope that you will notify me when I can be of any service to you here. I definitely count on your visiting me here once at Oxford, as promised. When do you think that this would be possible?– Once again, with best wishes for 1921, also from my father, I remain, yours very truly,

F. A. Lindemann.

## 241. From Edouard Guillaume

Berne, 23 December 1920

[Not selected for translation.]

## 242. From Albert G. Schmedeman

Christiania, December 23, 1920.

Dear Professor Einstein:

In reply to your letter of the 16th in which you inform me that the proposal made by the University of Wisconsin is acceptable to you providing satisfactory terms can be arranged by Mr. Paul Wahrburg or some one else,[1] I have to inform you that I have telegraphed this information to the regents of the University of Wisconsin.[2]

I quite agree with you that it would be impossible for you to arrange a series of lectures, and your suggestion that Mr. Wahrburg arrange this matter for you is an excellent idea and I hope that satisfactory arrangements can be made for you to come to the United States.[3] I am sure that you will receive a splendid welcome by the American people and that they are anxious to both see and hear you.

If arrangements are made for you to come to the University of Wisconsin, as I shall be at home at that time I hope to have the pleasure of calling on you and Mrs. Einstein.

With kindest regards to you and Mrs. Einstein, I am,

Sincerely yours,

A G Sch[medeman]
American Minister.

P. S. Since writing the foregoing letter I have received the following telegram from the regents of the University of Wisconsin:

"Will attempt to arrange summer conferences as per my letter of thirteenth[4] if Warburg has satisfactory terms."

A.G.[Sch].

# 243. From John G. Hibben

[Princeton,] December 24th, 1920.

My dear Professor Einstein:—

I have received your letter[1] and am greatly obliged to you for your kindly courtesy in replying to my inquiry concerning your possible visit to Princeton.

Unfortunately we are still struggling with the problems of reorganization, after the war, and we find ourselves financially straitened so that it will be impossible to consider the honorarium of $15,000 which you request for your lectures.[2] It is a great disappointment to me personally, as well as to the members of the Departments of Physics and Mathematics, that we shall not be able to avail ourselves of your presence for a while in the midst of our academic life.

With the assurance of my high regard, believe me

Faithfully yours,

John Grier Hibben

# 244. From Michele Besso

Berne, 7 Cäcilien Street, 24–27 December 1920

Dear Albert,

Today I read your speech of May 20 in Leyden—on the ether.[1] It was yet another one of those serene moments for me that you brought down from the stars. I do think, though, that you in fact gave that word the only possible meaning in the new domain, so that people who cling to it, Lorentz in particular, are not intimidated even further by apparent divergences—hence, something humane. Yet also something humanely beautiful.

Do you remember, when visiting you 8 years ago in Zurich, how I couldn't make head or tail of tensor calculus, despite all the fine sounding-board discussions about the new theory?[2] Now, *that* is how stuck I am in virtually everything. Financially, despite the exchange losses, it could have gone on like this for a few more years, even without good old Haller's kindness,[3] albeit with growing worry, of course; but that empty-headedness of those days has progressed strongly. And the order that was supposed to compensate for the resinification due to aging never set in.

May it do good. I'd like to continue with my lamentations, but it doesn't settle anything.

I was pleased recently, when I saw Maja for a quarter of an hour,[4] to hear the excellent news about your health. Also heard of the planned dollar-trip.[5] And that you complain about not being able to work because of all sorts of hindrances. Surely this has to be interpreted only relatively: perhaps also ⟨because the nuts that are now gathered, better said⟩ connected to that is the fact that the mountains towering before the mind of humanity, of which you are the embodiment, are immensely high again. Nevertheless, the problem of Weyl's theory still seems to be the most productive.[6] Do you remember the speculations about *non-Archimedean* geometry? (⟨the current⟩ Infinitely small, differing magnitudes are supposed to be set not equal.) ⟨—added to that⟩ We have often noted that the physical continuum could still play into this somewhere—comp[are] the second Riemannian hypothesis, space as something countable.–[7] Under *what conditions* could the various invariants of Weyl's theory remain indistinguishable in principle?—As frequently said, I should keep my mouth shut: I don't even know what kind of principally observable distinction exists between gravitational redshift and "redshift in an electric field,"[8] if such a thing really exists. That there is a gap here in my interpretation attempts is already evident from my inference, arising out of these interpretation attempts, that one would expect a red or blueshift between the respective ends of an electric line of force (for the content of a quasi stationary world—and what in the general case?), hence a plain potential effect, as with gravitation, whereas Weyl and you agree ⟨in this regard⟩ that a progressive change ought to be expected. Or should this change only come into consideration for fields not derivable from a potential? Could it be connected with the nonradiating rotating electron?—The more vacuous the brain, the wilder the speculations.

For today: Happy New Year for 1921! Yours,

Michele.

I might be able to receive a few marks from Germany in the form of books. I just want to order Born's *Theory of Relativity*, Nernst's ⟨little⟩ summarizing work on his heat theorem (exact title?), Sommerfeld's *Atomic Structure*, Born's atomic structure (?exact title?), Riemann/Weyl.[9] What other nice things to read (or to assign as reading . . .) could one order, I wonder?

Just in time before the end of the letter. Vero got engaged (with a good girl whom we've known for a long time already).[10]—The darkest spot is the financial issue, since he's in the middle of his studies for a long while yet. Otherwise we're happy.

# 245. To Ernest Pickworth Farrow[1]

[Berlin,] 28 December 1920

Highly est. Col.,

Your Christmas card is truly of exquisite humor and tickled me very much, the more so since I am just cramming English vocabulary, by the sweat of my brow, into my no longer so youthf. brain.

The kd. inquiry that you add to your letter filled me with utter joy, as a sign of affinity and high regard from that scientif. circle whose members I have always admired and about whom [a] friend told me many personally likeable things.[2]

The problems and attacks that fell to my share here[3] are not of such a serious nature as it may seem abroad. Given the great willingness on the part of our German colleagues in the profession and the authorities to oblige me, as I have continually experienced, it would be an act of [disdain] if during these times of need and degradation I were to turn my back on Germany. That I am Swiss and of thoroughly international persuasion changes nothing in this personal relationship.[4]

Thus I deem it as my duty to stay resolutely at my post as long as external circumstances do not make this practically impossible.[5]

With best wishes for the coming year, cordial regards from your

## 246. To Ayao Kuwaki[1]

Berlin, 28 December 1920

Highly esteemed Colleague,

I am extremely pleased that you and your colleague have translated my booklet into Japanese.[2] I still remember well your visit to Berne, especially since you were the first Japanese, indeed the first East Asian whose acquaintance I ever made. You astounded me then with your great theoretical knowledge.[3]

I am sending your letter to my publisher, Vieweg & Son, which is joint owner of the translation copyright.

Kind regards, yours,

A. Einstein.

## 247. From Hermann Anschütz-Kaempfe

Munich, 6 Leopold St., 28 December 1920

Esteemed Professor,

Like my wife and I, many others will regret that nothing should come of your visit to Munich. If it weren't for the Christmas holidays, I would suggest that you make a brief stop in Munich before the Vienna talk, around Jan. 5th. Around this time, there would be room for an internal colloquium too, of course; a select enthusiastic public would most heartily welcome you as well, which would probably be more agreeable to you than a more or less uncomprehending great mass of listeners in the main aud[itorium].

But I would not like to wait until next January or, as Sommerfeld hopes, until November[1] with my news about the metal sphere and the heat-rotation experiment.

As was to be expected, the sphere with 3 or more magnets (of 500 periods) is at an adequate distance and centered; but, unfortunately, the iron distribution within the sphere is asymmetrical (3 gyroscopes); hence preferred positions for the sphere, since the lines of force penetrate through the sphere's 3 m/m-thick aluminium wall.[2] An iron-sheet shield within the sphere still proved not to be sufficient. Even a larger number of magnets (more than 10 cannot be comfortably housed, probably for space reasons) will surely improve the situation but not solve it. There is simply an asymmetry left.

Building upon your suggestion, I have now suggested in Kiel to try a ring magnet of about 120 m/m diameter at a sphere size of 220 m/m. Thus something like this:[3]

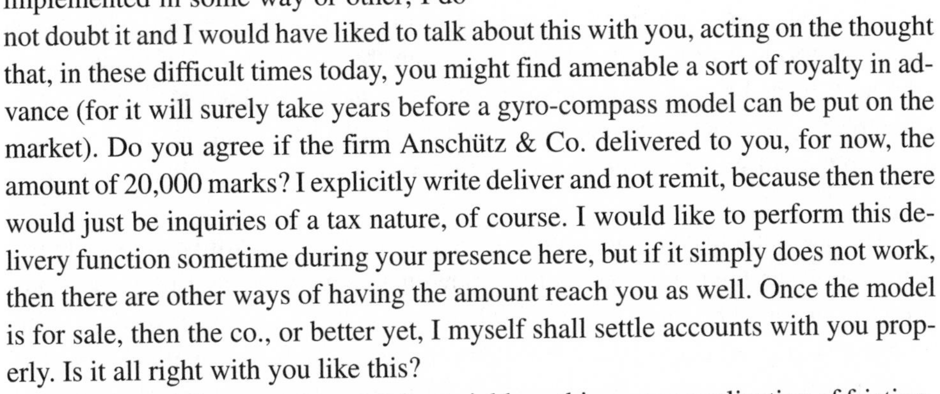


According to it, the construction would be independent of the iron distribution within the sphere. I do suspect, though, that the efficiency of this arrangement will be very bad.

Nevertheless, your suggestion will be implemented in some way or other; I do not doubt it and I would have liked to talk about this with you, acting on the thought that, in these difficult times today, you might find amenable a sort of royalty in advance (for it will surely take years before a gyro-compass model can be put on the market). Do you agree if the firm Anschütz & Co. delivered to you, for now, the amount of 20,000 marks? I explicitly write deliver and not remit, because then there would just be inquiries of a tax nature, of course. I would like to perform this delivery function sometime during your presence here, but if it simply does not work, then there are other ways of having the amount reach you as well. Once the model is for sale, then the co., or better yet, I myself shall settle accounts with you properly. Is it all right with you like this?

The heat-rotation experiment did not yield anything upon application of frictional heat.[4] Now I have set up a new experimental arrangement and send very hot oil (approx. 200°) through a thick-walled copper cylinder. You will immediately understand the enclosed diagram. I suspect that with this arrangement disturbances will occur less easily. I do still have some reservations about the use of a magnetic needle to measure the ev. magnetism, since at closer proximities of a magnet to the rotating copper cylinder, eddy currents must surely form, which might have a disturbing influence on the experiment. Finally, there are other control measurements as well that one can use once a positive result is in.

It goes without saying that I shall report to you further about the relevant experiments.

My wife and I send you our warmest wishes for the year 1921; in particular, my wife sends word that your agreement to come to Kiel in the summertime has been seriously taken note of and that we cannot release your from it so easily without completely compelling reasons. How you will wheedle yourself out of this one[5] is not clear to me yet, for I know my wife's pigheadedness; she never gives in.[6] And the sailboat or ship stands ready in the summer.

You see, here she is already. So please submit right away; it's much simpler.

Most cordially yours,

Anschütz.

Note by Reta Anschütz: "That's right!!! Your surrogate mother!"

## 248. From Carl Beck[1]

Chicago, 28 December 1920

My dear Professor Einstein,

I regretted very much during my visit in Berlin, I did not have the chance to see you. I telephoned to your home to make an appointment; but was told that you had left for Holland.

Professor Lecher of Vienna,[2] who is a friend of mine—and particularly Professor Ehrenfest,[3] another friend, promised to arrange with you a meeting; but unfortunately I had to return to the States. It was not the idle curiosity of many people who seek to make your acquaintance; but *other motives* which prompted me to want to see you. You, of course, are a persona grata with certain scientific circles, and also here in America, there is a great interest for you personally, and for your work.

I have no doubt that you are invited to this country by scientific bodies and men. Knowing American conditions as I do know them, and knowing European conditions as I do, and I do know them, I think, however, that such a visit while it offers on one side a great many possibilities—is also possibly fraught with certain dangers—not only for yourself but for your country, and I would very much like to discuss conditions with you on such a visit. America is not yet ready for experiments, and I think too highly of you and your country to make any experiments.

When I returned a couple of weeks ago from the other side, I had a long conference with Professor Vincent of the Rockefeller Foundation,[4] and with a number of others interested in Germany and German Science. I think I have accomplished a great deal by interesting the Rockefeller Foundation favorably, before I left for Europe through Professor Welch of John Hopkins University[5] and Professor Vincent, and I believe that my report coming from abroad, has clinched the argument in favor of appropriating $100,000 by the Rockefeller people for the aid of German and Austrian Universities.[6] Although my propositions as Professor Vincent writes me have not all been accepted, he says I will be satisfied for the present with what has been done.

There are in America a large number of serious men thinking well of Germany. There are, however, a large number of men who appear in that light; but in reality are not inclined that way, and the masses a[re] fickle and are easily swayed.

Will you be kind enough to write to me about your plans in regard to a visit to this country, and if I can do anything—as I enjoy the confidence of a great many of those who are today what one might call the leading men of science and progress in thought,—I shall be only too glad to put my services to your disposal.

Professor Ehren[haft] to whom I am sending a copy of this letter, wrote to me that you will be their guest in Vienna in January,[7] and it is possible he may give you this letter personally and discuss with you some phases of it.

With kind personal regards, I remain, Yours very truly,

Carl Beck

## 249. To Wilhelm Blaschke

Berlin, 29 December 1920

Dear Colleague,

An opinion on Laue is obviated.[1] Among the others under consideration, the achievements of Epstein seem to me to tower far above all the others. Lenz, Schrödinger, Thirring, and Flamm are all qualified theoreticians, each of whom is worthy of recommendation.[2] Among these I would prefer Lenz and Schrödinger, without being able to say which of these two I place above the other. Finally, I would not want to neglect pointing out Fritz Reiche,[3] who is an excellent theoretician and teacher, even if his accomplishments as regards originality perhaps rank behind those of Schrödinger and Lenz.

With best wishes for the New Year, yours,

## 250. To Edouard Guillaume

[29 December 1920]

Dear Guillaume,

Now I think I see what you are doing. You are observing a spherical wave. Observed from $K_1$, the world points taken at a time $u_1$ are characterized by a time $u_1$ = const. Observed from $K_2$, these world points lie on a surface that is characterized by the equation

$$u_2 = \frac{u_1}{\beta(1 + \cos\varphi_2)}.$$

*With reference to $K_2$ alone*, however, this surface has absolutely no physical meaning.[1]

I know, though, that we are dealing with some sort of idée fixe of yours and that all labors of love are futile. I am no angrier at you than at a sparrow for not singing like a nightingale. But it amuses me that you are apparently finding a faithful audience—but certainly no theoretical physicist of any standing will fall for this business of yours.[2] With Langevin, for inst., I am firmly persuaded that he will immediately see the whole picture.[3] I cannot imagine that Grossmann would have anything against you.[4] But he is probably embarrassed for Swiss physicists that the issue is not being challenged; one can't take offense at that. When asked, I cannot do otherwise than speak my mind. And I don't know what else to say to you than: do what you just cannot refrain from doing.

Best regards, yours,

A. Einstein.

# 251. To Mário Basto Wagner

Berlin, 29 December 1920

Esteemed Colleague,

Pardon my long silence.[1] I do not remember whether I ever published something about the problem of the entropy of solutions. This is not of interest, of course. As far as I know, I am in full agreement with Planck's treatment of the Nernst theorem, since he postulates the vanishing of entropy at absolute zero only for chemically homogeneous substances, not, however, for irregular mixtures of molecules of various sorts.[2] Unfortunately, it has been impossible for me to find the time to study your papers,[3] because I am currently not working in this field.

In utmost respect,

# 252. From Arnold Sommerfeld

curr. Garmisch, 29 December 1920

Dear Einstein,

That really is a shame! We would have been so pleased to have you.[1] But we won't give in. Note us down for the beginning of next *winter* term: *the 11th or 18th of November*. Anschütz is back in Munich then.[2] I propose definite dates only so that you do not shift the business to the familiar *calendas gräcas*[3] and I otherwise agree in advance to any counterproposal of yours.

I hear from Anschütz that you received 2,000 marks in Kiel;[4] we can give you that too. Now I will probably have to give the lecture at the Engineers Association myself, for better or for worse[5] (I will probably do it rather worse than better). When I see how irascibly I respond to unqualified questions regarding relativity links, then I can imagine the rage that you yourself eventually must feel at that.

That you are not angry at me about the citations from your letters, which I gather from your silence about that, is a relief for me. They simply fit too well in my flow of words.[6]

Will the new year bring the electron problem to maturity? That is probably the next and perhaps the largest fruit of rel. th.[7]

Herzfeld's idea is the following, very superficially described and without guarantee that I am relaying it to you exactly:[8] Upon magnetic reversal of a ferromagnetic atom, radiation is emitted into the ether; however, this emission carries with it (according to the general rules relating energy and momentum) just half of the change in momentum that corresponds to a change in the electrons' orbital sense; only the other half goes onto the magnet. Herzfeld wants to verify the presence of this radiation experimentally (frequency as for wirel[ess] telegr.). This can take a good while yet; so I ask you please not to breathe a word about it for the time being. It will be a big thing, if everything works.

I am in suspense about Bohr's new atom ideas.[9] In my opinion, he is a man of rare intellectual and personal charm.

With best regards, yours,

A. Sommerfeld.

Poor Rümelin has died of gastric ulcers, intrinsically harmless ones, but with a rupture into the abdominal membrane. How good that you took proper care of yourself in time, which Rümelin did not do.[10]

## 253. From Jewish Community of Berlin

Berlin N. 24, 2 Oranienburger St., 30 December 1920

To Professor Albert Einstein, Schöneberg, Berlin, 5 Haberland St.

To your obliging correspondence of the 22nd of this mo.[1] we reply that, according to local law, membership in a cultural community does not presuppose a declaration of membership. Rather, every Jew is obligated, by force of law, to be a taxpaying member of that Jewish community in whose district he resides. Pursuant to the statutes of the Jewish Community of Berlin, the amount of tax is set according to the national income tax, as is described in more detail in the commentary to the assessment being forwarded to you. The community is therefore not legally authorized to disregard your assessment pursuant to the rule. We permit ourselves to point out that a large part of the tax revenues are used for those purposes that your generous offer had in mind so that your intentions are also realized by a payment in the form of a tax.

Chairman of the Jewish Community. Certified:

Goldstein[2]

# INDEX

References are collected under the appropriate English heading. Certain institutions, organizations, and concepts that have no standard English translation are listed under their German designation (with cross references from an English translation). "Albert Einstein" is abbreviated to "AE" in subentries. Other abbreviations used in the index are: "DPG" for "Deutsche Physikalische Gesellschaft," "ETH" for "Eidgenössisch-Technische Hochschule," "GDNÄ" for "Gesellschaft deutscher Naturforscher und Ärzte," "KWIP" for "Kaiser-Wilhelm-Institut für Physik," "PAW" for "Preussische Akademie der Wissenschaften," "US" for the "United States of America."